# Structure Determination
# by X-Ray Crystallography

# Structure Determination by X-Ray Crystallography

**M. F. C. Ladd**

*University of Surrey*
*Guilford, England*

and

**R. A. Palmer**

*Birkbeck College*
*University of London*
*London, England*

**Plenum Press · New York and London**

Library of Congress Cataloging in Publication Data

Ladd, Marcus Frederick Charles.
  Structure determination by X-ray crystallography.

  Includes bibliographies and index.
  1. X-ray crystallography. I. Palmer, Rex Alfred, 1936-        joint
author. II. Title.
QD945.L32                    548'.83                    76-40229
ISBN 0-306-30844-4

© 1977 Plenum Press, New York
A Division of Plenum Publishing Corporation
227 West 17th Street, New York, N.Y. 10011

Printed in the United States of America

Technology & Science

# Preface

Crystallography may be described as the science of the structure of materials, using this word in its widest sense, and its ramifications are apparent over a broad front of current scientific endeavor. It is not surprising, therefore, to find that most universities offer some aspects of crystallography in their undergraduate courses in the physical sciences. It is the principal aim of this book to present an introduction to structure determination by X-ray crystallography that is appropriate mainly to both final-year undergraduate studies in crystallography, chemistry, and chemical physics, and introductory postgraduate work in this area of crystallography. We believe that the book will be of interest in other disciplines, such as physics, metallurgy, biochemistry, and geology, where crystallography has an important part to play.

In the space of one book, it is not possible either to cover all aspects of crystallography or to treat all the subject matter completely rigorously. In particular, certain mathematical results are assumed in order that their applications may be discussed. At the end of each chapter, a short bibliography is given, which may be used to extend the scope of the treatment given here. In addition, reference is made in the text to specific sources of information.

We have chosen not to discuss experimental methods extensively, as we consider that this aspect of crystallography is best learned through practical experience, but an attempt has been made to simulate the interpretive side of experimental crystallography in both examples and exercises.

During the preparation of this book, we have tried to keep in mind that students meeting crystallography for the first time are encountering a new discipline, and not merely extending a subject studied previously. In consequence, we have treated the geometry of crystals a little more fully than is usual at this level, for it is our experience that some of the difficulties which students meet in introductory crystallography lie in the unfamiliarity of its three-dimensional character.

We have limited the structure-determining techniques to the three that are used most extensively in present-day research, and we have described them in depth, particularly from a practical point of view. We hope that this treatment will indicate our belief that crystallographic methods can reasonably form part of the structural chemist's repertoire, like quantum mechanics and nmr spectroscopy.

Each chapter is provided with a set of problems, for which answers and notes are given. We recommend the reader to tackle these problems; they will provide a practical involvement which should be helpful to the understanding of the subject matter of the book. From experience in teaching this subject, the authors are aware of many of the difficulties encountered by students of crystallography, and have attempted to anticipate them both in these problems and the text. For any reader who has access to crystallographic computing facilities, the authors can supply copies of the data used to solve the structures described in Chapters 6 and 8. Certain problems have been marked with an asterisk. They are a little more difficult than the others and may be omitted at a first reading.

The Hermann–Mauguin system of symmetry notation is used in crystallography, but, unfortunately, this notation is not common to other disciplines. Consequently, we have written the Schoenflies symbols for point groups on some of the figures that depict point-group and molecular symmetry in three dimensions, in addition to the Hermann–Mauguin symbols. The Schoenflies notation is described in Appendix 3. General symbols and constants are listed in the Notation section.

We wish to acknowledge our colleague, Dr. P. F. Lindley, of Birkbeck College, London, who undertook a careful and critical reading of the manuscript and made many valuable suggestions. We acknowledge an unknown number of past students who have worked through many of the problems given in this book, to our advantage and, we hope, also to theirs. We are grateful to the various copyright holders for permission to reproduce those figures that carry appropriate acknowledgments. Finally, we thank the Plenum Publishing Company for both their interest in this book and their ready cooperation in bringing it to completion.

*University of Surrey*                                                M. F. C. Ladd
*Birkbeck College, London*                                      R. A. Palmer

# Contents

## Chapter 3

## Preliminary Examination of Crystals by Optical and X-Ray Methods

## Chapter 4

## Intensity of Scattering of X-Rays by Crystals . . . . . .    143

## Chapter 5

## Methods in X-Ray Structure Analysis. I . . . . . . . . .    183

## Chapter 6

## Methods in X-Ray Structure Analysis. II . . . . . . . . 201

## Chapter 7

## Some Further Topics . . . . . . . . . . . . . . . . . 271

## Chapter 8
# Examples of Crystal Structure Analysis . . . . . . . . 299

# Notation

These notes provide a key to the main symbols and constants used throughout the book. Inevitably, some symbols have more than one use. This feature arises partly from general usage in crystallography, and partly from a desire to preserve a mnemonic character in the notation wherever possible. It is our belief that, in context, no confusion will arise. Where several symbols are closely linked, they are listed together under the first member of the set.

$A'(hkl), B'(hkl)$ . . . . Components of the structure factor, measured along the real and imaginary axes, respectively, in the complex plane (Argand diagram)

$A(hkl), B(hkl)$ . . . Components of the geometric structure factor, measured along the real and imaginary axes, respectively, in the complex plane

$A$ . . . . . . $A$-face-centered unit cell; absorption correction factor

$Å$ . . . . . . Angstrom unit; $1\ Å = 10^{-8}\ cm = 10^{-10}\ m$

$a, b, c$ . . . . Unit-cell edges parallel to the $X$, $Y$, and $Z$ axes, respectively, of a crystal; intercepts made by the parametral plane on the $X$, $Y$, and $Z$ axes, respectively; glide planes with translational components of $a/2, b/2,$ and $c/2$, respectively

$\mathbf{a}, \mathbf{b}, \mathbf{c}$ . . . . Unit-cell edge vectors parallel to the $X$, $Y$, and $Z$ axes, respectively

$a^*, b^*, c^*$ . . . Edges in the reciprocal unit cell associated with the $X^*$, $Y^*$, and $Z^*$ axes, respectively

$\mathbf{a}^*, \mathbf{b}^*, \mathbf{c}^*$ . . . Reciprocal unit-cell vectors associated with the $X^*$, $Y^*$, and $Z^*$ axes, respectively

$B$ . . . . . . $B$-face-centered unit cell; overall isotropic temperature factor

$B_j$ . . . . . . Isotropic temperature factor for the $j$th atom

$C$ . . . . . . $C$-face-centered unit cell

| | |
|---|---|
| $c$ . . . . . . | Velocity of light ($2.9979 \times 10^{-8}$ m s$^{-1}$); as a subscript: calculated, as in $\lvert F_c \rvert$ |
| $D_m$ . . . . . | Experimentally measured crystal density |
| $D_c$ . . . . . | Calculated crystal density |
| $d$ . . . . . . | Interplanar spacing |
| $d(hkl)$ . . . . | Interplanar spacing of the $(hkl)$ family of planes |
| $d^*$ . . . . . | Distance in reciprocal space |
| $d^*(hkl)$ . . . | Distance from the origin to the $hkl$th reciprocal lattice point |
| $E, E(hkl)$ . . . | Normalized structure factor (centrosymmetric crystals) |
| $\mathscr{E}(hkl)$ . . . . | Total energy of the $hkl$th diffracted beam from one unit cell |
| $e$ . . . . . . | Electron charge ($1.6021 \times 10^{-19}$C); exponential factor |
| $\mathbf{F}(hkl)$ . . . . | Structure factor for the $hkl$th spectrum referred to one unit cell |
| $\mathbf{F}^*(hkl)$ . . . | Conjugate vector of $\mathbf{F}(hkl)$. |
| $\lvert F \rvert$ . . . . . | Modulus, or amplitude, of any vector $\mathbf{F}$ |
| $f$ . . . . . . | Atomic scattering factor |
| $f_{j,\theta}, f_j$ . . . . | Atomic scattering factor for the $j$th atom |
| $g$ . . . . . . | Glide line in two-dimensional space groups |
| $g_j$ . . . . . . | Atomic scattering factor for the $j$th atom, in a crystal, corrected for thermal vibrations |
| $H$ . . . . . . | Hexagonal (triply primitive) unit cell |
| $(hkl), (hkil)$ . . | Miller, Miller–Bravais, indices associated with the $X$, $Y$, and $Z$ axes or the $X$, $Y$, $U$, and $Z$ axes, respectively; any single index containing two digits has a comma placed *after* such an index |
| $\{hkl\}$ . . . . . | Form of $(hkl)$ planes |
| $hkl$ . . . . . | Reciprocal lattice point corresponding to the $(hkl)$ family of planes |
| $\mathbf{h}$ . . . . . . | Vector with components $h, k, l$ in reciprocal space. |
| $h$ . . . . . . | Planck's constant ($6.6256 \times 10^{-34}$ J s) |
| $I$ . . . . . . | Body-centered unit cell; intensity of reflection |
| $I(hkl)$ . . . . | Intensity of reflection from the $(hkl)$ planes referred to one unit cell |
| $\mathscr{I}$ . . . . . . | Imaginary axis in the complex plane |
| i . . . . . . | $\sqrt{-1}$; an operator that rotates a vector in the complex plane through 90° in a right-handed (anticlockwise) sense |

$J(hkl)$ . . . . .     Integrated reflection

$K$ . . . . . .     Reciprocal lattice constant; scale factor for $|F_o(hkl)|$ data

$L$ . . . . . .     Lorentz correction factor

$M$ . . . . . .     Relative molecular weight (mass)

$m$ . . . . . .     Mirror plane; atomic mass unit $(1.6604 \times 10^{-24}\,\text{g})$

$N$ . . . . . .     Number of atoms per unit cell

$n$ . . . . . .     Glide plane, with translational component of $(a+b)/2$, $(b+c)/2$, or $(c+a)/2$

$n_1, n_2, n_3$ . . .     Principal refractive indices in a biaxial crystal

$o$ . . . . . .     subscript: observed, as in $|F_o(hkl)|$

$P$ . . . . . .     Probability; Patterson function

$P(u, v, w)$ . . .     Patterson function at the fractional coordinates $u, v, w$ in the unit cell

$p$ . . . . . .     Polarization correction factor

$R$ . . . . . .     Rhombohedral unit cell; rotation axis (of degree $R$); reliability factor

$\bar{R}$ . . . . . .     Inversion axis

$\mathscr{R}$ . . . . . .     Real axis in the complex plane

$RU$ . . . . .     Reciprocal lattice unit

$s, s(hkl), s(\mathbf{h})$ .     Sign of a centric reflection

$T_{j,\theta}$ . . . . .     Thermal vibration parameter for the $j$th atom

$[UVW]$ . . .     Zone or direction symbol

$\langle UVW \rangle$ . . .     Form of zone axes or directions

$(u, v, w)$ . . .     Components of a vector in Patterson space

$\overline{U^2}$ . . . . .     Mean square amplitude of vibration

$V_c$ . . . . .     Volume of a unit cell

$w$ . . . . . .     Weight factor

$X, Y, Z;$     Spatial coordinates, in absolute measure, of a point,
  $X, Y, U, Z$ .     parallel to the $X, Y, (U)$, and $Z$ axes, respectively

$x, y, z$ . . . .     Spatial fractional coordinates in a unit cell

$x_j, y_j, z_j$ . . .     Spatial fractional coordinates of the $j$th atom in a unit cell

$[x, \beta, \gamma]$ . . .     Line parallel to the $x$ axis and intersecting the $y$ and $z$ axes at $\beta$ and $\gamma$, respectively

$(x, y, \gamma)$ . . .     Plane normal to the $z$ axis and intersecting at $\gamma$

$\pm\{x, y, z; \dots\}$ .     $x, y, z; \bar{x}, \bar{y}, \bar{z}; \dots$

$Z$ . . . . . .     Number of formula-entities of weight $M$ per unit cell

$Z_j$ . . . . . .     Atomic number of the $j$th atom in a unit cell

$\alpha, \beta, \gamma$ . . . .     Angles between the pairs of unit-cell edges $bc$, $ca$, and $ab$, respectively

$\alpha^*, \beta^*, \gamma^*$ . . .     Angles between the pairs of reciprocal unit-cell edges $b^*c^*$, $c^*a^*$, and $a^*b^*$, respectively

$\delta$ . . . . . .     Path difference

$\varepsilon, \varepsilon(hkl)$ . . .     Statistical weight of a reflection

$\varepsilon, \omega$ . . . . .     Principal refractive indices for a uniaxial crystal

$\theta$ . . . . . .     Bragg angle

$\lambda$ . . . . . .     Wavelength

$\mu$ . . . . . .     Linear absorption coefficient

$\nu$ . . . . . .     Frequency

$\nu_n$ . . . . . .     Spacing between the zeroth- and $n$th-layer lines

$\rho$ . . . . . .     Radius of stereographic projection

$\rho(x, y, z)$ . . .     Electron density at the point $x$, $y$, $z$

$\Phi$ . . . . . .     Interfacial (internormal) angle

$\phi(hkl), \phi(h), \varphi$ .     Phase angle associated with a structure factor

$\chi, \psi, \omega$ . . . .     $(\cos \chi, \cos \psi, \cos \omega)$ direction cosines of a line with respect to the $X$, $Y$, and $Z$ axes

$\omega$ . . . . . .     Angular frequency

$\Omega$ . . . . . .     Azimuthal angle in experimental methods

# Crystal Geometry. I

## 1.1 Introduction

Crystallography grew up as a branch of mineralogy, and involved mainly the recognition, description, and classification of naturally occurring crystal species. As a subject in its own right, crystallography is a relatively new discipline, dating from the discovery in 1912 of the diffraction of X-rays by crystals. This year marked the beginning of the experimental determination of crystal structures. Figure 1.1 illustrates the structure of sodium chloride, which was among the first crystals to be studied by the new X-ray techniques.

Nowadays, we can probe the internal structure of crystals by X-ray methods and determine with certainty the actual atomic arrangement in space. Figure 1.2 shows a three-dimensional contour map of the electron density in euphenyl iodoacetate, $C_{32}H_{53}IO_2$. The contour lines join points of equal electron density in the structure; hydrogen atoms are not revealed in this map because of their relatively small scattering power for X-rays. If we

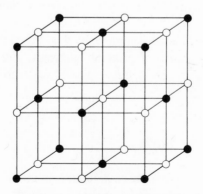

FIGURE 1.1 Crystal structure of sodium chloride: (●)$Na^+$, (○)$Cl^-$.

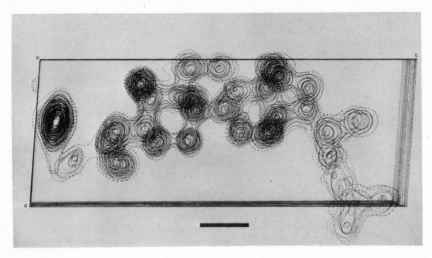

FIGURE 1.2. Three-dimensional electron density contour map for euphenyl iodoacetate as seen along the $b$ axis of the unit cell.

assume that the centers of atoms are located at the maxima in the electron density map, we can deduce the molecular model in Figure 1.3; the conventional chemical structural formula is shown for comparison. The iodine atom is represented by the large number of contours in the elongated peak at the extreme left of the figure. The carbon and oxygen atoms are depicted by approximately equal numbers of contours, except for the atoms in the side chain, shown on the extreme right of the figure. Thermal vibrations of the atoms are most severe in this portion of the molecule, and they have the effect of smearing out the electron density, so that its gradient, represented by the contour intervals, is less steep than in other parts of the molecule.

Molecules of much greater complexity than this example are now being investigated; the structures of proteins, enzymes, and nucleic acids—the "elements" of life itself—are being revealed by powerful X-ray diffraction techniques.

## 1.2  The Crystalline State

A crystalline substance may be defined as a homogeneous solid having an ordered internal atomic arrangement and a definite, though not necessarily stoichiometric, overall chemical composition. In addition to the more

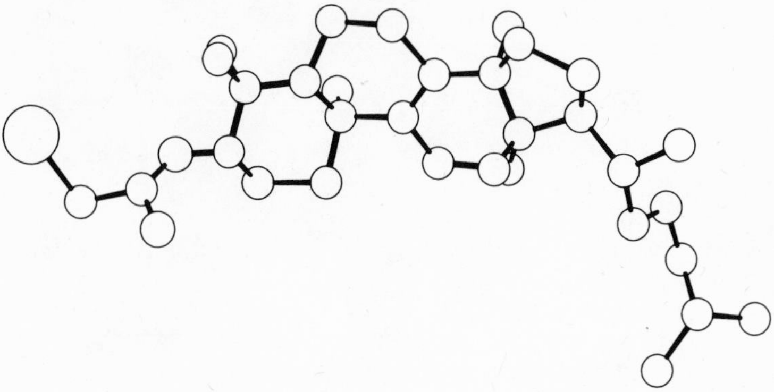

(a)

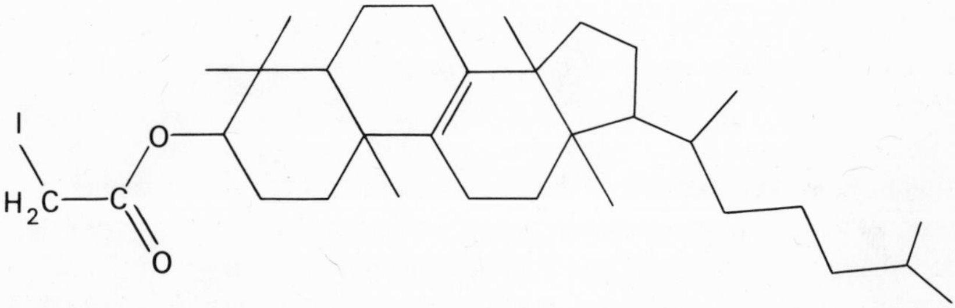

(b)

FIGURE 1.3. Euphenyl iodoacetate: (a) molecular model, excluding hydrogen atoms; (b) chemical structural formula.

obvious manifestations of crystalline material, other substances, such as cellophane sheet and fibrous asbestos, which reveal different degrees of long-range order (extending over many atomic dimensions), may be described as crystalline.

Fragments of glass and of quartz look similar to each other to the unaided eye, yet quartz is crystalline and glass is noncrystalline, or amorphous. Glass may be regarded as a supercooled liquid, with an atomic arrangement that displays only very short-range order (extending over few atomic dimensions). Figure 1.4 illustrates the structures of quartz and silica glass; both contain the same atomic group, the tetrahedral $SiO_4$ structural unit.

(a)

(b)

FIGURE 1.4. SiO$_4$ structural unit (the darker spheres represent Si): (a) $\alpha$-quartz, (b) silica glass. [Crown copyright. Reproduced from *NPL Mathematics Report Ma62* by R. J. Bell and P. Dean, with the permission of the Director, National Physical Laboratory, Teddington, Middlesex, England.]

A crystal may be defined as a substance that is crystalline in three dimensions and is bounded by plane faces. The word crystal is derived from the Greek κρυσταλλοσ, meaning *ice*, used to describe quartz, which once was thought to be water permanently congealed by intense cold. We have made the useful distinction that crystalline substances exhibit long-range order in three dimensions or less, whereas crystals have three-dimensional regularity and plane bounding faces.

## 1.2.1 Reference Axes

The description of crystals and their external features presents situations which are associated with coordinate geometry. In two dimensions, a straight line $AB$ may be referred to rectangular axes (Figure 1.5) and described by the equation

$$Y = mX + b \qquad (1.1)$$

where $m$ ($= \tan \Phi$) is the slope of the line and $b$ is the intercept made by $AB$ on the $Y$ axis. Any point $P(X, Y)$ on the line satisfies (1.1). If the line had been referred to oblique axes (Figure 1.6), its equation would have been

$$Y = MX + b \qquad (1.2)$$

where $b$ has the same value as before, and $M$ is given by

$$M = (\tan \Phi \sin \gamma - \cos \gamma) \qquad (1.3)$$

Evidently, oblique axes are less convenient in this case.

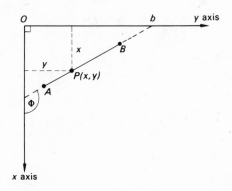

FIGURE 1.5. Line $AB$ referred to rectangular axes.

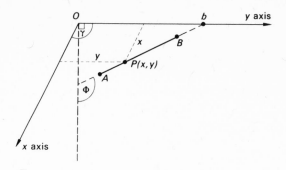

FIGURE 1.6.  Line *AB* referred to oblique axes.

We may describe the line in another way. Let *AB* intersect the *X* axis at *a* and the *Y* axis at *b* (Figure 1.7) and have slope *m*. At *X* = *a*, we have *Y* = 0, and, using (1.1),

$$ma + b = 0 \tag{1.4}$$

whence

$$Y = -(bX/a) + b \tag{1.5}$$

or

$$(X/a) + (Y/b) = 1 \tag{1.6}$$

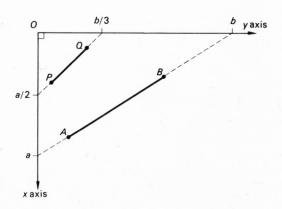

FIGURE 1.7.  Lines *AB* and *PQ* referred to rectangular axes.

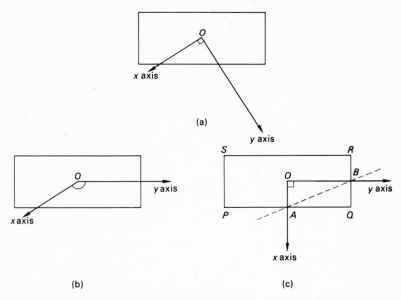

FIGURE 1.8. Rectangle referred to rectangular and oblique axes.

Equation (1.6) is the intercept form of the equation of the straight line $AB$. This line will be used as a reference, or parametral, line. Consider next any other line, such as $PQ$; let its intercepts on the $X$ and $Y$ axes be, for example, $a/2$ and $b/3$, respectively. This line may be identified by two numbers $h$ and $k$ defined such that $h$ is the ratio of the intercepts made on the $X$ axis by the parametral line and the line $PQ$, and $k$ is the corresponding ratio for the $Y$ axis. Thus

$$h = a/(a/2) = 2 \tag{1.7}$$

$$k = b/(b/3) = 3 \tag{1.8}$$

$PQ$ is described as the line (23)—two-three. It follows that $AB$ is (11). Although the values of $a$ and $b$ are not specified, once the parametral line is chosen, any other line can be defined uniquely by its indices $h$ and $k$.

   In the analysis of a plane figure, reference axes are again useful. In Figure 1.8, common sense (and convention, as we shall see) dictates the choice (c) for reference axes $X$ and $Y$; these lines are parallel to the perimeter lines, which are important features of the rectangle. If $AB$ is (11), then $PQ$, $QR$, $RS$, and $SP$ are (10), (01), ($\bar{1}$0),* and (0$\bar{1}$), respectively. A zero

* Read as "bar-one zero."

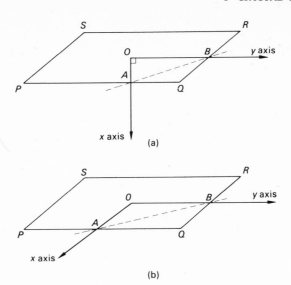

FIGURE 1.9. Parallelogram: (a) referred to rectangular
axes, (b) referred to oblique axes.

value for $h$ or $k$ indicates parallelism of the line with the corresponding axis
(its intercept is at infinity); a negative value for $h$ or $k$, indicated by a bar over
the symbol, implies an intercept on the negative side of the reference axis.

This simple description of the lines in $PQRS$ is not obtained with either
the orientation (a) or the oblique axes (b) in Figure 1.8. In considering a
parallelogram, however, oblique axes are the more convenient for our
purposes. It is left as an exercise for the reader to show that, if $AB$ in Figure
1.9 is (11), then $PQ$, $QR$, $RS$, and $SP$ are again (10), (01), ($\bar{1}$0), and (0$\bar{1}$),
respectively, provided that the reference axes are chosen parallel to the sides
of the figure.

### Crystallographic Axes

Three reference axes are needed for crystal description (Figure 1.10).
An extension of the above arguments leads to the adoption of $X$, $Y$, and $Z$
axes parallel to important directions in the crystal. We shall see later that
these directions (crystal edges or possible crystal edges) are related closely to
the symmetry of the crystal; in some cases, a choice of oblique axes then will
arise naturally.

It is usual to work with right-handed axes. In Figure 1.11, $+Y$ and $+Z$
are in the plane of the paper, as shown, and $+X$ is directed forwards; the

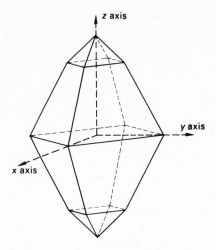

FIGURE   1.10. Idealized    tetragonal
crystal with the crystallographic axes
drawn in.

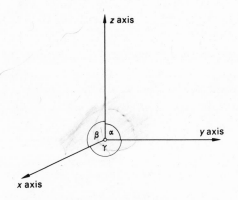

FIGURE   1.11. Right-handed    crystallographic axes and interaxial angles.

succession $X \to Y \to Z$ simulates a right-handed screw motion. Notice the
selection of the interaxial angles $\alpha$, $\beta$, and $\gamma$, and the mnemonic connection
between their positions and the directions of the $X$, $Y$, and $Z$ axes.

## 1.2.2   Equation of a Plane

In Figure 1.12, the plane $ABC$ intercepts the $X$, $Y$, and $Z$ axes at $A$, $B$,
and $C$, respectively. $ON$ is the perpendicular from the origin $O$ to the

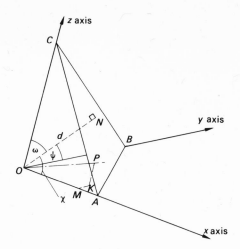

FIGURE 1.12. Plane *ABC* in three-dimensional space.

plane; it has the length $d$, and its direction cosines are $\cos \chi$, $\cos \psi$, and $\cos \omega$ with respect to *OA*, *OB*, and *OC*, respectively. *OA*, *OB*, and *OC* have the lengths $a$, $b$, and $c$, respectively, and *P* is any point *X*, *Y*, *Z* in the plane *ABC*. Let *PK* be parallel to *OC* and meet the plane *AOB* at *K*, and let *KM* be parallel to *OB* and meet *OA* at *M*. Then the lengths of *OM*, *MK*, and *KP* are *X*, *Y*, and *Z*, respectively. Since *ON* is the projection of *OP* onto *ON*, it is equal to the sum of the projections *OM*, *MK*, and *KP* all onto *ON*. Hence,

$$d = X \cos \chi + Y \cos \psi + Z \cos \omega \qquad (1.9)$$

In $\triangle OAN$, $d = OA \cos \chi = a \cos \chi$. Similarly, $d = b \cos \psi = c \cos \omega$, and, hence,

$$(X/a) + (Y/b) + (Z/c) = 1 \qquad (1.10)$$

Equation (1.10) is the intercept form of the equation of the plane *ABC*, and may be compared with (1.6).

### 1.2.3  Miller Indices

The faces of a crystal are planes in three dimensions. Once the crystallographic axes are chosen, a parametral plane may be defined and any other plane described in terms of three numbers $h$, $k$, and $l$. It is an experimental

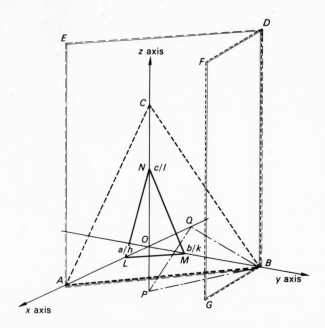

FIGURE 1.13. Miller indices of planes: $OA = a$, $OB = b$,
$OC = c$.

fact that, in crystals, if the parametral plane is designated by integral values
of $h$, $k$, and $l$, usually (111), then the indices of all other crystal faces are
integers. This notation for describing the faces of a crystal was introduced
first by Miller in 1839, and $h$, $k$, and $l$ are called Miller indices.

In Figure 1.13, let the parametral plane (111) be $ABC$, making intercepts $a$, $b$, and $c$ on the crystallographic axes. Another plane, $LMN$, makes
intercepts $a/h$, $b/k$, and $c/l$ along the $X$, $Y$, and $Z$ axes, respectively. Its
Miller indices are expressed by the ratios of the intercepts of the parametral
plane to those of the plane $LMN$. If in the figure, $a/h = a/4$, $b/k = b/3$, and
$c/l = c/2$, then $LMN$ is (432). If fractions occur in $h$, $k$, or $l$, they are
cleared by multiplication throughout by the lowest common denominator.
Conditions of parallelism to axes and intercepts on the negative sides of the
axes lead to zero or negative values for $h$, $k$, and $l$. Thus, $ABDE$ is (110),
$BDFG$ is (010), and $PBQ$ is ($\bar{2}1\bar{3}$). It may be noted that it has not been
necessary to assign numerical values to either $a$, $b$, and $c$ or $\alpha$, $\beta$, and $\gamma$ in
order to describe the crystal faces. In the next chapter we shall identify $a$, $b$,
and $c$ with the edges of the crystal unit cell in a lattice, but this relationship is
not needed at present.

The preferred choice of the parametral plane leads to small numerical values for the Miller indices of crystal faces. Rarely are $h$, $k$, and $l$ greater than 4. If $LMN$ had been chosen as (111), then $ABC$ would have been (346). Summarizing, we may say that the plane $(hkl)$ makes intercepts $a/h$, $b/k$, and $c/l$ along the crystallographic $X$, $Y$, and $Z$ axes, respectively, where $a$, $b$, and $c$ are the corresponding intercepts made by the parametral plane.

From (1.10), the intercept equation of the plane $(hkl)$ may be written as

$$(hX/a)+(kY/b)+(lZ/c) = 1 \qquad (1.11)$$

The equation of the parallel plane passing through the origin is

$$(hX/a)+(kY/b)+(lZ/c) = 0 \qquad (1.12)$$

since it must satisfy the condition $X = Y = Z = 0$.

Miller–Bravais Indices

In crystals that exhibit sixfold symmetry (see Table 1.3), four axes of reference are used. The axes are designated $X$, $Y$, $U$, and $Z$; the $X$, $Y$, and $U$ axes lie in one plane, at 120° to one another, and the $Z$ axis is perpendicular to the $XYU$ plane (Figure 1.14). As a consequence, planes in these crystals are described by four numbers, the Miller–Bravais indices $h$, $k$, $i$, and $l$. The

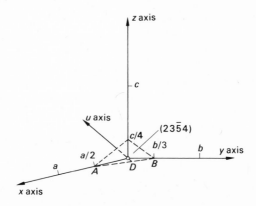

FIGURE 1.14. Miller–Bravais indices $(hkil)$. The crystallographic axes are labeled $X$, $Y$, $U$, $Z$, and the plane $(23\bar{5}4)$ is shown; the parametral plane is $(11\bar{2}1)$.

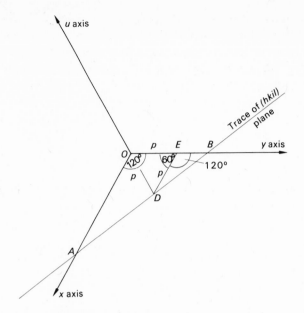

FIGURE 1.15.   Equivalence of $i$ and $-(h+k)$.

index $i$ is not independent of $h$ and $k$: thus, if the plane $ABC$ in Figure 1.14 intercepts the $X$ and $Y$ axes at $a/2$ and $b/3$, for example, then the $U$ axis is intercepted at $-u/5$. If the $Z$ axis is intercepted at $c/4$, the plane is designated $(23\bar{5}4)$. In general, $i = -(h+k)$, and the parametral plane is $(11\bar{2}1)$.

We can show that $i = -(h+k)$ with reference to Figure 1.15. From the definition of Miller indices,

$$OA = a/h \qquad\qquad (1.13)$$

$$OB = b/k \qquad\qquad (1.14)$$

Let the intercept of $(hkil)$ on the $U$ axis be $p$. Draw $DE$ parallel to $AO$. Since $OD$ bisects $\widehat{AOB}$,

$$\widehat{AOD} = 60° \qquad\qquad (1.15)$$

Hence, $\triangle ODE$ is equilateral, and

$$OD = DE = OE = p \qquad\qquad (1.16)$$

Triangles *EBD* and *OBA* are similar. Hence,

$$\frac{EB}{ED} = \frac{OB}{OA} = \frac{b/k}{a/h} \qquad (1.17)$$

But

$$EB = b/k - p \qquad (1.18)$$

Hence,

$$\frac{b/k - p}{p} = \frac{b/k}{a/h} \qquad (1.19)$$

or

$$p = \frac{ab}{ak + bh} \qquad (1.20)$$

Since $a = b = u$ in crystals with sixfold symmetry (hexagonal), and writing $p$ as $-u/i$, we have

$$i = -(h + k) \qquad (1.21)$$

An alternative, geometric, approach to this result consists in drawing the traces of any family of $(hkil)$ planes from the origin to $+u$, whence it will be clear that $i = -(h + k)$. This construction may be appreciated fully after a study of Chapter 2.

### 1.2.4   Axial Ratios

If both sides of (1.12) are multiplied by $b$, we obtain

$$\frac{hX}{a/b} + kY + \frac{lZ}{c/b} = 0 \qquad (1.22)$$

The quantities $a/b$ and $c/b$ are termed axial ratios; they can be deduced from the crystal morphology.

### 1.2.5   Zones

Most well-formed crystals have their faces arranged in groups of two or more with respect to certain directions in the crystal. In other words, crystals exhibit symmetry; this feature is an external manifestation of the ordered arrangement of atoms in the crystal. Figure 1.16 illustrates zircon, $ZrSiO_4$, an example of a highly symmetric crystal. It is evident that several faces have a given direction in common. Such faces are said to lie in a zone, and the common direction is called the zone axis. Two faces, $(h_1k_1l_1)$ and $(h_2k_2l_2)$, define a zone. The zone axis is the line of intersection of the two planes, and is given by the solution of the equations

$$(h_1X/a) + (k_1Y/b) + (l_1Z/c) = 0 \qquad (1.23)$$

and

$$(h_2X/a) + (k_2Y/b) + (l_2Z/c) = 0 \qquad (1.24)$$

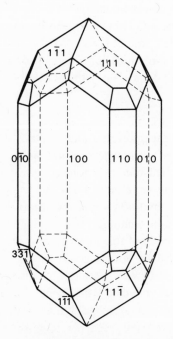

FIGURE 1.16.   A highly symmetric crystal (zircon, $ZrSiO_4$).

that is, by the line

$$\frac{X}{a(k_1 l_2 - k_2 l_1)} = \frac{Y}{b(l_1 h_2 - l_2 h_1)} = \frac{Z}{c(h_1 k_2 - h_2 k_1)} \tag{1.25}$$

which passes through the origin. It may be written as

$$X/(aU) = Y/(bV) = Z/(cW) \tag{1.26}$$

where $[UVW]$ is called the zone symbol.

If any other face $(hkl)$ lies in the same zone as $(h_1 k_1 l_1)$ and $(h_2 k_2 l_2)$, then it may be shown, from (1.12) and (1.26), that

$$hU + kV + lW = 0 \tag{1.27}$$

which is a symbolic expression of the Weiss zone law.

In the zircon crystal, the vertical (prism) faces lie in one zone. If the prism faces are indexed in the usual manner (see Figure 1.16), then, from (1.25) and (1.26), the zone symbol is $[001]$. From (1.27), we see that $(\bar{1}\bar{1}0)$ is a face in the same zone, but $(111)$ is not. In the manipulation of (1.25) and (1.26), it may be noted that the zone axis is described by $[UVW]$, the simplest symbol; the directions that may be described as $[nU, nV, nW]$ $(n = 0, \pm 1, \pm 2, \ldots)$ are coincident with $[UVW]$.

## Interfacial Angles

The law of constant interfacial angles states that in all crystals of the same substance, angles between corresponding faces have a constant value. Interfacial angles are measured by a goniometer, the simplest form of which is the contact goniometer (Figure 1.17). In using this instrument, large crystals are needed, a condition not easily obtainable in practice.

An improvement in technique was brought about by the reflecting goniometer. The principle of this instrument is shown in Figure 1.18, and forms the basis of modern optical goniometers. A crystal is arranged to rotate about a zone axis $O$, which is set perpendicular to a plane containing the incident and crystal-reflected light beams. Parallel light reflected from the face $AB$ is received by a telescope. If the crystal is rotated in a clockwise direction, a reflection from the face $BC$ is received next when the crystal has been turned through the angle $\Phi$ and the interfacial angle is $180 - \Phi$ degrees.

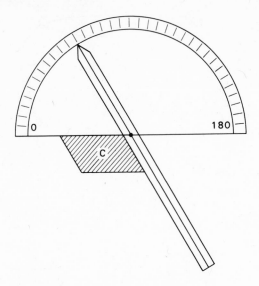

FIGURE 1.17. Contact goniometer with a
crystal (C) in the measuring position.

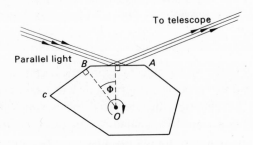

FIGURE 1.18. Principle of the reflecting
goniometer.

Accurate goniometry brought a quantitative significance to observable
angular relationships in crystals.

## 1.3  Stereographic Projection

The general study of the external features of crystals is called crystal
morphology. The analytical description of planes and zones given above is
inadequate for a simultaneous appreciation of the many faces exhibited by a

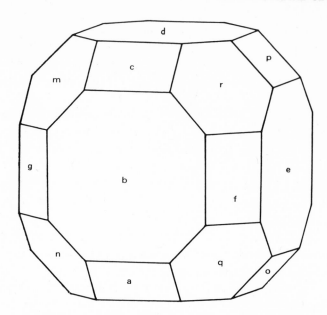

FIGURE 1.19. Cubic crystal showing three forms of planes: cube—*b*, *e*, *d*, and parallel faces; octahedron—*r*, *m*, *n*, *q*, and parallel faces; rhombic dodecahedron—*f*, *g*, *p*, *o*, *c*, *a*, and parallel faces.

crystal. It is necessary to be able to represent a crystal by means of a two-dimensional drawing, while preserving certain essential properties. For a study of crystal morphology, the interfacial angles, which are a fundamental feature of crystals, must be maintained in plane projection, and the stereographic projection is useful for this purpose. Furthermore, with imperfectly formed crystals, the true symmetry may not be apparent by inspection. In favorable cases, the symmetry may be completely revealed by a stereographic projection of the crystal. We shall develop this projection with reference to the crystal shown in Figure 1.19.

This crystal belongs to the cubic system (page 33): The crystallographic reference axes $X$, $Y$, and $Z$ are mutually perpendicular, and the parametral plane (111) makes equal intercepts $(a = b = c)$ on these axes. The crystal shows three forms of planes. In crystallography, a form of planes, represented by $\{hkl\}$, refers to the set of planes that are equivalent under the point-group symmetry of the crystal (see page 25*ff*). The crystal under discussion shows the cube form $\{100\}$—six faces (100), ($\bar{1}$00), (010), (0$\bar{1}$0), (001), and (00$\bar{1}$); the octahedron $\{111\}$—eight faces; and the rhombic dodecahedron {110}—12 faces. Each face on the crystal drawing has a related parallel face on the actual crystal, for example, $b$ (shown) and $b'$. The reader may care to list the sets of planes in the latter two cubic forms; the answer will evolve from the discussion of the stereographic projection of the crystal.

From a point within the crystal, lines are drawn normal to the faces of the crystal. A sphere of arbitrary radius is described about the crystal, its center being the point of intersection of the normals which are then produced to cut the surface of the sphere (Figure 1.20).

In Figure 1.21, the plane of projection is $ABCD$, and it intersects the sphere in the primitive circle. The portion of the plane of projection enclosed by the primitive circle is the primitive plane, or primitive. The point of intersection of each normal with the upper hemisphere is joined to the lowest point $P$ on the sphere. The intersection of each such line with the primitive is the stereographic projection, or pole, of the corresponding face on the crystal, and is indicated by a dot on the stereographic projection, or stereogram.

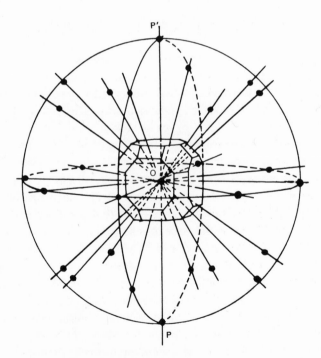

FIGURE 1.20. Spherical projection of the crystal in Figure 1.19.

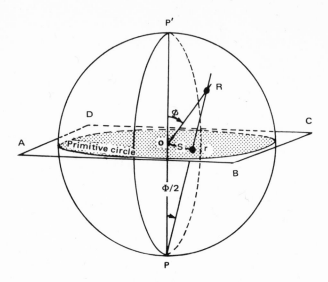

FIGURE 1.21. Development of the stereographic projection
(stereogram) from the spherical projection.

If the crystal is oriented such that the normal to face $d$ (and $d'$) coincides
with $PP'$ in the sphere, then the normals to the zone $e, f, b, \ldots, g'$ lie in the
plane of projection and intersect the sphere on the primitive circle. In order
to avoid increasing the size of the stereogram unduly, the intersections of the
face normals with the lower hemisphere are joined to the uppermost point $P'$
on the sphere and their poles are indicated on the stereogram by an open
circle.

The completed stereogram is illustrated by Figure 1.22. The poles now
should be compared with their corresponding faces on the crystal drawing. A
fundamental property of the stereogram is that all circles drawn on the
sphere project as circles. Thus, the curve $G_1 G_1'$ is an arc of a circle;
specifically, it is the projection of a great circle that is inclined to the plane of
projection. A great circle is the trace, on the sphere, of a plane that passes
through the center of the sphere; it may be likened to a meridian on the
globe of the world. Limiting cases of inclined great circles are the primitive
circle, which lies in the plane of projection, and straight lines, such as $G_2 G_2'$,
which are projections of great circles lying normal to the plane of projection.
All poles on one great circle represent faces lying in one and the same zone.

Circles formed on the surface of the sphere by planes that do not pass
through the center of the sphere are called small circles; they may be likened
to parallels of latitude on the globe.

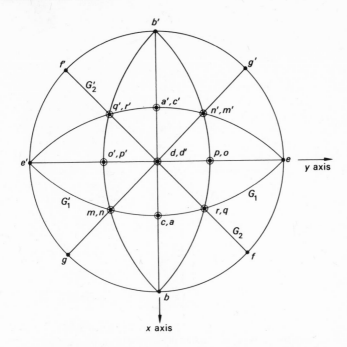

FIGURE 1.22. Stereogram of the crystal in Figure 1.19. The zone circle (great circle) $G_1 G_1'$ passes through $e, q, a, n, e', q', a', n'$; the zone circle $G_2 G_2'$ passes through $f, r, d, q', f', r', d', q$.

In order to construct Figure 1.22, the following practical principles must be followed. The interfacial angles are measured in zones. If an optical goniometer is used, the angle $\Phi$ (see Figure 1.18) is plotted directly on the stereogram. Although $\Phi$ is the angle between the normals to planes, it is often called the interfacial angle in this context. Next, the crystal orientation with respect to the sphere is chosen: for example, let zone $b, f, e, \ldots$ be on the primitive circle, and zone $b, c, d, \ldots$ run from bottom to top of the projection. Since $\widehat{bf}$, the angle between faces $b$ and $f$, is $45°$, zone $f, r, d, \ldots$ can be located on the stereogram.

The distance $S$ of the pole $r$ from the center of the stereogram is given by

$$S = \rho \tan(\Phi/2) \tag{1.28}$$

where $\rho$ is the radius of the stereogram and $\Phi$ is the interfacial angle $\widehat{dr}$. This equation may be examined by means of Figure 1.23.

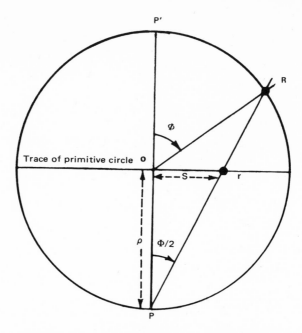

FIGURE 1.23. Evaluation of the stereographic distance $S$
for the pole $r$.

A simple graphical method, employing a Wulff net* (Figure 1.24), is
often sufficiently accurate to locate poles on a stereogram. On this net, the
curves running from top to bottom are projected great circles; curves
running from left to right are projected small circles. In use, the center of the
net is pivoted at the center of the stereogram, the interfacial angle measured
along the appropriate great circle, and the pole plotted. The pole of face $r$
lies at the intersection of two zone circles, $e, r, c, \ldots$ and $f, r, d, \ldots$; if $\widehat{dc}$ and
$\widehat{bf}$, for example, are known, $r$ can be located. Interfacial angles may be
measured on a stereogram by aligning a Wulff net in the manner described,
rotating it until the poles in question lie on the same great circle, and then
reading directly the angle.

The completed stereogram (Figure 1.22) may now be indexed. The
parametral plane is chosen as face $r$ (the parametral plane must intersect all
three crystallographic axes), and the remaining faces are then allocated $h, k$,
and $l$ values (Figure 1.25). It is not necessary to write the indices for both

* Obtainable from one of the authors (M.F.C.L.).

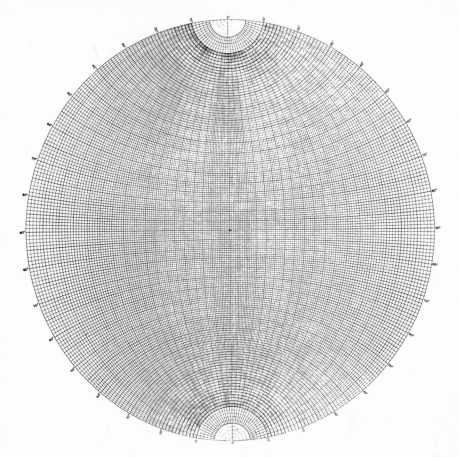

FIGURE 1.24. Wulff net.

poles at the same point on the stereogram. If the dot is *hkl*, then we know that the open circle is, in general, $hk\bar{l}$. Figure 1.26 shows the crystal of Figure 1.19 again, but with the Miller indices inserted for direct comparison with its stereogram.

We shall not be concerned here with any further development of the stereogram.* The angular truth of the stereographic projection makes it very suitable for representing not only interfacial angles, but also symmetry directions, point groups, and bond directions in molecules and ions.

---

* For a fuller discussion, see Bibliography.

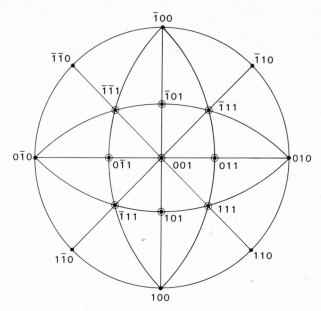

FIGURE 1.25. Stereogram in Figure 1.22 indexed, taking $r$ as 111. The zone containing (100) and (111) is [01$\bar{1}$], and that containing (010) and (001) is [100]; the face $p$ common to these two zones is (011)—see (1.25) to (1.27).

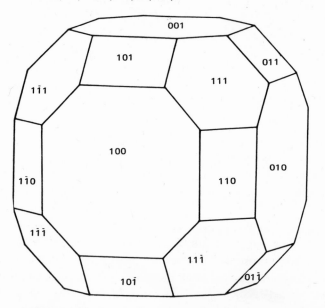

FIGURE 1.26. Crystal of Figure 1.19 with Miller indices inserted.

# 1.4   External Symmetry of Crystals

Often, the faces on a crystal are arranged in groups of two or more in a similar orientation with respect to some line or plane in the crystal. In other words, the crystal exhibits symmetry. The crystal drawing of zircon in Figure 1.16 shows several sets of symmetrically arranged faces.

Symmetry is a property by means of which an object is brought into self-coincidence by a certain operation. For our purposes, the operation will be considered to take place in space and to represent an action about a symmetry element. A symmetry element may be thought of as a geometric entity which generates symmetry operations, that is, a point, line, or plane with respect to which symmetry operations may be performed. The object under examination may contain more than one symmetry element. A collection of interacting symmetry elements is called a point group, which may be defined as a set of symmetry elements, the operation about which leaves at least one point unmoved. This point is the origin, and all symmetry elements in a point group pass through this point. We shall introduce the ideas of symmetry elements first in two dimensions. Self-coincidence will be judged by appearance, by performing or imagining the symmetry operation, or by measuring interfacial angles.

It should be noted that the symmetry of a crystal may be different with respect to different physical properties, such as optical refraction, magnetism, or photoelasticity. We are concerned with the symmetry of crystals as revealed by optical or X-ray goniometry. It may be argued that, because of imperfections in real crystals, self-coincidence can be obtained *only* by a 360° rotation about any line in the crystal, the equivalent of doing nothing, and that when we speak of self-coincidence, we mean an apparent self-coincidence judged by the measuring property. Rather than enter the hypothetical realm of conceptual, geometrically perfect crystals, we shall note that for most practical purposes, the effects of crystal imperfections on symmetry observations are of a very small order, and we shall employ the term self-coincidence with this understanding.

## 1.4.1   Two-Dimensional Point Groups

If we examine various two-dimensional objects (Figure 1.27), we can discover two types of symmetry element that can bring an object into self-coincidence: Parts (a)–(e) of Figure 1.27 depict rotational symmetry, whereas (f) shows reflection symmetry.

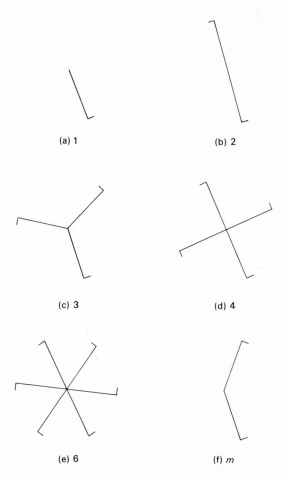

FIGURE 1.27. Two-dimensional objects and their point groups. The pattern is generated from the asymmetric unit (a) by operating on it with the symmetry elements of the point group.

## Rotation Symmetry

An object possesses rotational symmetry of degree $R$ (or $R$-fold symmetry) if it is brought into self-coincidence for each rotation of $(360/R)$ degrees about the symmetry point, or rotation point. Figures 1.27a–e illustrate the rotational symmetry elements $R$ of 1, 2, 3, 4, and 6, respectively. The onefold element is the identity element, and is trivial; every object has onefold symmetry.

Reflection Symmetry

An object possesses reflection symmetry, symbol $m$, in two dimensions if it is brought into self-coincidence by reflection across a line. Reflection (mirror) symmetry is a nonperformable operation. We can make an object like Figure 1.27d and rotate it into self-coincidence. On the other hand, we cannot physically reflect Figure 1.27f across a symmetry line bisecting the figure. We can imagine, however, that this line divides the figure into its asymmetric unit, ∨, and the mirror image or enantiomorph of this part, ∧, which situation is characteristic of reflection symmetry.

Each of the objects in Figure 1.27 has a symmetry pattern which can be described by a two-dimensional point group, and it is convenient to illustrate these point groups by stereograms. Figure 1.28 shows stereograms for the

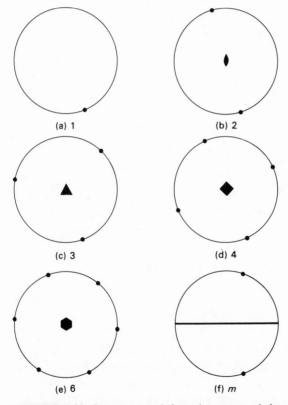

(a) 1          (b) 2

(c) 3          (d) 4

(e) 6          (f) $m$

FIGURE 1.28. Stereograms of the point groups of the objects in Figure 1.27; the conventional graphic symbols for $R$ and $m$ are shown.

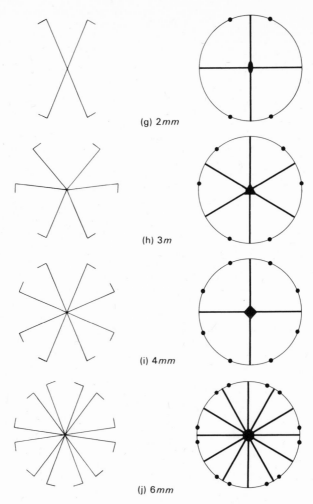

FIGURE 1.29. Other two-dimensional objects with their
stereograms and point groups.

two-dimensional point groups 1, 2, 3, 4, 6, and *m*. It should be noted that in
using stereogram-like drawings to illustrate two-dimensional symmetry, the
representative points (poles) are placed on the perimeter; such situations
may represent special forms (page 37) on stereograms of three-dimensional
objects.

Combinations of *R* and *m* lead to four more point groups; they are
illustrated in Figure 1.29. We have deliberately omitted point groups in
which *R* = 5 and *R* ≥ 7; the reason for this choice will be discussed later.

TABLE 1.1  Two-Dimensional Point Groups

| System | Point groups | Symbol meaning, appropriate to position occupied | | |
| --- | --- | --- | --- | --- |
| | | First position | Second position | Third position |
| Oblique | 1, 2 | Rotation about a point | — | — |
| Rectangular | $1m^a$ | { Rotation about | $m \perp X$ | — |
| | $2mm$ | a point } | $m \perp X$ | $m \perp Y$ |
| Square | 4 | { Rotation about | — | — |
| | $4mm$ | a point } | $m \perp X, Y$ | $m$ at 45° to $X, Y$ |
| Hexagonal | 3 | | — | — |
| | $3m$ | { Rotation about | $m \perp X, Y, U$ | — |
| | 6 | a point } | — | — |
| | $6mm$ | | $m \perp X, Y, U$ | $m$ at 30° to $X, Y, U$ |

$^a$ The full symbol is given here in order to clarify the location of the positions.

It is convenient to allocate the ten two-dimensional point groups to two-dimensional systems, and to choose reference axes in close relation to the directions of the symmetry elements. Table 1.1 lists these systems, together with the meanings of the three positions in the point-group symbols. It should be noted that combinations of $m$ with $R$ ($R \geqslant 2$) introduce additional reflection lines of a different crystallographic form. In the case of $3m$ however, these additional $m$ lines are coincident with the first set; the symbol $3mm$ is not used.

It is important to *remember* the relative orientations of the symmetry elements in the point groups, and the variations in the meanings of the positions in the different systems. In the two-dimensional hexagonal system, three axes are chosen in the plane; this selection corresponds with the use of Miller–Bravais indices in three dimensions (page 12).

## 1.4.2  Three-Dimensional Point Groups

The symmetry elements encountered in three dimensions are rotation axes ($R$), inversion axes ($\bar{R}$), and a reflection (mirror) plane ($m$). A center of symmetry can be invoked also, although neither this symmetry element nor the $m$ plane is independent of $\bar{R}$.

The operations of rotation and reflection are similar to those in two dimensions, except that the geometric extensions of the operations are now increased to rotation about a line and reflection across a plane.

Inversion Axes

An object is said to possess an inversion axis $\bar{R}$ (read as bar-$R$), if it is brought into self-coincidence by the combination of a rotation of $(360/R)$ degrees and inversion through the origin. Like the mirror plane, the inversion axis depicts a nonperformable symmetry operation, and it may be represented conveniently on a stereogram. It is a little more difficult to envisage this operation than those of rotation and reflection. Figure 1.30 illustrates a crystal having a vertical $\bar{4}$ axis: the stereoscopic effect can be created by using a stereoviewer or, with practice, by the unaided eyes (see Appendix A.1).

In representing three-dimensional point groups, it is helpful to indicate the third dimension on the stereogram, and, in addition, to illustrate the change-of-hand relationship that occurs with $\bar{R}$ (including $m$) symmetry elements. For example, referring to Figure 1.31, the element 2 lying in the plane of projection, and the element $\bar{4}$ normal to the plane of projection, when acting on a point derived from the upper hemisphere (symbol ●) both move the point into the lower hemisphere region (symbol ○). Both operations involve a reversal of the sign of the vertical coordinate, but only $\bar{4}$ involves also a change of hand, and this distinction is not clear from the conventional notation. Consequently, we shall adopt a symbolism, common to three-dimensional space groups, which will effect the necessary distinction.

A representative point in the positive hemisphere will be shown by $\bigcirc^+$, signifying, for example, the face $(hkl)$. A change of hemisphere, to $(h'k'\bar{l})^*$, will be indicated by $\bigcirc^-$, and a change of hand by reflection or inversion by $\odot^+$ or $\odot^-$ (see Figure 1.32). This notation may appear to nullify the purpose of a stereogram. However, although the stereogram is a two-dimensional diagram, it should convey a three-dimensional impression, and the notation is used as an aid to this purpose.

Crystal Classes

There are 32 crystal symmetry classes, each characterized by a point group. They comprise the symmetry elements $R$ and $\bar{R}$, taken either singly or in combination, with $R$ restricted to the values 1, 2, 3, 4, and 6. A simple explanation for this restriction is that figures based only on these rotational symmetries can be stacked together to fill space completely, as Figure 1.33 shows.

\* $h'$, $k'$ may or may not be the same as $h$, $k$.

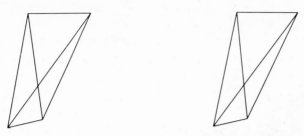

FIGURE 1.30. Stereoscopic pair of idealized tetragonal crystals. The vertical direction in the crystal is the $\bar{4}$ axis.

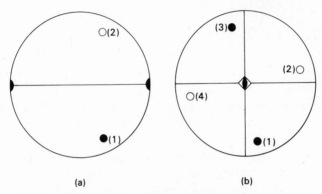

FIGURE 1.31. Stereograms of general forms: (a) point group 2 (axis horizontal and in the plane of the stereogram), (b) point group $\bar{4}$ (axis normal to the plane of the stereogram). In (a), the point ● is rotated through 180° to ○: (1) → (2). In (b), the point ● is rotated through 90° and then inverted through the origin to ○; this combined operation generates, in all, four symmetry-equivalent points: (1) → (4) → (3) → (2).

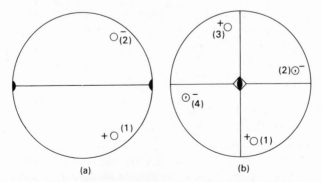

FIGURE 1.32. Stereograms from Figure 1.31 in the revised notation; the different natures of points (2) in (a) and (2) and (4) in (b) with respect to point (1) are clear.

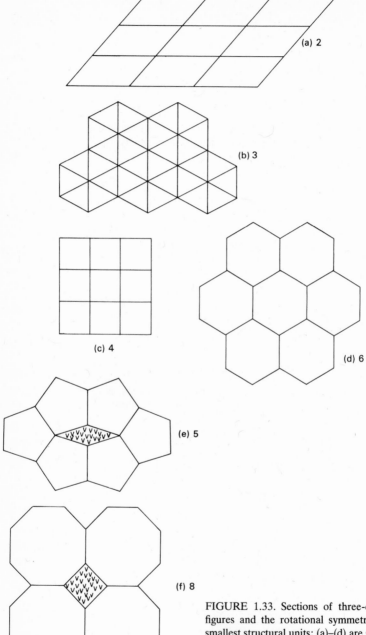

FIGURE 1.33. Sections of three-dimensional figures and the rotational symmetries of their smallest structural units; (a)–(d) are space-filling patterns. In (e) and (f) the v-marks represent voids in the pattern (see also Problem 1.11).

TABLE 1.2. Three-Dimensional Symmetry Symbols

| Symbol | Name | Action for self-coincidence | Graphic symbol |
|---|---|---|---|
| 1 | Monad | 360° rotation; identity | None |
| 2 | Diad | 180° rotation | ● ⊥ projection    ( ‖ projection |
| 3 | Triad | 120° rotation | ▲ ⊥ or inclined to projection |
| 4 | Tetrad | 90° rotation | ◆ ⊥ projection    ■ ‖ projection |
| 6 | Hexad | 60° rotation | ⬢ ⊥ projection |
| $\bar{1}$ | Inverse monad | Inversion[a] | ○ |
| $\bar{3}$ | Inverse triad | 120° rotation + inversion | △ ⊥ or inclined to projection |
| $\bar{4}$ | Inverse tetrad | 90° rotation + inversion | ◈ ⊥ projection    ▯ ‖ projection |
| $\bar{6}$ | Inverse hexad | 60° rotation + inversion | ⬡ ⊥ projection |
| $m$ | Mirror plane[b] | Reflection across plane | — ⊥ projection    ( ‖ projection |

[a] $R$ is equivalent to $R$ plus $\bar{1}$ only where $R$ is an odd number: $\bar{1}$ represents the center of symmetry, but $\bar{2}, \bar{4}$, and $\bar{6}$ are not centrosymmetric point groups.
[b] The symmetry elements $m$ and $\bar{2}$ produce an equivalent operation.

We shall not be concerned here to derive the crystallographic point groups—and there are several ways in which it can be done—but to give, instead, a scheme which allows them to be worked through simply and adequately for present purposes.

The symbols for rotation and reflection are similar to those used in two dimensions. Certain additional symbols are required in three dimensions, and Table 1.2 lists them all.

Figure 1.34a shows a stereogram for point group $m$. The inverse diad is lying normal to the $m$ plane. A consideration of the two operations in the given relative orientations shows that they produce equivalent actions. It is conventional to use the symbol $m$ for this operation, although sometimes it is helpful to employ $\bar{2}$ instead. Potassium tetrathionate (Figure 1.34b) crystallizes in point group $m$.

Crystal Systems and Point-Group Scheme

The broadest classification of crystals is carried out in terms of rotation axes and inversion axes. Crystals are grouped into seven systems according

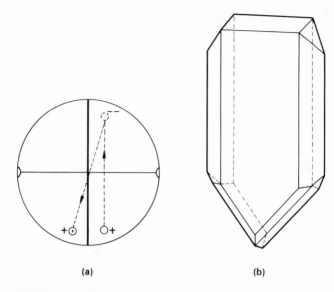

(a)                                             (b)

FIGURE 1.34. Point group $m$: (a) stereogram showing equivalence of $m$ and $\bar{2}$ (the symbol $()$ for $\bar{2}$ is not conventional); (b) crystal of potassium tetrathionate ($K_2S_4O_6$), point group $m$.

TABLE 1.3.  The Seven Crystal Systems

| System | Characteristic symmetry axes, with their orientation | Parametral plane intercepts and interaxial angles, assuming the simplest indexing of faces[a,b] |
|---|---|---|
| Triclinic | None | $a \neq b \neq c$; $\alpha \neq \beta \neq \gamma \neq 90°$, $120°$ |
| Monoclinic | One 2 or $\bar{2}$ axis along $Y$ | $a \neq b \neq c$; $\alpha = \gamma = 90°$; $\beta \neq 90°$, $120°$ |
| Orthorhombic | Three mutually perpendicular 2 or $\bar{2}$ axes[c] along $X$, $Y$, and $Z$ | $a \neq b \neq c$; $\alpha = \beta = \gamma = 90°$ |
| Tetragonal | One 4 or $\bar{4}$ axis along $Z$ | $a = b \neq c$; $\alpha = \beta = \gamma = 90°$ |
| Trigonal[d] | One 3 axis along $Z$ | $a = b \neq c$; $\alpha = \beta = 90°$; |
| Hexagonal | One 6 or $\bar{6}$ axis along $Z$ | $\gamma = 120°$ |
| Cubic | Four 3 axes inclined at $54°44'$ ($\cos^{-1}\frac{1}{\sqrt{3}}$) to $X$, $Y$, and $Z$ | $a = b = c$; $\alpha = \beta = \gamma = 90°$ |

[a] We shall see in Chapter 2 that the same constraints apply to conventional unit cells in lattices.

[b] The symbol $\neq$ should be read as "not constrained by symmetry to equal."

[c] It must be remembered that $\bar{2}$ is equivalent to an $m$ plane normal to the $\bar{2}$ axis.

[d] For convenience, the trigonal system is referred to hexagonal axes.

TABLE 1.4. Point-Group Scheme[a]

| Type | Triclinic | Monoclinic | Trigonal | Tetragonal | Hexagonal | Cubic[b] |
|---|---|---|---|---|---|---|
| $R$ | 1 | 2 | 3 | 4 | 6 | 23 |
| $\bar{R}$ | $\bar{1}$ | $m$ | $\bar{3}$ | $\bar{4}$ | $\bar{6}$ | $m3$ |
| $R$ + center | | $2/m$ | | $4/m$ | $6/m$ | |
| | | Orthorhombic | | | | |
| $R2$ | | 222 | 32 | 422 | 622 | 432 |
| $Rm$ | | $mm2$ | $3m$ | $4mm$ | $6mm$ | $\bar{4}3m$ |
| $\bar{R}m$ | | | $\bar{3}m$ | $\bar{4}2m$ | $\bar{6}m2$ | $m3m$ |
| $R2$ + center | | $mmm$ | | $\frac{4}{m}mm$ | $\frac{6}{m}mm$ | |

[a] The reader should consider the implication of the unfilled spaces in this table.
[b] The cubic system is characterized by threefold axes; $R$ refers to the element 2, 4, or $\bar{4}$ here, but 3 is always present.

to characteristic symmetry, as listed in Table 1.3. The characteristic symmetry refers to the minimum necessary for classification of a crystal in a system; a given crystal may contain more than the characteristic symmetry of its system. The conventional choice of crystallographic reference axes leads to special relationships between the intercepts of the parametral plane (111) and between the interaxial angles $\alpha$, $\beta$, and $\gamma$.

A crystallographic point-group scheme is given in Table 1.4, under the seven crystal systems as headings. The main difficulty in understanding point groups lies not in knowing the action of the individual symmetry elements, but rather in appreciating both the relative orientation of the different elements in a point-group symbol and the fact that this orientation changes among the crystal systems according to the principal symmetry axis.* These orientations must be learned: They are the key to point-group and space-group studies.

Table 1.5 lists the meanings of the three positions in the three-dimensional point-group symbols. Tables 1.4 and 1.5 should be studied carefully in conjunction with Figure 1.39.

The reader should not be discouraged by the wealth of convention which surrounds this part of the subject. It arises for two main reasons. There are many different, equally correct ways of describing crystal geometry. For example, the unique axis in the monoclinic system could be chosen along $X$ or $Z$ instead of $Y$, or even in none of these directions. Second, a strict system of notation is desirable for the purposes of concise,

* Rotational axis of highest degree, $R$.

TABLE 1.5. Three-Dimensional Point Groups

| System | Point groups | Symbol meaning, appropriate to position occupied | | |
|---|---|---|---|---|
| | | First position | Second position | Third position |
| Triclinic | $1, \bar{1}$ | One symbol position only, denoting all directions in the crystal | | |
| Monoclinic[a,b] | $2, m, 2/m$ | One symbol position only: 2 or $\bar{2}$ along $Y$ | | |
| Orthorhombic | $222, mm2,$ $mmm$ | 2 and/or $\bar{2}$ along $X$ | 2 and/or $\bar{2}$ along $Y$ | 2 and/or $\bar{2}$ along $Z$ |
| Tetragonal | $4, \bar{4}, 4/m$ $422, 4mm,$ $\bar{4}2m, \dfrac{4}{m}mm$ | 4 and/or $\bar{4}$ along $Z$ | — 2 and/or $\bar{2}$ along $X, Y$ | — 2 and/or $\bar{2}$ at 45° to $X, Y$ in the $XY$ plane |
| Trigonal[c] | $3, \bar{3}$ $32, 3m, \bar{3}m$ | 3 or $\bar{3}$ along $Z$ | — 2 and/or $\bar{2}$ along $X, Y, U$ | — |
| Hexagonal | $6, \bar{6}, 6/m$ $622, 6mm,$ $\bar{6}m2, \dfrac{6}{m}mm$ | 6 and/or $\bar{6}$ along $Z$ | — 2 and/or $\bar{2}$ along $X, Y, U$ | — 2 and/or $\bar{2}$ at 30° to $X, Y, U$ in the $XYU$ plane |
| Cubic | $23, m3$ | 2 and/or $\bar{2}$ along $X, Y, Z$ | 3 or $\bar{3}$ at 54°44′[d] to $X, Y, Z$ | — |
| | $432, \bar{4}3m,$ $m3m$ | 4 and/or $\bar{4}$ along $X, Y, Z$ | | 2 and/or $\bar{2}$ at 45° to $X, Y, Z$ in $XY, YZ$, and $ZX$ planes |

[a] In the monoclinic system, the $Y$ axis is taken as the unique 2 or $\bar{2}$ axis. Since $\bar{2} \equiv m$, then if $\bar{2}$ is along $Y$, the $m$ plane represented by the same position in the point-group symbol is perpendicular to $Y$. The latter comment applies *mutatis mutandis* in other crystal systems. (It is best to specify the orientation of a plane by that of its normal.)

[b] $R/m$ occupies a single position in a point-group symbol.

[c] For convenience, the trigonal system is referred to hexagonal axes.

[d] Actually $\cos^{-1}(1/\sqrt{3})$.

unambiguous communication of crystallographic material. With familiarity, the conventions cease to be a problem.

We shall now consider two point groups in a little more detail in order to elaborate the topics discussed so far.

*Point Group mm2.* We shall see that once we fix the orientations of two of the symmetry elements in this point group, the third is introduced.

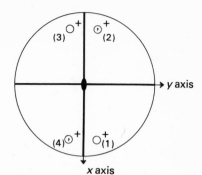

FIGURE 1.35. Stereogram of point group *mm*2.

Referring to Figure 1.35, we start with *mm* as shown. Point (1), in a general position, is reflected across the *m* plane perpendicular to the $X$ axis ($m_X$) to give point (2). This point is now reflected across the second *m* plane to (3). Then either (3) across $m_X$ or (1) across $m_Y$ produces (4). It is evident now that (1) and (3), and (2) and (4), are related by the twofold rotation axis along $Z$.

*Point Group 4mm.* If we start with 4 along $Z$ and *m* perpendicular to $X$, we see straightaway that another *m* plane (perpendicular to $Y$) is required (Figures 1.36a and b); the fourfold axis acts on other symmetry elements in the crystal as well as faces. A general point operated on by the symmetry 4*m* produces eight points in all (Figure 1.36c). The stereogram shows that a second form of *m* planes, lying at 45° to the first set,* is introduced (Figure 1.36d). No further points are introduced by the second set of *m* planes: a fourfold rotation, (1) → (2), followed by reflection across the mirror plane normal to the $X$ axis, (2) → (3), is equivalent to reflection of the original point across the mirror at 45° to $X$, (1) → (3). The reader should now look again at Table 1.5 for the relationship between the positions of the symmetry elements and the point-group symbols, particularly for the tetragonal and orthorhombic systems, from which these detailed examples have been drawn.

In this discussion, we have used a general form, which we may think of as $\{hkl\}$, to represent the point group. Each symmetry-equivalent point lies in a general position (point-group symmetry 1) on the stereogram. Certain crystal planes may coincide with symmetry planes or lie normal to symmetry axes. These planes constitute special forms, and their poles lie in special positions on the stereogram; the forms $\{110\}$ and $\{001\}$ in 4*mm* are examples of special forms. The need for the general form in a correct description of a

---

* This description is not strict; more fully, we may say that the normals to the two forms of *m* planes are at 45° to one another.

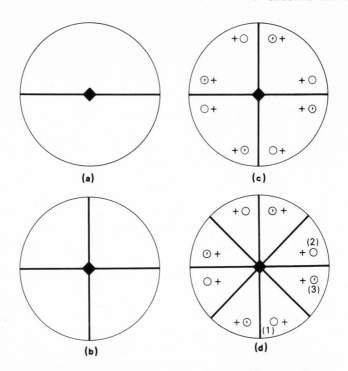

FIGURE 1.36. Intersecting symmetry elements: (a) one *m* plane is inconsistent with intersecting 4; (b) consistent; (c) points generated by 4*m*; (d) complete stereogram (4*mm*).

point group is illustrated in Figure 1.37. The poles of the faces on each of the two stereograms shown are identical, although they may be derived for crystals in different classes, 4*mm* and $\overline{4}2m$ in this example. Figure 1.38 shows crystals of these two classes with the {110} form, among others, developed. In Figure 1.38b, the presence of only special forms led originally to an incorrect deduction of the point group of this crystal.

The stereograms for the 32 crystallographic point groups are shown in Figure 1.39. The conventional crystallographic axes are drawn once for each system. Two comments on the nomenclature are necessary at this stage. The symbol $^{-}\oplus^{+}$ indicates two points, $\bigcirc^{+}$ and $\odot^{-}$, related by a mirror plane in the plane of projection. In the cubic system, the four points related by a fourfold axis in the plane of the stereogram lie on a small circle (Figure 1.40). In general, two of the points are projected from the upper hemisphere and the other two points from the lower hemisphere. We can distinguish them readily by remembering that 2 is a subgroup (page 45) of both 4 and $\overline{4}$.

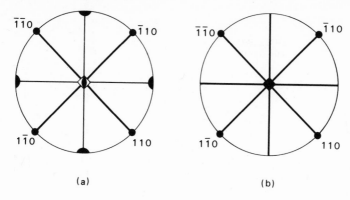

FIGURE 1.37. {110} form in tetragonal point groups: (a) point group $\bar{4}2m$, (b) point group $4mm$.

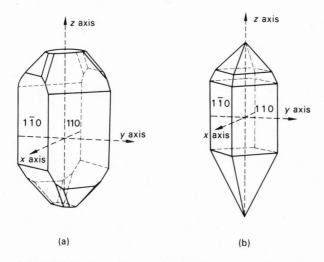

FIGURE 1.38. Tetragonal crystals showing, among others, the {110} form: (a) copper pyrites ($\bar{4}2m$); (b) iodosuccinimide ($4mm$)—X-ray photographs revealed that the true point group is 4.

Appendix A.2 describes a scheme for the study and recognition of the crystallographic point groups. Appendix A.3 discusses the Schoenflies symmetry notation for point groups. Because this system is still in use, we have written the Schoenflies symbols in Figure 1.39, in parentheses, after the Hermann–Mauguin symbol.

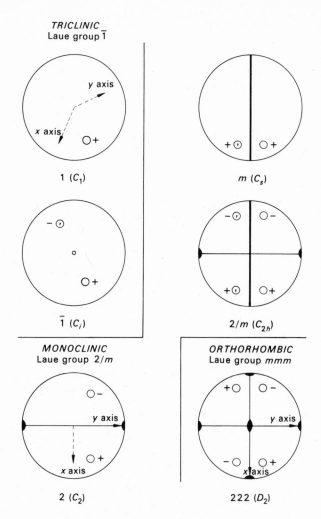

FIGURE 1.39.  Stereograms showing both the symmetry elements and the general form $\{hkl\}$ in the 32 crystallographic point groups. The arrangement is by system and common Laue group. The crystallographic axes are named once for each system and the $Z$ axis is chosen normal to the stereogram. The Schoenflies symbols are given in parentheses.

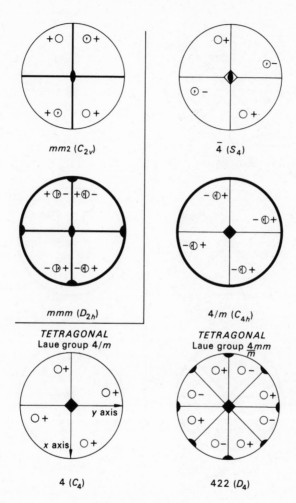

$mm2\ (C_{2v})$

$\bar{4}\ (S_4)$

$mmm\ (D_{2h})$

$4/m\ (C_{4h})$

*TETRAGONAL*
Laue group $4/m$

*TETRAGONAL*
Laue group $\dfrac{4mm}{m}$

$4\ (C_4)$

$422\ (D_4)$

FIGURE 1.39.—*cont.*

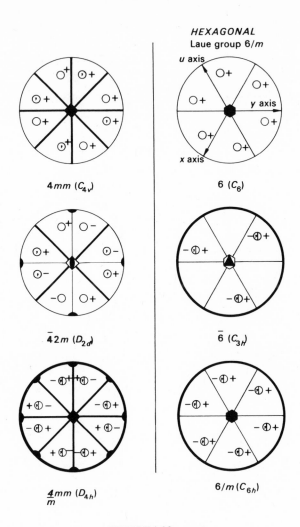

FIGURE 1.39.—*cont.*

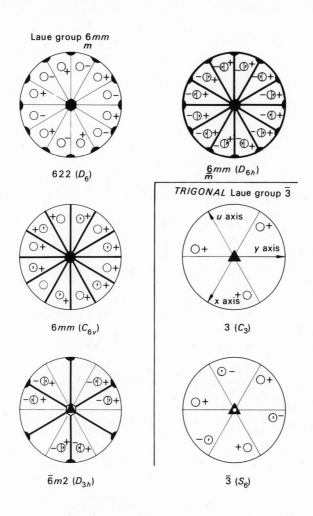

FIGURE 1.39.—*cont.*

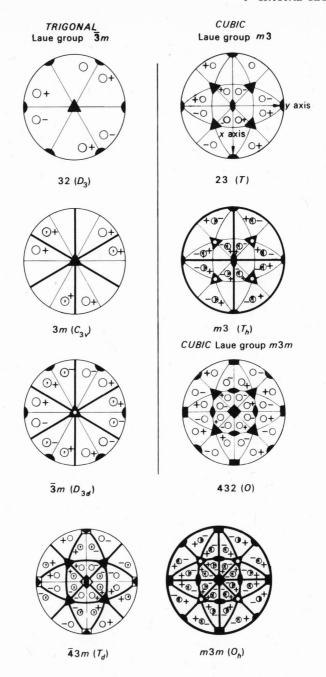

FIGURE 1.39.—*cont.*

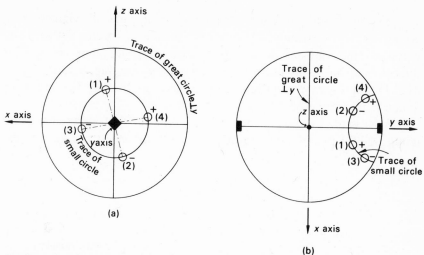

FIGURE 1.40.   Stereogram nomenclature for points related by a fourfold axis ($Y$) lying in the plane of the stereogram. The ± signs refer to the $Z$-axis direction: (a) vertical section normal to the $Y$ axis, (b) corresponding stereogram; the pairs of points (1)–(2) and (3)–(4) are related by twofold symmetry (subgroup of 4).

## Subgroups and Laue Groups

A subgroup of a given point group is a point group of lower symmetry than the given group, contained within it and capable of separate existence as a point group. For example, 32 is a subgroup of $\bar{3}m$, 622, $\bar{6}m2$, $\frac{6}{m}\,mm$, 432, and $m3m$, whereas $\bar{4}$ is a subgroup of $\frac{4}{m}$, $\bar{4}2m$, $\frac{4}{m}\,mm$, $\bar{4}3m$, and $m3m$. The subgroup principle provides a rationale for some of the graphic symbols for symmetry elements. Thus, $\bar{4}$ is shown by a square (fourfold rotation), unshaded (to distinguish it from 4), and with a twofold rotation symbol inscribed (2 is a subgroup of $\bar{4}$).

Point group $\bar{1}$ and point groups that have $\bar{1}$ as a subgroup are centrosymmetric. Since, as we shall see, X-ray diffraction effects are, in general, centrosymmetric, the symmetry pattern of X-ray diffraction spots on a flat-plate film, obtained from any crystal, can exhibit only the symmetry that would be obtained from a crystal having the corresponding centrosymmetric point group.

There are 11 such point groups; they are called Laue groups, since symmetry is often investigated by the Laue X-ray method (page 122). Neither the Laue photograph, however, nor any other X-ray photograph can show directly the presence (or absence) of a center of symmetry in a crystal. In Table 1.6, the point groups are classified according to their Laue group, and the symmetry of the Laue flat-plate film photographs is given for directions of the X-ray beam normal to the crystallographic forms listed. The Laue-projection symmetry corresponds to one of the ten two-dimensional point groups.

What is the Laue-projection symmetry on {120} of a crystal of point group 422? This question can be answered with the stereogram of the

TABLE 1.6. Laue Groups and Laue-Projection Symmetry

| System | Point groups | Laue group | Laue-projection symmetry normal to the given form | | |
|---|---|---|---|---|---|
| | | | {100} | {010} | {001} |
| Triclinic | $1, \bar{1}$ | $\bar{1}$ | 1 | 1 | 1 |
| Monoclinic | $2, m, 2/m$ | $2/m$ | $m$ | 2 | $m$ |
| Orthorhombic | $222, mm2,$ $mmm$ | $mmm$ | $2mm$ | $2mm$ | $2mm$ |
| | | | {001} | {100} | {110} |
| Tetragonal | $4, \bar{4}, 4/m$ | $4/m$ | 4 | $m$ | $m$ |
| | $422, 4mm,$ $\bar{4}2m, \frac{4}{m}mm$ | $\frac{4}{m}mm$ | $4mm$ | $2mm$ | $2mm$ |
| | | | {0001} | {10$\bar{1}$0} | {11$\bar{2}$0} |
| Trigonal[a] | $3, \bar{3}$ | $\bar{3}$ | 3 | 1 | 1 |
| | $32, 3m, \bar{3}m$ | $\bar{3}m$ | $3m$ | $m$ | 2 |
| Hexagonal | $6, \bar{6}, 6/m$ | $6/m$ | 6 | $m$ | $m$ |
| | $622, 6mm,$ $\bar{6}m2, \frac{6}{m}mm$ | $\frac{6}{m}mm$ | $6mm$ | $2mm$ | $2mm$ |
| | | | {100} | {111} | {110} |
| Cubic | $23, m3$ | $m3$ | $2mm$ | 3 | $m$ |
| | $432, \bar{4}3,$ $m3m$ | $m3m$ | $4mm$ | $3m$ | $2mm$ |

[a] Referred to hexagonal axes.

corresponding Laue group, $\frac{4}{m}$ *mm*. Reference to the appropriate diagram in Figure 1.39 shows that an X-ray beam traveling normal to $\{hk0\}$ ($h \neq k$) encounters only *m* symmetry. The entries in Table 1.6 can be deduced in this way. The reader should refer again to Table 1.5 and compare it with Table 1.6.

## Noncrystallographic Point Groups

We have seen that in crystals the elements $R$ and $\bar{R}$ are limited to the numerical values 1, 2, 3, 4, and 6. However, there are molecules that exhibit symmetries other than those of the crystallographic point groups. Indeed, $R$ can, in principle, take any integer value between one and infinity. The statement $R = \infty$ implies cylindrical symmetry; the molecule of carbon monoxide has an $\infty$-axis along the C—O bond, if we assume spherical atoms.

In biscyclopentadienyl ruthenium (Figure 1.41) a fivefold symmetry axis is present, and the point-group symbol may be written as $\overline{10}m2$, or $\frac{5}{m}$ *m*. The stereogram of this point group is shown in Figure 1.42; the symbol for $\frac{5}{m}$ ($\overline{10}$) is not standard.

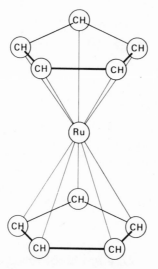

FIGURE 1.41. Biscyclopenta-dienyl ruthenium, $(C_5H_5)_2Ru$.

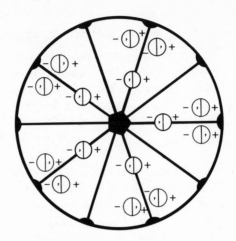

FIGURE 1.42. Stereogram of the noncrystallo-
graphic point group $\overline{10}m2$ ($D_{5h}$) showing the general
form (20 poles) and a special form of 10 poles, lying
on the $m$ planes, which can be used to represent the
CH groups in $(C_5H_5)_2Ru$.

Other examples of noncrystallographic point groups will be encountered among chemical molecules, and always a stereogram can be used to
represent the point-group symmetry. In every such example, however, the
substance will crystallize in one of the seven crystal systems and the crystals
will belong to one of the 32 crystal classes.

## Bibliography

### General and Historical Study of Crystallography

BRAGG, W. L., *A General Survey* (*The Crystalline State*, Vol. I), London,
Bell.

EWALD, P. P. (Editor), *Fifty Years of X-Ray Diffraction*, Utrecht, Oosthoek.

### Crystal Morphology and Stereographic Projection

PHILLIPS, F. C., *An Introduction to Crystallography*, London, Longmans.

## Crystal Symmetry and Point Groups

HENRY, N. F. M., and LONSDALE, K. (Editors), *International Tables for X-Ray Crystallography*, Vol. I, Birmingham, Kynoch Press.

## Problems

**1.1.** The line $AC$ (Figure 1P.1) may be indexed as (12) with respect to the rectangular axes $X$ and $Y$. What are the "indices" of the same line with respect to the axes $X'$ and $Y$, where the angle $\widehat{X'OY} = 120°$? $PQ$ is the parametral line for both sets of axes, and $OB/OA = 2$.

**1.2.** Write the Miller indices for planes that make the intercepts given below:

   (a)   $a, -b/2, \|c.$       (b)   $2a, b/3, c/2.$
   (c)   $\|a, \|b, -c.$        (d)   $a, -b, 3c/4.$
   (e)   $\|a, -b/4, c/3.$    (f)   $-a/4, b/2, -c/3.$

**1.3.** Evaluate zone symbols for the pairs of planes given below:

   (a)   $(123), (0\bar{1}1).$     (b)   $(20\bar{3}), (111).$
   (c)   $(41\bar{5}), (1\bar{1}0).$     (d)   $(\bar{1}1\bar{2}), (001).$

**1.4.** What are the Miller indices of the plane that lies in both the zones [123] and [$\bar{1}1\bar{1}$]? Why are there, apparently, two answers?

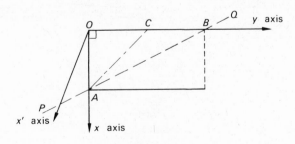

FIGURE 1P.1

**\*1.5.**    Anatase, $TiO_2$, is tetragonal, and the following goniometric measurements have been recorded for the zones shown.

| Zones $a, m, a, \ldots$ | | Zones $m, s, p, \ldots$ | | Zones $a, r, r, \ldots$ | |
|---|---|---|---|---|---|
| $\widehat{am}$ | 45°00′ | $\widehat{ms}$ | 11°14′ | $\widehat{ar}$ | 53°52′ |
| $\widehat{aa}$ | 90°00′ | $\widehat{sp}$ | 10°24′ | $\widehat{rr}$ | 72°15′ |
| | | $\widehat{pr}$ | 11°52′ | | |
| Zones $a, s, s, \ldots$ | | $\widehat{rz}$ | 16°31′ | Zones $a, e, a, \ldots$ | |
| | | $\widehat{zz}$ | 79°53′ | | |
| $\widehat{as}$ | 46°06′ | | | $\widehat{ae}$ | 29°22′ |
| $\widehat{ss}$ | 87°48′ | | | $\widehat{ee}$ | 121°15′ |
| | | Zones $a, p, p, \ldots$ | | | |
| Zones $a, z, z, \ldots$ | | $\widehat{ap}$ | 48°55′ | Zones $m, e, e, \ldots$ | |
| | | $\widehat{pp}$ | 82°10′ | | |
| $\widehat{az}$ | 63°00′ | | | $\widehat{me}$ | 51°56′ |
| $\widehat{zz}$ | 54°00′ | | | $\widehat{ee}$ | 76°07′ |

(a)  Each result is the mean of eight measurements, except for $aa$ where four results have been averaged. Study the morphology of the anatase crystal illustrated in Figure 1P.2. Plot the goniometric measurements on a stereogram, radius 3 in. Let zones $a$, $m$, $a$, ... and $a$, $e$, $e$, ... be on the primitive circle and running from bottom to top, respectively.

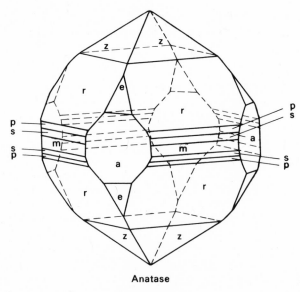

**Anatase**

FIGURE 1P.2

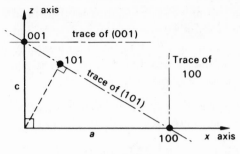

FIGURE 1P.3

(b) From a consideration of the construction in Figure 1P.3, determine the axial ratio $c/a$.

(c) Index the poles on the stereogram.

(d) List the indices of the general forms and the special forms present on this crystal.

(e) Write the point group of anatase.

**1.6.** Take the cover of a matchbox (Figure 1P.4a).

(a) Ignore the label, and write down its point group.

Squash it diagonally (Figure 1P.4b).

(b) What is the point group now?

(c) In each case, what is the point group if the label is not ignored?

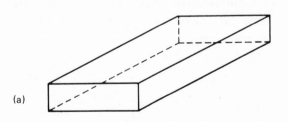

(a)

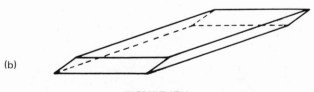

(b)

FIGURE 1P.4

**1.7.** Draw stereograms to show the general form in each of the point groups deduced in Problems 1.6a and 1.6b. Satisfy yourself that in 1.6a three, and in 1.6b two, symmetry operations carried out in sequence produce a resultant action that is equivalent to another operation in the group.

**1.8.** How many planes are there in the forms $\{010\}$, $\{\overline{1}10\}$, and $\{11\overline{3}\}$ in each of the point groups $2/m$, $\overline{4}2m$, and $m3$?

**1.9.** What symmetry would be revealed by Laue flat-film photographs where the X-ray beam is normal to a plane in the form given in each of the examples below?

| | Point group | Orientation | | Point group | Orientation |
|---|---|---|---|---|---|
| (a) | $\overline{1}$ | $\{100\}$ | (f) | $3m$ | $\{11\overline{2}0\}$ |
| (b) | $mmm$ | $\{011\}$ | (g) | $\overline{6}$ | $\{0001\}$ |
| (c) | $m$ | $\{010\}$ | (h) | $\overline{6}m2$ | $\{0001\}$ |
| (d) | 422 | $\{120\}$ | (i) | 23 | $\{111\}$ |
| (e) | 3 | $\{10\overline{1}0\}$ | (j) | 432 | $\{110\}$ |

In some examples, it may help to draw stereograms.

**1.10.** Name each species of molecule or ion in the 10 drawings of Figure 1.P.5 and write down its point-group symbol in both the Hermann–Mauguin and Schoenflies nomenclatures. Study the stereograms in Figure 1.39, and suggest the form $\{hkl\}$, the normals to faces of which may be identified with the following bond directions: (a) Pt—Cl (six bonds), (b) C—O (three bonds), (d) C—H (four bonds), (h) C—S (four bonds). In some examples, it may help to make ball-and-spoke models.*

---

* A simple apparatus for this purpose is marketed by Morris Laboratory Instruments, 480 Bath Road, Slough, Bucks., England. [See also M. F. C. Ladd, Crystal Structure Models, *Education in Chemistry* **5**, 186 (1968).]

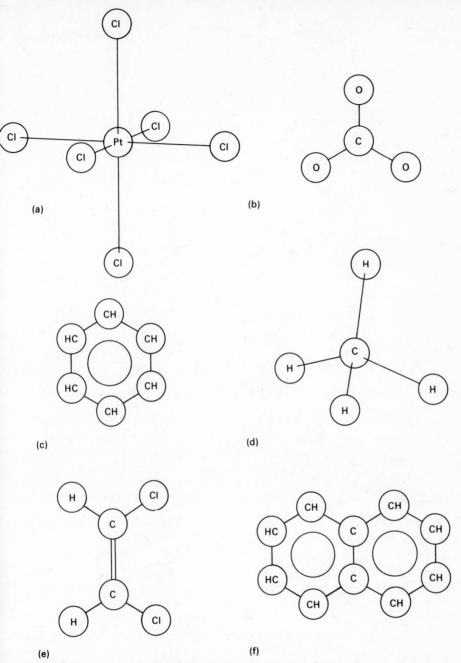

FIGURE 1P.5. (a) $[PtCl_6]^{2-}$. All Cl–Pt–Cl angles are 90°. (b) $CO_3^{2-}$. Planar. (c) $C_6H_6$. Planar. (d) $CH_4$. All H–C–H angles are 109° 28′. (e) CHCl–CHCl. Planar. (f) $C_{10}H_8$. Planar. (g) $SO_4^{2-}$. All O–S–O angles are 109° 28′. (h) $C(SCH_3)_4$. All S–C–S angles are 109° 28′ and all C–S–CH$_3$ angles are 118° 30′. (i) $C_3N_3(N_3)_3$. Planar. (j) CHBrClF. All $X$–C–$Y$ angles are approximately 109° but are not equal.

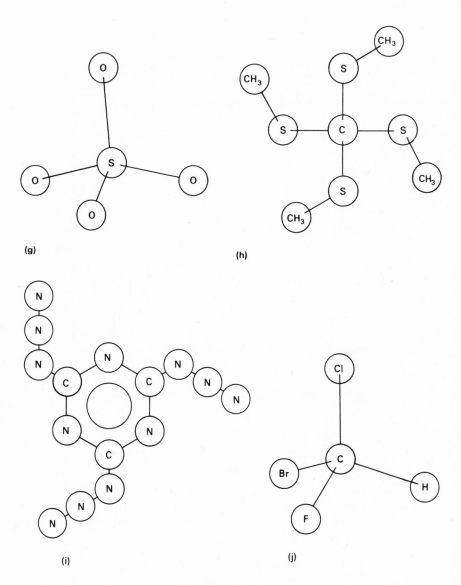

(g)

(h)

(i)

(j)

FIGURE 1P.5—*cont.*

**1.11.** What is the nontrivial symmetry of the figure obtained by packing a number of equivalent but irregular quadrilaterals in one plane?

# 2

# Crystal Geometry. II

## 2.1 Introduction

In this chapter, we continue our study of crystal geometry by investigating the internal arrangements of crystalline materials. Crystals are characterized by periodicities in three dimensions. An atomic grouping is repeated over and over again by a certain symmetry mechanism so as to build up a crystal, and thus we are led to a consideration of space-group symmetry. A space-group pattern in its simplest form may be considered to be derived by repeating a motif having a crystallographic point-group symmetry by the translations of a lattice, and so it is appropriate to examine lattices next.

## 2.2 Lattices

Every crystal has a lattice as its geometric basis. A lattice may be described as a regular, infinite arrangement of points in which every point has the same environment as any other point. This description is applicable, equally, in one-, two-, and three-dimensional space.

Crystal lattice geometry is described in relation to three basic repeat, or translation, vectors $\mathbf{a}$, $\mathbf{b}$, and $\mathbf{c}$. Any point in the lattice may be chosen as an origin, whence a vector $\mathbf{r}$ to any other lattice point is given by

$$\mathbf{r} = U\mathbf{a} + V\mathbf{b} + W\mathbf{c} \tag{2.1}$$

where $U$, $V$, and $W$ are positive or negative integers or zero, and represent the coordinates of the lattice point. The direction, or directed line, joining the origin to the points $U$, $V$, $W$; $2U$, $2V$, $2W$; . . . ; $nU$, $nV$, $nW$ defines the row $[UVW]$. A set of such rows, or directions, related by the crystal symmetry constitutes a form of directions $\langle UVW \rangle$ (compare zone symbols, page 16).

## 2.2.1  Two-Dimensional Lattices

We begin our study of lattices in two dimensions rather than three. A two-dimensional lattice is called a net; it may be imagined as being formed by aligning, in a regular manner, rows of equally spaced points (Figure 2.1). The net is the array of points; the connecting lines are a convenience, drawn to aid our appreciation of the lattice geometry.

Since nets exhibit symmetry, they can be allocated to the two-dimensional systems (page 29). The most general net is shown in Figure 2.1b. A sufficient representative portion of the lattice is the unit cell, outlined by the vectors **a** and **b**; an infinite number of such unit cells stacked side by side builds up the net.

The net under consideration exhibits twofold rotational symmetry about each point; consequently, it is placed in the oblique system. The chosen unit cell is primitive (symbol $p$), which means that one lattice point is associated with the area of the unit cell; each point is shared equally by four adjacent unit cells. In the oblique unit cell, $a \neq b$, and $\gamma \neq 90°$ or $120°$; angles of $90°$ and $120°$ may imply symmetry higher than 2.

Consider next the stacking of unit cells in which $a \neq b$ but $\gamma = 90°$ (Figure 2.2). The symmetry at every point is $2mm$, and this net belongs to the rectangular system. The net in Figure 2.3 may be described by a unit cell in which $a' = b'$ and $\gamma' \neq 90°$ or $120°$. It may seem at first that this net is oblique, but careful inspection shows that each point has $2mm$ symmetry, and so this net, too, is allocated to the rectangular system.

In order to display this fact clearly, a centered (symbol $c$) unit cell is chosen, shown in Figure 2.3 by the vectors **a** and **b**. This cell has two lattice points per unit cell area. It is left as an exercise to the reader to show that a

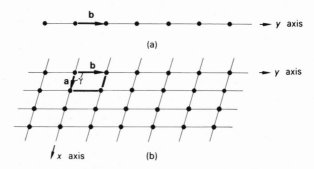

FIGURE 2.1.  Formation of a net: (a) row of equally spaced points, (b) regular stack of rows.

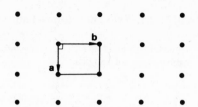

FIGURE 2.2. Rectangular net with a
*p* unit cell drawn in.

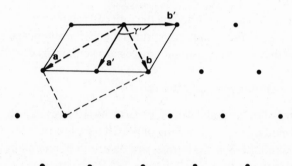

FIGURE 2.3. Rectangular net with *p* and *c* unit cells drawn in.

centered, oblique unit cell does not represent a net which is fundamentally
different from that in Figure 2.1.

### 2.2.2 Choice of Unit Cell

From the foregoing discussion, it will be evident that the choice of unit
cell is somewhat arbitrary. We shall follow a universal crystallographic
convention in choosing a unit cell: The unit cell is the smallest repeat unit for
which its delineating vectors are parallel to, or coincide with, important
symmetry directions in the lattice. Returning to Figure 2.3, the centered cell
is preferred because **a** and **b** coincide with the symmetry (*m*) lines in the net.
The primitive unit cell (**a′**, **b′**) is, of course, a possible unit cell, but it does not,
in isolation, reveal the lattice symmetry clearly. The symmetry is still there;
it is invariant under choice of unit cell, as is shown by the following
equations:

$$a'^2 = a^2/4 + b^2/4 \tag{2.2}$$

$$b'^2 = a^2/4 + b^2/4 \tag{2.3}$$

the value of $\gamma'$ depends only on the ratio $a'/b'$.

TABLE 2.1.  Two-Dimensional Lattices

| System | Unit cell symbol(s) | Point group of unit cell | Unit cell edges and angles |
|---|---|---|---|
| Oblique | $p$ | 2 | $a \neq b,\ \gamma \neq 90°,\ 120°$ |
| Rectangular | $p, c$ | $2mm$ | $a \neq b,\ \gamma = 90°$ |
| Square | $p$ | $4mm$ | $a = b,\ \gamma = 90°$ |
| Hexagonal | $p$ | $6mm$ | $a = b,\ \gamma = 120°$ |

Two other nets are governed by the unit cell relationships $a = b$, $\gamma = 90°$ and $a = b$, $\gamma = 120°$; their study constitutes the first problem at the end of this chapter. The five two-dimensional lattices are summarized in Table 2.1.

## 2.2.3  Three-Dimensional Lattices

The three-dimensional lattices, or Bravais lattices, may be imagined as being developed by regular stacking of nets. There are 14 ways in which this can be done, and the Bravais lattices are distributed, unequally, among the seven crystal systems, as shown in Figure 2.4. Each lattice is represented by a unit cell, outlined by three noncoplanar vectors[*] **a**, **b**, and **c**. In accordance with convention, these vectors are chosen so that they both form a parallelepipedon of smallest volume in the lattice and are parallel to, or coincide with, important symmetry directions in the lattice. In three dimensions, we encounter unit cells centered on a pair of opposite faces, body-centered, or centered on all faces. Table 2.2 lists the unit cell types and their nomenclature.

### Triclinic Lattice

If oblique nets are stacked in a general, but regular, manner, a triclinic lattice is obtained (Figure 2.5). The unit cell is characterized by the conditions $a \neq b \neq c$ and $\alpha \neq \beta \neq \gamma \neq 90°$, $120°$, and corresponds to point group $\bar{1}$. This unit cell is primitive (symbol $P$),[†] which means that one lattice point is associated with the unit cell volume; each point is shared equally by eight adjacent unit cells in three dimensions (see Figure 2.6). There is no symmetry direction to constrain the choice of the unit cell vectors, and a parallelepipedon of smallest volume can always be chosen conventionally. The angles $\alpha$, $\beta$, and $\gamma$ are selected to be oblique, as far as is practicable.

[*] The magnitudes of the edges $a$, $b$, and $c$ and the angles $\alpha$, $\beta$, and $\gamma$ provide an equivalent description.
[†] Capital letters are used in three dimensions.

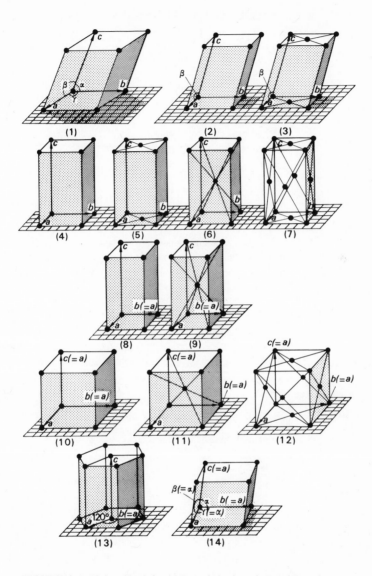

FIGURE 2.4. Unit cells of the 14 Bravais lattices; interaxial angles are 90° unless indicated otherwise by a numerical value or symbol: (1) triclinic *P*, (2) monoclinic *P*, (3) monoclinic *C*, (4) orthorhombic *P*, (5) orthorhombic *C*, (6) orthorhombic *I*, (7) orthorhombic *F*, (8) tetragonal *P*, (9) tetragonal *I*, (10) cubic *P*, (11) cubic *I*, (12) cubic *F*, (13) hexagonal *P*, (14) trigonal *R*.

TABLE 2.2.   Unit Cell Nomenclature

| Centering site(s) | Symbol | Miller indices of centered faces of the unit cell | Fractional coordinates[a] of centered sites |
|---|---|---|---|
| None | $P$ | — | — |
| $bc$ faces | $A$ | 100 | $0, \frac{1}{2}, \frac{1}{2}$ |
| $ca$ faces | $B$ | 010 | $\frac{1}{2}, 0, \frac{1}{2}$ |
| $ab$ faces | $C$ | 001 | $\frac{1}{2}, \frac{1}{2}, 0$ |
| Body center | $I$ | — | $\frac{1}{2}, \frac{1}{2}, \frac{1}{2}$ |
| All faces | $F$ | $\begin{cases} 100 \\ 010 \\ 001 \end{cases}$ | $\begin{cases} 0, \frac{1}{2}, \frac{1}{2} \\ \frac{1}{2}, 0, \frac{1}{2} \\ \frac{1}{2}, \frac{1}{2}, 0 \end{cases}$ |

[a] A fractional coordinate $x$ is given by $X/a$, where $X$ is the coordinate in absolute measure (Å) and $a$ is the unit-cell repeat distance in the same direction and in the same units.

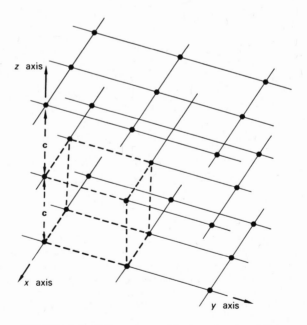

FIGURE  2.5.  Oblique  nets  stacked  to  form  a  triclinic lattice.

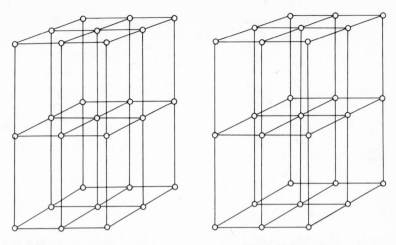

FIGURE 2.6. Stereoscopic pair of drawings showing eight adjacent $P$ unit cells in a monoclinic lattice. The sharing of corner points can be seen readily by focusing attention on the central lattice point in the drawings.

## Monoclinic Lattices

The monoclinic system is characterized by one diad (rotation or inversion), which is chosen to be parallel to the $Y$ axis (and $b$). The conventional unit cell is specified by the conditions $a \neq b \neq c$, $\alpha = \gamma = 90°$, and $\beta \neq 90°$, $120°$. Figure 2.6 illustrates a stereoscopic pair of drawings of a monoclinic lattice, showing eight unit cells; according to convention, the $\beta$ angle is chosen to be oblique.

Reference to Figure 2.4 shows that there are two conventional monoclinic lattices, symbolized by the unit cell types $P$ and $C$.

A monoclinic unit cell centered on the $A$ faces is equivalent to that described as $C$; the choice of the $b$ axis* is governed by symmetry, but $a$ and $c$ are interchangeable labels.

The centering of the $B$ faces is illustrated in Figure 2.7. In this situation a new unit cell, $\mathbf{a}'$, $\mathbf{b}'$, $\mathbf{c}'$, can be defined by the following equations:

$$\mathbf{a}' = \mathbf{a} \tag{2.4}$$

$$\mathbf{b}' = \mathbf{b} \tag{2.5}$$

$$\mathbf{c}' = \mathbf{a}/2 + \mathbf{c}/2 \tag{2.6}$$

---

* We often speak of the $b$ axis (to mean $Y$ axis) because our attention is usually confined to the unit cell.

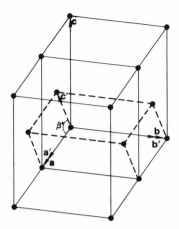

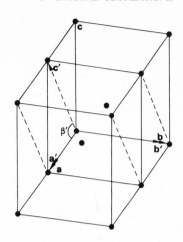

FIGURE 2.7. Monoclinic lattice
showing that $B \equiv P$.

FIGURE 2.8. Monoclinic lattice
showing that $I \equiv C$.

*Since c′ lies in the ac* plane, $\alpha' = \gamma' = 90°$, but $\beta' \neq 90°$ or 120°. The new monoclinic cell is primitive; symbolically we may write $B \equiv P$. Similarly, it may be shown that $I \equiv F \equiv C$ (Figures 2.8 and 2.9).

If the $C$ cell (Figure 2.10) is reduced to primitive, it no longer displays the characteristic monoclinic symmetry clearly (see Table 2.3), and we may conclude that there are two monoclinic lattices, described by the unit cell types $P$ and $C$.

It may be necessary to calculate the new dimensions of a transformed

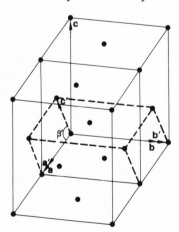

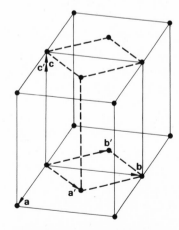

FIGURE 2.9. Monoclinic lattice
showing that $F \equiv C$.

FIGURE 2.10. Monoclinic lattice
showing that $C \not\equiv P$.

unit cell. Consider the example $B \to P$, (2.4)–(2.6). Clearly, $a' = a$ and $b' = b$. Taking the scalar product* of (2.6) with itself, we obtain

$$\mathbf{c'} \cdot \mathbf{c'} = (\mathbf{a}/2 + \mathbf{c}/2) \cdot (\mathbf{a}/2 + \mathbf{c}/2) \tag{2.7}$$

Hence,

$$c'^2 = a^2/4 + c^2/4 + ac(\cos\beta)/2 \tag{2.8}$$

The new angle $\beta'$ is given by†

$$\cos\beta' = \mathbf{a'} \cdot \mathbf{c'}/a'c' \tag{2.9}$$

Using (2.6), we obtain

$$\cos\beta' = [a/2 + c(\cos\beta)/2]/c' \tag{2.10}$$

where $c'$ is given by (2.8). These calculations can be applied to all crystal systems, giving due consideration to any nontrivial relationships between $a$, $b$, and $c$ and between $\alpha$, $\beta$, and $\gamma$.

## Orthorhombic Lattices

The monoclinic system was treated in some detail. It will not be necessary to give such an extensive discussion for either this system or the remaining crystal systems.

The orthorhombic system is characterized by three mutually perpendicular diads (rotation and/or inversion); the unit cell vectors are chosen to be parallel to, or coincide with, these symmetry axes. The orthorhombic unit cell is specified by the relationships $a \neq b \neq c$ and $\alpha = \beta = \gamma = 90°$. It will not be difficult for the reader to verify that the descriptions $P$, $C$, $I$, and $F$ are necessary and sufficient in this system. One way in which this exercise may be carried out is as follows. After centering the $P$ unit cell, three questions must be asked, in the following order:

1. Does the centered cell represent a true lattice?
2. If it is a lattice, is the symmetry changed?
3. If the symmetry is unchanged, does it represent a new lattice, and has the unit cell been chosen correctly?

We answered these questions implicitly in discussing the monoclinic lattices.

---

*The scalar (dot) product of two vectors $\mathbf{p}$ and $\mathbf{q}$ is denoted by $\mathbf{p} \cdot \mathbf{q}$ and is equal to $pq(\cos\hat{pq})$, where $\hat{pq}$ represents the angle between the positive directions of $\mathbf{p}$ and $\mathbf{q}$.
†To make $\beta'$ obtuse, it may be necessary to use $-\mathbf{a}/2$ in (2.6) and (2.7).

It should be noted that the descriptions $A$, $B$, and $C$ do not remain equivalent for orthorhombic space groups in the class $mm2$; it is necessary to distinguish $C$ from $A$ (or $B$). The reader may like to consider now, or later, why this distinction is necessary.

## Tetragonal Lattices

The tetragonal system is characterized by one tetrad (rotation or inversion along $Z$ (and $c$); the unit cell conditions are $a = b \neq c$ and $\alpha = \beta = \gamma = 90°$. There are two tetragonal lattices, specified by the unit cell symbols $P$ and $I$ (Figure 2.4); $C$ and $F$ tetragonal unit cells may be transformed to $P$ and $I$, respectively.

## Cubic Lattices

The symmetry of the cubic system is characterized by four triad axes at angles of $\cos^{-1}(1/3)$ to one another; they are the body diagonals $\langle 111 \rangle$ of a cube. The threefold axes in this orientation introduce twofold axes along $\langle 100 \rangle$. There are three cubic Bravais lattices (Figure 2.4) with conventional unit cells $P$, $I$, and $F$.

## Hexagonal Lattice

The basic feature of a hexagonal lattice is that it should be able to accommodate a sixfold symmetry axis. This requirement is achieved by a lattice based on a $P$ unit cell, with $a = b \neq c$, $\alpha = \beta = 90°$, and $\gamma = 120°$, the $c$ direction being parallel to a sixfold axis in the lattice.

## Lattices in the Trigonal System

A two-dimensional unit cell in which $a = b$ and $\gamma = 120°$ is compatible with either sixfold or threefold symmetry (see Figure 2.22, plane groups $p6$ and $p3$). For this reason, the hexagonal lattice ($P$ unit cell) is used for certain crystals which belong to the trigonal system. However, as shown in Figure 2.11, the presence of two threefold axes within a unit cell, with $x, y$ coordinates of $\frac{2}{3}, \frac{1}{3}$ and $\frac{1}{3}, \frac{2}{3}$, respectively, and parallel to the $Z$ axis, introduces the possibility of a crystal structure which belongs to the trigonal system but has a triply primitive hexagonal unit cell $R_{hex}$ with centering points at $\frac{2}{3}, \frac{1}{3}, \frac{1}{3}$ and $\frac{1}{3}, \frac{2}{3}, \frac{2}{3}$ in the unit cell.

Thus for some trigonal crystals the unit cell will be $P$, and for others it will be $R_{hex}$, the latter being distinguished by systematically absent X-ray

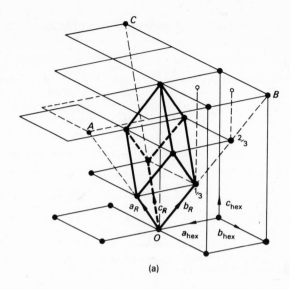

(a)

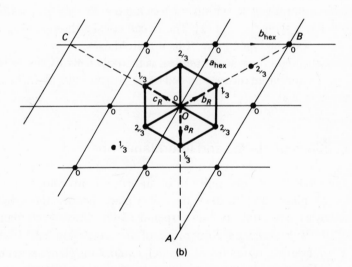

(b)

FIGURE 2.11. Trigonal lattice (the fractions refer to values of $c_{hex}$): (a) rhombohedral ($R$) unit cell (obverse setting) developed from a triply primitive hexagonal ($R_{hex}$) unit cell. (In the *reverse* setting, the rhombohedral lattice and unit cell are rotated about [111] 60° clockwise with respect to the $R_{hex}$ axes.) The ratio of the volumes of any two unit cells in one and the same lattice is equal to the ratio of the numbers of lattice points in the two unit cell volumes. (b) Plan view of (a) as seen along $c_{hex}$.

TABLE 2.3.  Fourteen Bravais Lattices

| System | Unit cell(s) | Point group of unit cell | Axial relationships[a] |
|---|---|---|---|
| Triclinic | $P$ | $\bar{1}$ | $a \neq b \neq c$; $\alpha \neq \beta \neq \gamma \neq 90°, 120°$ |
| Monoclinic | $P, C$ | $2/m$ | $a \neq b \neq c$; $\alpha = \gamma = 90°$; $\beta \neq 90°, 120°$ |
| Orthorhombic | $P, C, I, F$ | $mmm$ | $a \neq b \neq c$; $\alpha = \beta = \gamma = 90°$ |
| Tetragonal | $P, I$ | $\dfrac{4}{m}mm$ | $a = b \neq c$; $\alpha = \beta = \gamma = 90°$ |
| Cubic | $P, I, F$ | $m3m$ | $a = b = c$; $\alpha = \beta = \gamma = 90°$ |
| Hexagonal | $P$ | $\dfrac{6}{m}mm$ | $a = b \neq c$; $\alpha = \beta = 90°$; $\gamma = 120°$ |
| Trigonal | $R^b$ or $P^c$ | $\bar{3}m$ | $a = b = c$; $\alpha = \beta = \gamma \neq 90°, <120°$ |

[a] Remember to read $\neq$ as "not constrained by symmetry to equal."
[b] On hexagonal axes, column 4 is the same as for the hexagonal system.
[c] Column 4 is the same as for the hexagonal system.

reflections (Table 4.1). The $R_{hex}$ cell can be transformed to a primitive rhombohedral unit cell $R$, with $a = b = c$ and $\alpha = \beta = \gamma \neq 90°$, $<120°$; the threefold axis is then along [111]. The $R$ cell may be thought of as a cube extended or squashed along one of its threefold axes.

The lattice based on an $R$ unit cell is the only truly exclusive trigonal lattice, the lattice based on a $P$ unit cell being borrowed from the hexagonal system (Table 2.3).

## 2.3　Families of Planes and Interplanar Spacings

Figure 2.12 shows one unit cell of an orthorhombic lattice projected onto the $ab$ plane. The trace of the (110) plane nearest the origin O is indicated by a dashed line, and the perpendicular distance of this plane from O is $d(110)$. By repeating the operation of the translation $\pm \mathbf{d}(110)$ on the plane (110), a series, or family, of parallel, equidistant planes is generated, as shown in Figure 2.13. Our discussion of the external symmetry of crystals led to a description of the external faces of crystals by Miller indices, which are by definition prime to one another. In discussing X-ray diffraction effects, however, it is necessary to consider planes for which the indices $h$, $k$, and $l$ may contain a common factor while still making intercepts $a/h$, $b/k$, and $c/l$ on the $X$, $Y$, and $Z$ axes, respectively, as required by the definition of Miller indices. It follows that the plane with indices ($nh$, $nk$, $nl$) makes

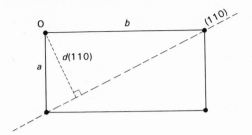

FIGURE 2.12. One unit cell in an orthorhombic
lattice as seen in projection along $c$.

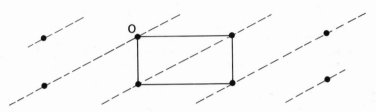

FIGURE 2.13. Family of (110) planes.

intercepts $a/nh$, $b/nk$, and $c/nl$ along $X$, $Y$, and $Z$, respectively, and that this plane is nearer to the origin by a factor of $1/n$ than is the plane $(hkl)$. In other words, $d(nh, nk, nl) = d(hkl)/n$. In general, we denote a family of planes as $(hkl)$ where $h$, $k$, and $l$ may contain a common factor. For example, the (220) family of planes is shown in Figure 2.14 with interplanar spacing $d(220) = d(110)/2$; alternate (220) planes therefore coincide with (110) planes. It should be noted that an external crystal face normal to $d(hh0)$ would always be designated (110), since external observations reveal the shape but not the size of the unit cell.

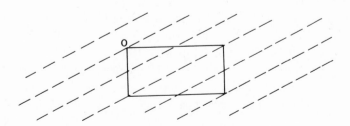

FIGURE 2.14. Family of (220) planes.

## 2.4  Reciprocal Lattice

We introduce the reciprocal lattice concept in this chapter; it will be needed in the study of X-ray diffraction from crystals. For each direct (Bravais) lattice, a corresponding reciprocal lattice may be postulated. It has the same symmetry as the direct lattice, and may be derived from it graphically. Let Figure 2.15a represent a monoclinic direct lattice as seen in a direction normal to the (010) plane. From the origin O of the $P$ unit cell, lines are drawn normal to families of planes ($hkl$) in direct space. It may be noted in passing that the normal to a plane ($hkl$) does not, in general, coincide with the direction of the same indices [$hkl$]. Along each line, reciprocal lattice points $hkl$ (no parentheses) are marked off such that the distance from the origin to the first point in any line is inversely proportional to the corresponding interplanar spacing $d(hkl)$.†

In three dimensions, we refer to $d^*(100)$, $d^*(010)$, and $d^*(001)$ as $a^*$, $b^*$, and $c^*$, respectively, and so define a unit cell in the reciprocal lattice. In general,

$$d^*(hkl) = K/d(hkl) \qquad (2.11)$$

where $K$ is a constant. Hence, for the monoclinic system,

$$a^* = K/d(100) = K/(a \sin \beta) \qquad (2.12)$$

From Figure 2.15a, the scalar product $\mathbf{a} \cdot \mathbf{a}^*$ is given by

$$\mathbf{a} \cdot \mathbf{a}^* = aa^* \cos(\beta - 90°) = aK\frac{\cos(\beta - 90°)}{a \sin \beta} = K \qquad (2.13)$$

The mixed scalar products, such as $\mathbf{a} \cdot \mathbf{b}^*$ are identically zero, because the angle between $a$ and $b^*$ is 90° (see Figure 2.15a).

The reciprocal lattice points form a true lattice with a representative unit cell outlined by $\mathbf{a}^*$, $\mathbf{b}^*$, and $\mathbf{c}^*$, and, therefore, involving six reciprocal cell parameters in the most general case—three sides $a^*$, $b^*$, and $c^*$, and three angles $\alpha^*$, $\beta^*$, and $\gamma^*$. The size of the reciprocal cell is governed by the choice of the constant $K$. In practice, $K$ is frequently taken as the wavelength $\lambda$ of the X-radiation used; reciprocal lattice units are then dimensionless.

† The correspondence between directions of reciprocal lattice vectors containing no common factor and those in a spherical projection (see Figure 1.20) should be noted.

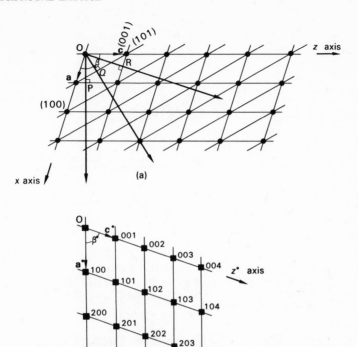

FIGURE 2.15.  Direct and reciprocal lattices: (a) monoclinic $P$,
as seen in projection along $b$, showing three families of planes;
(b) corresponding reciprocal lattice.

A reciprocal lattice row $hkl$; $2h,2k,2l$; ... may be considered to be
derived from the families of planes $(nh, nk, nl)$ with $n = 1, 2, \ldots$, since $d(nh, nk, nl) = d(hkl)/n$. Hence,

$$d^*(nh, nk, nl) = nd^*(hkl) \tag{2.14}$$

where $d^*(hkl)$ is the distance of the reciprocal lattice point $hkl$ from the
origin, expressed in reciprocal lattice units (RU). The vector $\mathbf{d}^*(hkl)$ is given
by

$$\mathbf{d}^*(hkl) = h\mathbf{a}^* + k\mathbf{b}^* + l\mathbf{c}^* \tag{2.15}$$

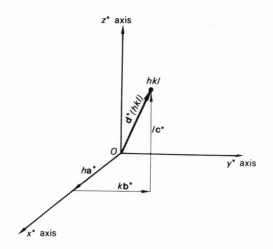

FIGURE 2.16. Vector components of $\mathbf{d}^*(hkl)$ in a reciprocal lattice. Following the appropriate vector paths parallel to the three reciprocal lattice axes $X^*$, $Y^*$, and $Z^*$ gives the result, $\mathbf{d}^*(hkl) = h\mathbf{a}^* + k\mathbf{b}^* + l\mathbf{c}^*$.

This equation is illustrated in Figure 2.16. Equation (2.15) provides a straightforward method for deriving expressions for $d^*$ and $d$. Thus, generally, from (2.15)

$$\mathbf{d}^*(hkl) \cdot \mathbf{d}^*(hkl) = d^{*2}(hkl) = h^2a^{*2} + k^2b^{*2} + l^2c^{*2} + 2klb^*c^* \cos \alpha^* \\ + 2lhc^*a^* \cos \beta^* + 2hka^*b^* \cos \gamma^*$$

$$(2.16)$$

Now $d(hkl)$ may be obtained from (2.11) and (2.16). Simplifications of (2.16) arise through symmetry constraints on the unit cell vectors in different crystal systems. The reader should check the entries in Table 2.4, starting with Table 2.3 and equation (2.16).

It should be noted that the relationship between the direct and reciprocal lattices requires that $\mathbf{a}^*$ be perpendicular to both $\mathbf{b}$ and $\mathbf{c}$, and, conversely, that $\mathbf{a}$ be perpendicular to both $\mathbf{b}^*$ and $\mathbf{c}^*$, and so on for the other cell constants. Hence, in converting from direct to reciprocal space and vice versa, the crystal system is preserved. For an orthogonal lattice, the direct and reciprocal axes are coincident.

We give, without proof, the following general formulae for the triclinic system:

$$a^* = Kbc(\sin \alpha)/V_c \qquad (2.17)$$

## TABLE 2.4. Expressions for $d^{*2}(hkl)$ and $d^2(hkl)$ in the Seven Crystal Systems[a]

| System | $d^{*2}(hkl)$ | $d^2(hkl)$ |
|---|---|---|
| Triclinic | $h^2a^{*2}+k^2b^{*2}+l^2c^{*2}+2klb^*c^*\cos\alpha^* + 2lhc^*a^*\cos\beta^*+2hka^*b^*\cos\gamma^*$ | $K^2/d^{*2}(hkl)$ |
| Monoclinic | $h^2a^{*2}+k^2b^{*2}+l^2c^{*2}+2hla^*c^*\cos\beta^*$ | $\left\{\dfrac{1}{\sin^2\beta}\left[\dfrac{h^2}{a^2}+\dfrac{l^2}{c^2}-\dfrac{2hl\cos\beta}{ac}\right]+\dfrac{k^2}{b^2}\right\}^{-1}$ |
| Orthorhombic | $h^2a^{*2}+k^2b^{*2}+l^2c^{*2}$ | $\left\{\dfrac{h^2}{a^2}+\dfrac{k^2}{b^2}+\dfrac{l^2}{c^2}\right\}^{-1}$ |
| Tetragonal | $(h^2+k^2)a^{*2}+l^2c^{*2}$ | $\left\{\dfrac{h^2+k^2}{a^2}+\dfrac{l^2}{c^2}\right\}^{-1}$ |
| Hexagonal and trigonal $(P)$ | $(h^2+k^2+hk)a^{*2}+l^2c^{*2}$ | $\left\{\dfrac{4(h^2+k^2+hk)}{3a^2}+\dfrac{l^2}{c^2}\right\}^{-1}$ |
| Trigonal $(R)$ (rhombohedral) | $[h^2+k^2+l^2+2(hk+kl+hl)(\cos\alpha^*)]a^{*2}$ | $a^2(TR)^{-1}$, where $T=h^2+k^2+l^2+2(hk+kl+hl)[(\cos^2\alpha-\cos\alpha)/\sin^2\alpha]$ and $R=(\sin^2\alpha)/(1-3\cos^2\alpha+2\cos^3\alpha)$ $\left\{\dfrac{h^2+k^2+l^2}{a^2}\right\}^{-1}$ |
| Cubic | $(h^2+k^2+l^2)a^{*2}$ | $\dfrac{a^2}{h^2+k^2+l^2}$ |

[a] In the monoclinic system, $d(100)=a\sin\beta$, $d(001)=c\sin\beta$, and hence $a=K/(a^*\sin\beta^*)$ and $c=K/(c^*\sin\beta^*)$.
In the hexagonal system (and trigonal $P$), $a=b=K/(a^*\sin\gamma^*)=K/(a^*\sqrt{3}/2)$.
In general, the expressions for $d^{*2}$ are simpler in form than the corresponding expressions for $d^2$.

and similarly for $b^*$ and $c^*$;

$$\cos \alpha^* = (\cos \beta \cos \gamma - \cos \alpha)/(\sin \beta \sin \gamma) \qquad (2.18)$$

and similarly for $\cos \beta^*$ and $\cos \gamma^*$; $V_c$ is the unit cell volume, given by

$$V_c = abc(1 - \cos^2\alpha - \cos^2\beta - \cos^2\gamma + 2 \cos \alpha \cos \beta \cos \gamma)^{1/2} \qquad (2.19)$$

Hence, the volume of the reciprocal unit cell $V^*$ is given by

$$V^* = K^3/V_c \qquad (2.20)$$

Simplifications of the expressions (2.17–2.19) arise for other crystal systems.

## 2.5   Rotational Symmetries of Lattices

We can now discuss analytically the permissible rotational symmetries in lattices, already stated to be of degrees 1, 2, 3, 4, and 6. In Figure 2.17, let $A$ and $B$ represent two adjacent lattice points, of repeat distance $t$, in any row. An $R$-fold rotation axis is imagined to act at each point and to lie normal to the plane of the diagram. An anticlockwise rotation of $\Phi$ about $A$ maps $B$ onto $B'$, and a clockwise rotation of $\Phi$ about $B$ maps $A$ onto $A'$. Lines $AB'$ and $BA'$ are produced to meet in $Q$.

Since triangles $ABQ$ and $A'B'Q$ are similar, $A'B'$ is parallel to $AB$. From the properties of lattices, it follows that $A'B' = Jt$, where $J$ is an integer.

Lines $A'S$ and $B'T$ are drawn perpendicular to $AB$, as shown. Hence,

$$A'B' = TS = AB - (AT + BS) \qquad (2.21)$$

or

$$Jt = t - 2t \cos \Phi \qquad (2.22)$$

whence

$$\cos \Phi = (1 - J)/2 = M/2 \qquad (2.23)$$

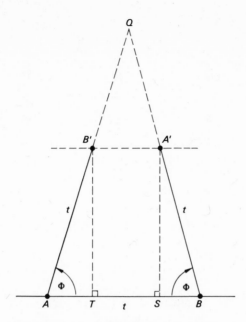

FIGURE 2.17. Rotational symmetry in crystal
lattices.

where $M$ is another integer. Since $-1 \leqslant \cos\Phi \leqslant 1$, and, from (2.23), the only admissible values for $M$ are $0, \pm 1, \pm 2$, these values give rise to the rotational symmetries already discussed. This treatment gives a quantitative aspect to the packing considerations mentioned previously (page 30).

## 2.6  Space Groups

In order to extend our study of crystal geometry into the realm of atomic arrangements, we must consider now the symmetry of extended, ideally infinite, patterns in space. We recall that a point group describes the symmetry of a finite body, and that a lattice constitutes a mechanism for repetition, to an infinite extent, by translations parallel to three noncoplanar directions. We may ask, therefore, what is the result of repeating a point-group pattern by the translations of a Bravais lattice. We shall see that it is like an arrangement of atoms in a crystal.

A space group may be described as a set of symmetry elements, the operation of any of which brings the infinite array of points to which they

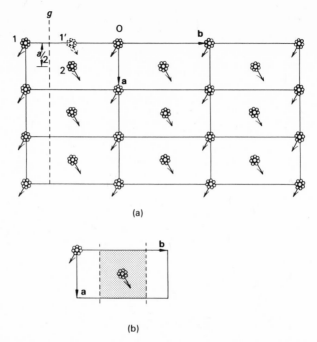

(a)

(b)

FIGURE 2.18. Wallpaper pattern: (a) extended pattern, (b) asymmetric unit.

belong into self-coincidence. We may apply space-group rules to crystals because the dimensions of crystals used in experimental investigations are large with respect to the repeat distances of the pattern.

A space group may be considered to be made up of two parts, a pattern unit and a repeat mechanism. An analogy may be drawn with a wallpaper (Victorian style), a simple example of which is shown in Figure 2.18a. We shall analyze this pattern.

The conventional unit cell for this pattern is indicated by the vectors **a** and **b**. If we choose a pattern unit consisting of two flowers (Figure 2.18b) and continue it indefinitely by the repeat vectors **a** and **b**, the plane pattern is generated. However, we have ignored the symmetry between the two flowers in the chosen pattern unit. If one flower (1) is reflected across the dashed line ($g$) to ($1'$) and then translated by **a**/2, it then occupies the position of the second flower (2), or the pattern represented by Figure 2.18a is brought into self-coincidence by the symmetry operation. This operation takes place across a glide line, a symmetry element that occurs in some extended two-dimensional patterns.

We say that the necessary and sufficient pattern unit is a single flower, occupying the asymmetric unit—the shaded (or unshaded) portion of Figure 2.18b. If the single flower is repeated by both the glide-line symmetry and the unit cell translations, then the extended pattern is again generated. Thus, to use our analogy, if we know the asymmetric unit of a crystal structure, which need not be the whole unit cell contents, and the space-group symbol for the crystal, we can generate the whole structure.

### 2.6.1 Two-Dimensional Space Groups

Our discussion leads naturally into two-dimensional space groups, or plane groups. Consider a pattern motif showing twofold symmetry (Figure 2.19a)—the point-group symbolism is continued into the realm of space groups. Next, consider a primitive oblique net (Figure 2.19b); it is of infinite extent in the plane, and the framework of lines divides the field into a number of identical primitive ($p$) unit cells. An origin is chosen at any lattice point.

Now, let the motif be repeated around each point in the net, and in the same relationship, with the twofold rotation points of the motif and the net in coincidence (Figure 2.19c). It will be seen that additional twofold rotation points are introduced, at the fractional coordinates (see footnote to Table 2.2) $0, \frac{1}{2}; \frac{1}{2}, 0;$ and $\frac{1}{2}, \frac{1}{2}$ in each unit cell. We must always look for such "additional" symmetry elements after the point-group motif has been operated on by the unit cell repeats. This plane group is given the symbol $p2$.

In general, we shall not need to draw several unit cells; one cell will suffice provided that the pattern motif is completed around all lattice points intercepted by the given unit cell. Figure 2.20 illustrates the standard drawing of $p2$: The origin is taken on a twofold point, the $X$ axis runs from top to bottom, and the $Y$ axis runs from left to right. Thus, the origin is considered to be in the top left-hand corner of the cell, but each corner is an equivalent position; we must remember that the drawing is a representative portion of an infinite array, whether in two or three dimensions.

The asymmetric unit, represented by $\bigcirc$, may be placed anywhere in the unit cell (for convenience, near the origin), and then repeated by the symmetry $p2$ to build up the picture, taking care to complete the arrangement around each corner. The additional twofold points can then be identified. The reader should now carry out this construction.

The list of fractional coordinates in Figure 2.20 refers to symmetry-related sites in the unit cell. The maximum number of sites generated by

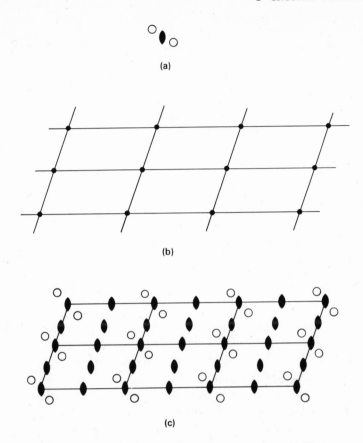

FIGURE 2.19  Plane group $p2$: (a) twofold symmetry motif, (b) oblique
net with $p$ unit cells, (c) extended pattern of plane group $p2$.

the space-group symmetry are called the general equivalent positions. In $p2$
they are given the coordinates $x, y$ and $\bar{x}, \bar{y}$. We could use $1-x, 1-y$ instead
of $\bar{x}, \bar{y}$, but it is more usual to work with a set of coordinates near one and the
same origin.

Each coordinate line in the space-group description lists, in order from
left to right, the number of positions in each set, the Wyckoff* notation (for
reference purposes only), the symmetry at each site in the set, and the
coordinates of all sites in the set.

In a conceptual two-dimensional crystal, or projected real atomic
arrangement, the asymmetric unit may contain either a single atom or a

* See Bibliography.

Origin at 2

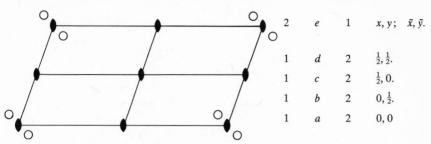

| 2 | $e$ | 1 | $x, y;$  $\bar{x}, \bar{y}.$ |
|---|---|---|---|
| 1 | $d$ | 2 | $\frac{1}{2}, \frac{1}{2}.$ |
| 1 | $c$ | 2 | $\frac{1}{2}, 0.$ |
| 1 | $b$ | 2 | $0, \frac{1}{2}.$ |
| 1 | $a$ | 2 | $0, 0$ |

FIGURE 2.20. Standard drawing and description of plane group $p2$. The lines which divide the unit cell into four quadrants are drawn for convenience only.

group of atoms. If it consists of part (half, in this plane group) of one molecule, then the whole molecule, as seen in projection at least, must contain twofold rotational symmetry, or a symmetry of which 2 is a subgroup. There are four unique twofold points in the unit cell; in the Wyckoff notation they are the sets (a), (b), (c), and (d), and they constitute the sets of special equivalent positions in this plane group. Notice that general positions have symmetry 1, whereas special positions have a higher crystallographic point-group symmetry. Where the unit cell contains fewer (an integral submultiple) of a species than the number of general equivalent positions in its space group, then it may be assumed that the species are occupying special equivalent positions and have the symmetry of the special site, at least. Exceptions to this rule arise in disordered structures, but this topic will not be discussed in this book.

We move now to the rectangular system, which includes point groups $m$ and $2mm$, and both $p$ and $c$ unit cells. We shall consider first plane groups $pm$ and $cm$.

The formation of these plane groups may be considered along the lines already described for $p2$, and we refer immediately to Figure 2.21. The origin is chosen on $m$, but its $y$ coordinate is not defined by this symmetry element. In $pm$, the general equivalent positions are two in number, and there are two sets of special equivalent positions on $m$ lines.

Plane group $cm$ introduces several new features. The coordinate list is headed by the expression $(0, 0; \frac{1}{2}, \frac{1}{2}) +$; this means that two translations—$0, 0$ and $\frac{1}{2}, \frac{1}{2}$—are added to all the listed coordinates. Hence, the full list of general positions reads

$$x, y;  \quad \bar{x}, y;  \quad \tfrac{1}{2}+x, \tfrac{1}{2}+y;  \quad \tfrac{1}{2}-x, \tfrac{1}{2}+y$$

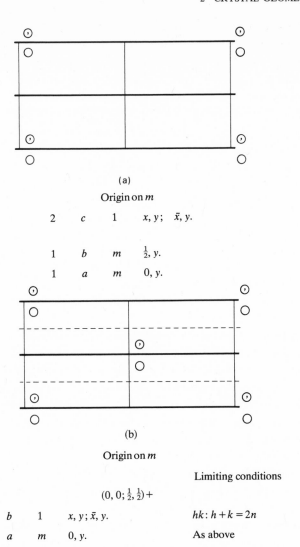

FIGURE 2.21. Plane groups in the rectangular system: (a) $pm$, (b) $cm$.

Given $x$, the distance $\frac{1}{2} - x$, for example, is found by first moving $\frac{1}{2}$ along the $a$ axis from the origin and then moving back along the same direction by the amount $x$.

The centering of the unit cell in conjunction with the $m$ lines introduces the glide-line symmetry element (symbol $g$ and graphic symbol - - -). The glide lines interleave the mirror lines, and their action is a combination of reflection and translation. The translational component is one-half of the repeat distance in the direction of the glide line. Thus, the pair of general positions $x, y$ and $\frac{1}{2} - x, \frac{1}{2} + y$ are related by the $g$ line. We shall encounter glide lines in any centered unit cell where $m$ lines are present, and in certain other plane groups. For example, we may ask if there is any meaning to the symbol $pg$, a glide-symmetry motif repeated by the lattice translations. The answer is that $pg$ is a possible plane group; in fact, it is the symmetry of the pattern in Figure 2.18.

There is only one set of special positions in $cm$, in contrast to two sets in $pm$. This situation arises because the centering condition in $cm$ requires that both mirror lines in the unit cell be included in one and the same set. If we try to postulate two sets, by analogy with $pm$, we obtain

$$0, y; \quad \tfrac{1}{2}, \tfrac{1}{2} + y \tag{2.24}$$

and

$$\tfrac{1}{2}, y; \quad 0 \,(\text{or } 1), \tfrac{1}{2} + y \tag{2.25}$$

Expressions (2.24) and (2.25) differ only in the value of the variable $y$ and therefore do not constitute two different sets of special equivalent positions.

We could refer to plane group $cm$ by the symbol $cg$. If we begin with the origin on $g$ and draw the general positions as before, we should find the glide lines interleaved with $m$ lines. Two patterns that differ only in the choice of origin or the values attached to the coordinates of the equivalent positions do not constitute different space groups. The reader can illustrate this statement by drawing $cg$, and also, by drawing $pg$, can show that $pm$ and $pg$ are different. The glide line, or, indeed, any translational symmetry element is not encountered in point groups; it is a property of infinite patterns. The 17 plane groups are illustrated in Figure 2.22. The asymmetric unit is represented by a scalene triangle instead of the usual circle.

## 2.6.2 Limiting Conditions Governing X-Ray Reflection

Our main reason for studying space-group symmetry is that it provides information about the repeat patterns of atoms in crystal structures. X-ray diffraction spectra are characterized partly by the indices of the families of planes from which, in the Bragg treatment of diffraction (page 116), the X-rays are considered to be reflected. The pattern of indices reveals information about the space group of the crystal. Where a space group

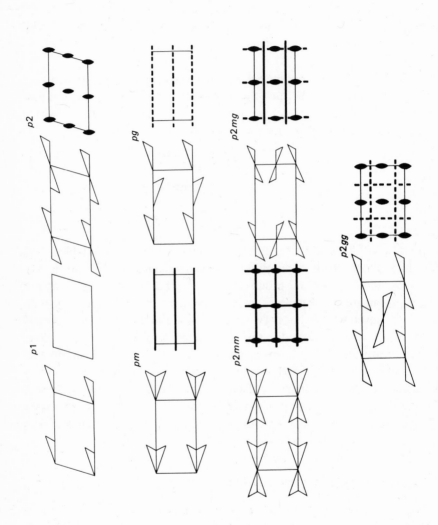

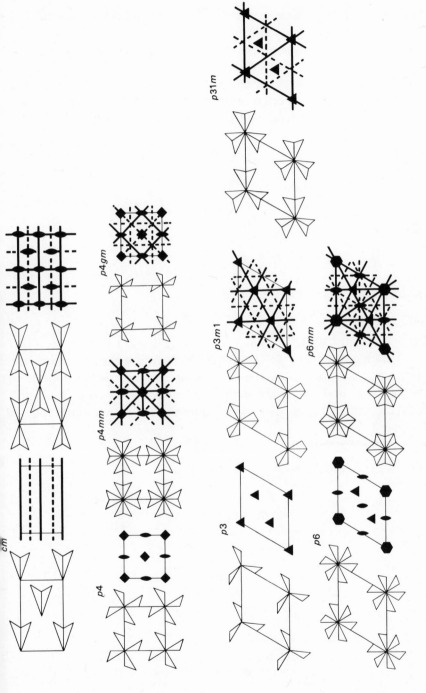

FIGURE 2.22.  Unit cells of the 17 plane groups.

contains translational symmetry, certain sets of reflections will be systematically absent from the experimental diffraction data record. We meet this
situation for the first time in *cm* (Figure 2.21b); a reflection *hk* is absent
unless the sum $h + k$ is an even number.

Figure 2.23 illustrates a rectangular lattice. Two unit cells are depicted
on this lattice, a centered cell with vectors **A** and **B**, and a primitive cell with
vectors **a** and **b**. The relationship between them is summarized by the
equations

$$\mathbf{A} = \mathbf{a} - \mathbf{b} \tag{2.26}$$

$$\mathbf{B} = \mathbf{a} + \mathbf{b} \tag{2.27}$$

It is shown in Appendix A.6 that Miller indices of planes transform in the
same way as unit cell vectors. Hence,

$$H = h - k \tag{2.28}$$

$$K = h + k \tag{2.29}$$

Adding (2.28) and (2.29), we obtain

$$H + K = 2h \tag{2.30}$$

which is even for all values of $h$.

Limiting conditions describe circumstances in which reflections can
occur; systematic absences refer to conditions under which reflections
cannot arise. Both terms are in common use, and we must distinguish
between them carefully. Limiting conditions are discussed more fully in
Chapter 4.

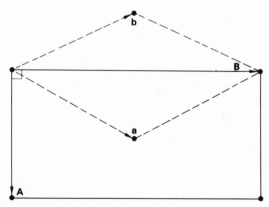

FIGURE 2.23. Centered rectangular unit cell (**A**, **B**)
and primitive unit cell (**a**, **b**) within the same lattice.

Point group $2mm$ belongs to the rectangular system, and, as a final example in two dimensions, we shall study plane group $p2gg$. It is often helpful to recall the parent point group of any space group. All that we need to do is to ignore the unit cell symbol, and replace any, translational symmetry elements by the corresponding nontranslational symmetry elements. Thus, $pg$ is derived from point group $m$, and $p2gg$ from $2mm$.

In $p2gg$, we are not at liberty to choose the actual positions of 2 and the two $g$ lines freely. In studying point groups, we saw that the symmetry elements in a given symbol have a definite relative orientation with respect to the crystallographic axes; this is preserved in the corresponding space groups, Thus, we know that the $g$ lines are normal to the $X$ and $Y$ axes, and we can take an origin, initially, at their intersection (Figure 2.24a). In Figure 2.24b the general equivalent positions have been inserted; this diagram reveals the positions of the twofold points, inserted in Figure 2.24c, together with the additional $g$ lines in the unit cell. The standard orientation of $p2gg$ places the twofold point at the origin; Figure 2.24d shows this setting and the description of this plane group. We see again that two interacting symmetry elements lead to a combined action which is equivalent to that of a third symmetry element, but their positions must be chosen correctly. This question did not arise in point groups because, by definition, all symmetry elements pass through a point—the origin.

There are two sets of special equivalent positions in $p2gg$; the pairs of twofold rotation points must be selected correctly. One way of ensuring proper selection is by inserting the coordinate values of the point-group symmetry element constituting a special position into the coordinates of the general positions. Thus, by taking $x = y = 0$ for one of the twofold points, we obtain a set of special positions with coordinates 0, 0 and $\frac{1}{2}, \frac{1}{2}$. If we had chosen 0, 0 and $0, \frac{1}{2}$ as a set, the resulting pattern would not have conformed to $p2gg$ symmetry, but to $pm$, as Figure 2.25 shows. Special positions form a subset of the general positions, under the same space-group symmetry.

The general equivalent positions give rise to two limiting conditions, because the structure is "halved" with respect to $b$ for the reflections $0k$, and with respect to $a$ for the reflections $h0$. The special positions take both of these conditions, and the extra conditions shown because occupancy of the special positions* in this plane group gives rise to centered arrangements (Figure 2.26). After the development of the structure factor (page 153), some of the different limiting conditions will be derived analytically.

---

* The entities occupying special positions must, themselves, conform to the space group symmetry.

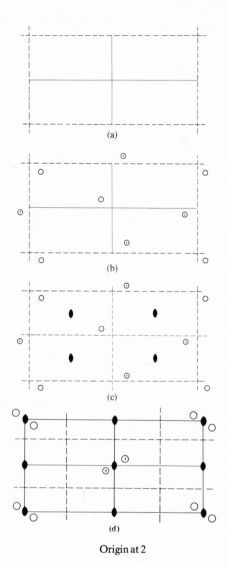

Origin at 2

Limiting conditions

| | | | | | |
|---|---|---|---|---|---|
| 4 | $c$ | 1 | $x,y;$ $\bar{x},\bar{y};$ $\frac{1}{2}+x,\frac{1}{2}-y;$ $\frac{1}{2}-x,\frac{1}{2}+y.$ | | $hk$: None |
| | | | | | $h0$: $h = 2n$ |
| | | | | | $0k$: $k = 2n$ |
| 2 | $b$ | 2 | $\frac{1}{2},0;$ $0,\frac{1}{2}.$ | | As above + |
| 2 | $a$ | 2 | $0,0;$ $\frac{1}{2},\frac{1}{2}.$ | | $hk$: $h + k = 2n$ |

FIGURE 2.24.  Formation and description of $p2gg$.

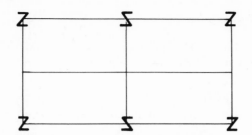

FIGURE 2.25. Occupation of the special positions 0, 0 and 0, $\frac{1}{2}$ leads to *pm* symmetry.

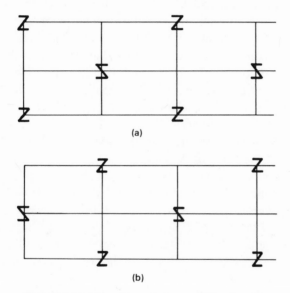

FIGURE 2.26. Special equivalent positions in *p2gg*; both sets (a) and (b) give rise to centered arrangements of the entity at the center of the Z symbol.

### 2.6.3 Three-Dimensional Space Groups

The principles which have emerged from the discussion on plane groups can be extended to three dimensions. Whereas the plane groups are limited to 17 in number, there are 230 space groups. We shall limit our discussion to a few space groups in the monoclinic and orthorhombic systems. We believe this will prove adequate because most of the important principles will evolve and, from a practical point of view, about 90% of crystals belong to these two systems.

## Monoclinic Space Groups

In the monoclinic system, the lattices are characterized by the *P* and *C* unit cell descriptions, and the point groups are 2, *m*, and 2/*m*. We consider first space groups *P*2 and *C*2.

As with plane groups, we may begin with a motif which has twofold symmetry, but now about a line (axis) in three-dimensional space. This motif is arranged in a fixed orientation with respect to the points of a monoclinic lattice. Figure 2.27 shows a stereoscopic pair of illustrations for the unit cell of *C*2, drawn with respect to the conventional right-handed axes (page 9).

In Figure 2.28, *P*2 and *C*2 are shown in projection. The standard drawing of space-group diagrams is on the *ab* plane of the unit cell, with +*X* running from top to bottom, +*Y* from left to right, both in the plane of the paper, and +*Z* coming up from the paper. The positive or negative signs attached to the representative points indicate the *z* coordinates, that is, in $O^+$ and $O^-$, the signs stand for *z* and $\bar{z}$, respectively. The relationship with the chosen stereogram nomenclature will be evident here.

In both *P*2 and *C*2, the origin is chosen on 2, and is, thus, defined with respect to the *X* and *Z* axes, but not with respect to *Y*. The graphic symbol for a diad axis in the plane of the diagram is→ .

In space group *P*2, the general and special equivalent positions may be derived quite readily. The special sets (b) and (d) should be noted carefully; they are sometimes forgotten by the beginner because symmetry elements distant *c*/2 from those drawn in the *ab* plane are not indicated on the conventional diagrams. The diad along *Y* and at $x = 0$, $z = \frac{1}{2}$, for example, relates *x*, *y*, *z* to a point at $1 - x$, $1 - y$, $1 - z$; its presence, and that of the diad

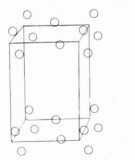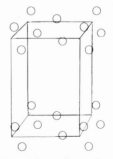

FIGURE 2.27. Stereoscopic pair of illustrations of the environs of one unit cell of space group *C*2; general equivalent positions are shown.

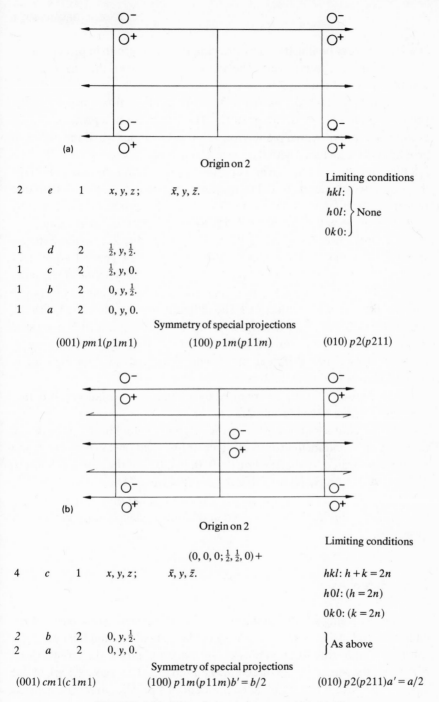

(a)

Origin on 2

| | | | | Limiting conditions |
|---|---|---|---|---|
| 2 | e | 1 | $x, y, z$;    $\bar{x}, y, \bar{z}.$ | $hkl:$ |
| | | | | $h0l:$ } None |
| | | | | $0k0:$ |

| | | | |
|---|---|---|---|
| 1 | d | 2 | $\frac{1}{2}, y, \frac{1}{2}.$ |
| 1 | c | 2 | $\frac{1}{2}, y, 0.$ |
| 1 | b | 2 | $0, y, \frac{1}{2}.$ |
| 1 | a | 2 | $0, y, 0.$ |

Symmetry of special projections

(001) $pm1(p1m1)$      (100) $p1m(p11m)$      (010) $p2(p211)$

(b)

Origin on 2

$(0, 0, 0; \frac{1}{2}, \frac{1}{2}, 0) +$

| | | | | Limiting conditions |
|---|---|---|---|---|
| 4 | c | 1 | $x, y, z$;    $\bar{x}, y, \bar{z}.$ | $hkl: h + k = 2n$ |
| | | | | $h0l: (h = 2n)$ |
| | | | | $0k0: (k = 2n)$ |

| | | | | |
|---|---|---|---|---|
| 2 | b | 2 | $0, y, \frac{1}{2}.$ | |
| 2 | a | 2 | $0, y, 0.$ | } As above |

Symmetry of special projections

(001) $cm1(c1m1)$      (100) $p1m(p11m)b' = b/2$      (010) $p2(p211)a' = a/2$

FIGURE 2.28. Monoclinic space groups: (a) $P2$, (b) $C2$.

at $x = z = \frac{1}{2}$, may be illustrated by drawing the space group in projection on the $ac$ plane of the unit cell. The reader should make this drawing and compare it with Figure 2.28a.

It is often useful to consider a structure in projection onto one of the principal planes (100), (010), or (001). The symmetry of a projected space group corresponds with a plane group, and the symmetries of the principal projections are included with the space-group description (Figure 2.28). The full plane-group symbols, given in parentheses, indicate the orientations of the symmetry elements. In $C2$, certain projections produce more than one repeat in certain directions; the projected cell dimensions, represented by $a'$, $b'$, and $c'$, may then be halved with respect to their original values.

The projection of $C2$ onto (100) is shown by Figure 2.29 in three stages, starting from the $y$ and $z$ coordinates of the set of general equivalent positions. The question is sometimes asked, how do two points of the same hand, such as $x, y, z$ and $\bar{x}, y, \bar{z}$ in $C2$, become of opposite hand, such as $y, z$ and $y, \bar{z}$ in $p1m$, after projection? The difficulty may be associated with the use of the highly symmetric circle as a representative point. It is suggested that the reader make a drawing of $C2$, and of the stages of the projection on to (100), using either $\bf 9$ instead of $\bigcirc^{+}$, and $\textit{9}$ instead of $\bigcirc^{-}$, or the scalene triangle shown in Figure 2.22.

Space group $C2$ may be obtained by adding the translation $\frac{1}{2}, \frac{1}{2}, 0$, that associated with a $C$ cell (Table 2.2), to the equivalent positions of $P2$. This operation is equivalent to repeating the original twofold motif at the lattice points of the $C$ monoclinic unit cell. This simple relationship between $P$ and $C$ cells is indicated by the heading $(0, 0, 0; \frac{1}{2}, \frac{1}{2}, 0)+$ of the coordinate list in $C2$; it may be compared with that for $cm$ (Figure 2.21b).

There are four sets of special positions in $P2$, but only two sets in $C2$; the reason for this has been discussed in relation to plane groups $pm$ and $cm$ (page 79).

### 2.6.4  Screw Axes

The centering of the unit cell in $C2$ introduces screw axes which interleave the diads. A screw axis may be designated $R_p$, and the operation consists of an $R$-fold rotation plus a translation parallel to the screw axis of $p/R$ times the repeat in that direction. Thus, in $C2$, the screw axes are of the type $2_1$ and have a translational component of $\frac{1}{2}$ parallel to $b$. The general equivalent positions $x, y, z$ and $\frac{1}{2} - x, \frac{1}{2} + y, \bar{z}$ are related by a $2_1$ axis along

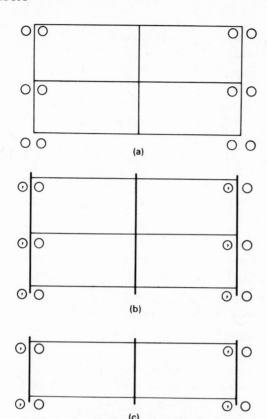

FIGURE 2.29.  Projection of $C2$ onto (100): (a) $y, z$ positions from $C2$ ($Z$ axis left to right), (b) two-dimensional symmetry elements added, (c) one unit cell—$p1m$ ($p11m$), $b' = b/2$, $c' = c$. (Plane groups $p11m$ and $p1m1$ are equivalent because they correspond only to an interchange of the $X$ and $Y$ axes; 1 is a trivial symmetry element.)

$[\frac{1}{4}, y, 0]$.* Screw axes are present in the positions shown by their graphic symbol $\rightarrow$.

## Limiting Conditions in $C2$

The limiting conditions for this space group are given in Figure 2.28b. Two of them are placed in parentheses; this notation is used to indicate that

* We use this nomenclature to describe, in this example, the line parallel to the $Y$ axis through $x = \frac{1}{4}$, $z = 0$.

they are dependent upon a more general condition. Thus, since the *hkl* reflections are limited by the condition $h + k = 2n$ (even), because the cell is *C*-centered, it follows that $h0l$ are limited by $h = 2n$ (0 is an even number). There are several other nonindependent conditions which could have been listed. For example, $0kl$: $k = 2n$ and $h00$: $h = 2n$. However, in the monoclinic system, in addition to the *hkl* reflections, we are concerned particularly only with $h0l$ and $0k0$, because the symmetry plane is parallel to (010) and the symmetry axis is parallel to [010]. This feature is discussed more fully in Chapter 4.

Space Group $P2_1$

Space groups $C2$ and $C2_1$ are equivalent (compare *cm* and *cg*). On the other hand, $P2$ contains no translational symmetry, so $P2_1$ is a new space group (Figure 2.30). There are no special positions in $P2_1$. Special positions cannot exist on a single translational symmetry element, since it would mean that the entity placed on such an element consisted of an infinite repeating pattern.

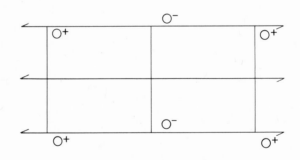

Origin on $2_1$

|  |  |  |  |  | Limiting conditions |
|---|---|---|---|---|---|
| 2 | *a* | 1 | $x, y, z$; | $\bar{x}, \tfrac{1}{2}+y, \bar{z}.$ | $hkl$: None |
|  |  |  |  |  | $h0l$: None |
|  |  |  |  |  | $0k0: k = 2n$ |

Symmetry of special projections

(001) $pg1(p1g1)$      (100) $p1g(p11g)$      (010) $p2(p211)$

FIGURE 2.30.  Space group $P2_1$.

TABLE 2.5. Glide-Plane Notation

| Symbol | Orientation | Graphic symbol | | Translational component |
|--------|-------------|----------------|---|------------------------|
| | | ‖ to projection | ⊥ to projection | |
| $a$ | (010) or (001) | - - - - - | | $a/2$ |
| $b$ | (100) or (001) | - - - - - | | $b/2$ |
| $c$ | (100) or (010) | · · · · · · · · · | None | $c/2$ |
| $n$ | (100) | | | $(b+c)/2$ |
| | (010) | | | $(a+c)/2$ |
| | (001) | | | $(a+b)/2$ |

## 2.6.5 Glide Planes

If a space group is formed from the combination of a point group with $m$ planes and a lattice of centered unit cells, glide planes are introduced into the space group. They are the three-dimensional analog of glide lines. The glide-plane operation consists of reflection across the plane plus a translation parallel to the plane. The direction of translation is indicated by the glide-plane symbol (Table 2.5). Other types of glide planes exist, particularly in the higher symmetry systems.

As an example of a space group with a glide plane, we shall study $P2_1/c$, a space group encountered frequently in practice. This space group is derived from point group $2/m$, and must, therefore, be centrosymmetric. However, the center of symmetry does not lie at the intersection of $2_1$ and $c$. It is convenient to take the origin on a center of symmetry* in centrosymmetric space groups, and, in this example, we must determine the correct positions of the symmetry elements in the unit cell. We shall approach the solution of this problem in two ways, the first of which is similar to our treatment of plane group $p2gg$.

Since the screw axis must intersect the glide plane, the point of intersection will be taken as an origin and the space group drawn (Figure 2.31). We see now that the centers of symmetry lie at points such as $0, \frac{1}{4}, \frac{1}{4}$. This point may be taken as a new origin, and the space group redrawn (Figure 2.32); a fraction ($\frac{1}{4}$, for example) placed next to a symmetry element indicates the position of that symmetry element with respect to the $ab$ plane.

---

* Sometimes the origin will have a point symmetry higher than $\bar{1}$, for example, $2/m$ or $mmm$, but $\bar{1}$ is a subgroup of such point symmetries.

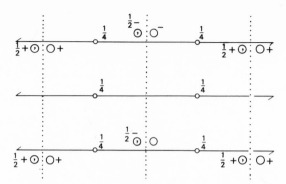

FIGURE 2.31.  Space group $P2_1/c$ with the origin at an
intersection of $2_1$ and $c$.

It is desirable, however, to be able to draw the standard space-group
illustration at the outset. From a choice of origin, and using the full meaning
of the space-group symbol, we can obtain the positions of the symmetry
elements by means of a simple scheme.

Let the symmetry elements be placed as follows:

$\bar{1}$ at 0, 0, 0 (choice of origin)
$2_1$ parallel to $[p, y, r]$ (parallel to the $Y$ axis)
$c$ parallel to $(x, q, z)$ (normal to the $Y$ axis)

It is important to note that we have employed only the standard choice of
origin and the information contained in the space-group symbol. Next, we

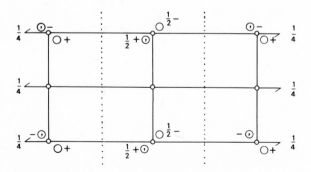

FIGURE 2.32.  Space group $P2_1/c$ with the origin on $\bar{1}$ (standard setting).

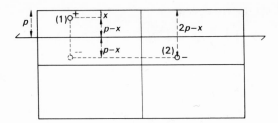

FIGURE 2.33. Operation about a $2_1$ axis along the line $[p, y, 0]$: $x \rightarrow 2p - x$. A similar construction may be used for the $y$ coordinate in a $c$-glide operation, for example.

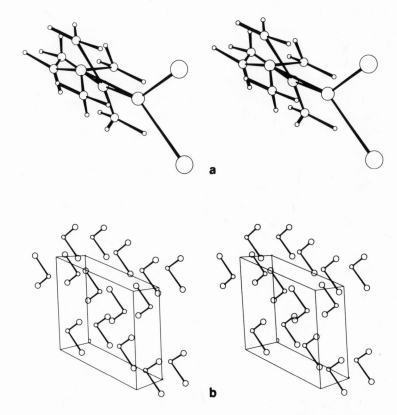

FIGURE 2.34. Stereoviews of the structure of diiodo-$(N,N,N',N'$-tetramethylethylenediamine)/zinc(II): (a) Structural formula; the circles, in decreasing order of size, represent I, Zn, N, C, and H. (b) Unit cell; for clarity, only the I and Zn atoms are shown.

carry out the symmetry operations:

$$(1)\quad x, y, z \xrightarrow{\;2_1\;} 2p - x, \tfrac{1}{2} + y, 2r - z \quad (2)$$

$$-c$$

$$2p - x, 2q - \tfrac{1}{2} - y, -\tfrac{1}{2} + 2r - z \quad (3)$$

$$\xrightarrow{\;\bar{1}\;} -x, -y, -z \quad (4)$$

The symbol $-c$ is used to indicate that the $c$-glide translation of $\tfrac{1}{2}$ is subtracted, which is crystallographically equivalent to being added.

We now use the fact that the combined effect of two operations is equivalent to a third operation, starting from the original point (1). Symbolically, $c.2_1 \equiv \bar{1}$, or $2_1$ followed by $c$ is equivalent to $\bar{1}$. Thus, points (3) and (4) are one and the same, whence, by comparing coordinates, $p = 0$ and $q = r = \tfrac{1}{4}$. Comparison with Figure 2.32 shows that these conditions lead to the desired positions of the three symmetry elements in $P2_1/c$.

The change in the $x$ coordinate in the operation (1) $\rightarrow$ (2) is illustrated in Figure 2.33; the argument can be applied to any similar situation in monoclinic and orthorhombic space groups, and we can always consider one coordinate at a time. The completion of the details of this space group forms the basis of a problem at the end of this chapter.

We shall not discuss centered monoclinic space groups, but they do not present difficulty once the primitive space groups have been mastered. Figure 2.34 shows a stereoscopic pair of illustrations of the unit cell of diiodo-($N,N,N',N'$-tetramethylethylenediamine)/zinc(II), $I_2[(CH_3)_2NCH_2CH_2N(CH_3)_2]Zn$, which crystallizes in space group $C2/c$ with four molecules in the unit cell; the zinc atoms lie on twofold axes.*

## Orthorhombic Space Groups

We shall consider two orthorhombic space groups, $P2_12_12_1$ and *Pnma*. The first is illustrated in Figure 2.35; it should be noted that the three mutually perpendicular $2_1$ axes do *not* intersect one another in this space group. Although $P2_12_12_1$ is a noncentrosymmetric space group, the three principal projections are centrosymmetric; each has the two-dimensional space group $p2gg$.

* S. Htoon and M. F. C. Ladd, *Journal of Crystal and Molecular Structure* **4**, 357 (1974).

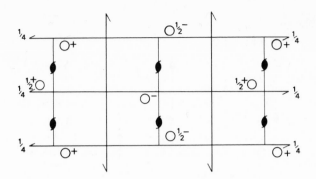

Origin halfway between three pairs of non-intersecting screw axes

Limiting conditions

$4 \quad a \quad 1 \quad x, y, z; \quad \frac{1}{2}-x, \bar{y}, \frac{1}{2}+z; \quad \frac{1}{2}+x, \frac{1}{2}-y, \bar{z}; \quad \bar{x}, \frac{1}{2}+y, \frac{1}{2}-z.$

$hkl:$
$0kl:$
$\left. \begin{array}{l} \\ \\ \end{array} \right\}$ None
$h0l:$
$hk0:$

$h00: h = 2n$

$0k0: k = 2n$

$00l: l = 2n$

Symmetry of special projections

$(001)\, p2gg$          $(100)\, p2gg$          $(010)\, p2gg$

*FIGURE* 2.35. Space group $P2_12_12_1$: in space-group diagrams, ⚭ represents a $2_1$ axis normal to the plane of projection.

*Change of Origin.* Considering the projection of $P2_12_12_1$ onto (001), we obtain from the general equivalent positions the two-dimensional set

$$x, y; \quad \frac{1}{2}-x, \bar{y}; \quad \frac{1}{2}+x, \frac{1}{2}-y; \quad \bar{x}, \frac{1}{2}+y$$

It is convenient to change the origin to a twofold rotation point, say at $\frac{1}{4}$, 0. To carry out this transformation, the coordinates of the new origin are subtracted from the original coordinates:

$$x-\tfrac{1}{4}, y; \quad \tfrac{1}{4}-x, \bar{y}; \quad \tfrac{1}{4}+x, \tfrac{1}{2}-y; \quad -x-\tfrac{1}{4}, \tfrac{1}{2}+y$$

Next, new variables $x_0$ and $y_0$ are chosen such that, for example, $x_0 = x - \frac{1}{4}$ and $y_0 = y$. Then, by substituting, we obtain

$$x_0, y_0; \quad \bar{x}_0, \bar{y}_0; \quad \tfrac{1}{2}+x_0, \tfrac{1}{2}-y_0; \quad \tfrac{1}{2}-x_0, \tfrac{1}{2}+y_0$$

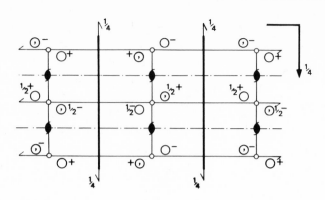

Origin at $\bar{1}$

8  $d$  1  $x, y, z;$  $\frac{1}{2}+x, \frac{1}{2}-y, \frac{1}{2}-z;$  $\bar{x}, \frac{1}{2}+y, \bar{z};$  $\frac{1}{2}-x, \bar{y}, \frac{1}{2}+z;$

          $\bar{x}, \bar{y}, \bar{z};$  $\frac{1}{2}-x, \frac{1}{2}+y, \frac{1}{2}+z;$  $x, \frac{1}{2}-y, z; \frac{1}{2}+x, y, \frac{1}{2}-z.$

Limiting conditions

$hkl$: None

$0kl$: $k+l = 2n$

$h0l$: None

$hk0$: $h = 2n$

$h00$: $(h = 2n)$

$0k0$: $(k = 2n)$

$00l$: $(l = 2n)$

4  $c$  $m$  $x, \frac{1}{4}, z;$  $\bar{x}, \frac{3}{4}, \bar{z};$  $\frac{1}{2}-x, \frac{3}{4}, \frac{1}{2}+z;$  $\frac{1}{2}+x, \frac{1}{4}, \frac{1}{2}-z.$    As above

4  $b$  $\bar{1}$  $0, 0, \frac{1}{2};$  $0, \frac{1}{2}, \frac{1}{2};$  $\frac{1}{2}, 0, 0;$  $\frac{1}{2}, \frac{1}{2}, 0.$    As above +

4  $a$  $\bar{1}$  $0, 0, 0;$  $0, \frac{1}{2}, 0;$  $\frac{1}{2}, 0, \frac{1}{2};$  $\frac{1}{2}, \frac{1}{2}, \frac{1}{2}.$    $hkl$: $h+l = 2n; k = 2n$

Symmetry of special projections

(001) $p2gm$          (100) $c2mm$          (010) $p2gg$

FIGURE 2.36.  Space group $Pnma$.

If the subscript is dropped, these coordinates are exactly those given already for $p2gg$ (Figure 2.24d). This type of change of origin is useful when studying projections. In this example, the reverse transformation takes us back to the standard setting in $P2_12_12_1$.

Space group $Pnma$ is shown with the origin on $\bar{1}$ (Figure 2.36). The symbol tells us that the unit cell is primitive, with an $n$-glide plane normal to the $X$ axis (see Table 2.5), an $m$ plane normal to $Y$, and an $a$-glide plane normal to $Z$. Although this space group is derived from point group $mmm$, we cannot assume that the three planes in *Pnma* intersect in a center of symmetry. We are, therefore, faced with a problem similar to that discussed with $P2_1/c$. The solution of this problem depends upon the fact that $m.m.m \equiv \bar{1}$, and is illustrated fully in Problem 12 at the end of this chapter.

The coordinates of the general and the special equivalent positions can be derived easily from the diagram. The translational symmetry elements $n$ and $a$ give rise to the limiting conditions shown. Nonindependent conditions are shown in parentheses; in the orthorhombic system, all of the classes of reflection listed should be considered, as will be discussed in Chapter 4.

It is useful to remember that in the triclinic, monoclinic, and orthorhombic space groups, at least, pairs of coordinates which have one *sign* change of $x$, $y$, or $z$ indicate a symmetry plane normal to the axis of the coordinate with the changed sign. If two sign changes exist, a symmetry axis lies parallel to the axis of the coordinate that has *not* changed sign. Three sign changes indicate a center of symmetry. In these three systems, where any coordinate, say $x$, is related by symmetry to another at $t - x$, the symmetry element intersects the $X$ axis at $t/2$.

### 2.6.6 Analysis of the Space-Group Symbol

In this section we consider the general interrelationship between space-group symbols and point-group symbols. On encountering a space-group symbol, the first problem is to determine the parent point group. This process has been discussed (page 83); here are a few more examples. It is not necessary to have explored all space groups in order to carry out this exercise:

$$P2_1/c \rightarrow (2_1/c) \rightarrow (2/c) \rightarrow 2/m$$
$$Ibca \rightarrow mmm$$
$$P4_12_12 \rightarrow 422$$
$$F\bar{4}3c \rightarrow \bar{4}3m$$

Next we must identify a crystal system for each point group:

$$2/m \rightarrow \text{monoclinic}$$
$$mmm \rightarrow \text{orthorhombic}$$
$$422 \rightarrow \text{tetragonal}$$
$$\bar{4}3m \rightarrow \text{cubic}$$

Now, from Table 1.5, we can associate certain crystallographic directions with each symmetry element in the space group symbol:

$P2_1/c$:   Primitive, monoclinic unit cell; $c$-glide plane $\perp b$; $2_1$ axis $\| b$; centrosymmetric.

$Ibca$:   Body-centered, orthorhombic unit cell; $b$-glide plane $\perp a$; $c$-glide plane $\perp b$; $a$-glide plane $\perp c$; centrosymmetric.

$P4_12_12$:   Primitive, tetragonal unit cell; $4_1$ axis $\| c$; $2_1$ axes $\| a$ and $b$; twofold axes at 45° to $a$ and $b$; noncentrosymmetric.

$F\bar{4}3c$:   Face-centered, cubic unit cell; $\bar{4}$ axes $\| a$, $b$, and $c$; threefold axes $\| \langle 111 \rangle$; $c$-glide planes $\perp \langle 110 \rangle$; noncentrosymmetric.

It should be noted carefully that the unique symmetry elements (where there are more than two present) given in a space-group symbol may not intersect, and the origin must always be selected with care. Appropriate procedures for the monoclinic and orthorhombic systems have been discussed; in working with higher symmetry space groups, similar rules can be evaluated.

Because of the similarities between space groups and their parent point groups, a reflection symmetry, for example, in the same orientation with respect to the crystallographic axes always produces the same changes in the *signs* of the coordinates. Thus, the $m$ plane perpendicular to $Z$ in point group $mmm$ changes $x$, $y$, $z$ to $x$, $y$, $\bar{z}$. The $a$-glide plane in *Pnma* changes $x$, $y$, $z$ to $\frac{1}{2}+x$, $y$, $\frac{1}{2}-z$; the translational components of $\frac{1}{2}$ are a feature of this space group, but the signs of $x$, $y$, and $z$ are still $+$, $+$, and $-$ after the operation (see also Appendix A7).

## Bibliography

### Lattices and Space Groups

HENRY, N. F. M., and LONSDALE, K. (Editors), *International Tables for X-Ray Crystallography*, Vol. I, Birmingham, Kynoch Press.

## Problems

**2.1.**   Two nets are described by the unit cells (i) $a = b$, $\gamma = 90°$ and (ii) $a = b$, $\gamma = 120°$. In each case (a) what is the symmetry at each net point, (b) to which two-dimensional system does the net belong, and (c) what are the results of centering the unit cell?

**2.2.** A monoclinic $F$ unit cell has the dimensions $a = 6.000$, $b = 7.000$, $c = 8.000$ Å and $\beta = 110.0°$. Show that an equivalent monoclinic $C$ unit cell, with an *obtuse* $\beta$ angle, can represent the same lattice, and calculate its dimensions. What is the ratio of the volume of the $C$ cell to that of the $F$ cell?

**2.3.** Carry out the following exercises with drawings of a tetragonal $P$ unit cell.

(a) Center the $B$ faces. Comment on the result.
(b) Center the $A$ and $B$ faces. Comment on the result.
(c) Center all faces. What conclusions can you draw now?

**2.4.** Calculate the length of $[31\bar{2}]$ (see page 55) for both unit cells in Problem 2.2.

**2.5.** The relationships $a \neq b \neq c$, $\alpha \neq \beta \neq 90°$, $120°$, and $\gamma = 90°$ may be said to define a diclinic system. Is this a new system? Give reasons for your answer.

**2.6.** (a) Draw a diagram to show the symmetry elements and general equivalent positions in $c2mm$ (origin on $2mm$). Write the coordinates and point symmetry of the general and special positions, in their correct sets, and give the conditions limiting X-ray reflection in this plane group. (b) Draw a diagram of the symmetry elements in plane group $p2mg$ (origin on 2); take care not to put the twofold point at the intersection of $m$ and $g$ (why?). On the diagram, insert each of the motifs P, V, and Z in turn, using the *minimum* number of motifs consistent with the space-group symmetry.

**2.7.** (a) Continue the study of space group $P2_1/c$ (page 94). Write the coordinates of the general and special positions, in their correct sets. Give the limiting conditions for all sets of positions, and write the plane-group symbols for the three principal projections. Draw a diagram of the space group as seen along the $b$ axis. (b) Biphenyl, ⟨O⟩-⟨O⟩, crystallizes in space group $P2_1/c$, with two molecules per unit cell. What can be deduced about both the positions of the molecules in the unit cell and the molecular conformation? (The planarity of each benzene ring in the molecule may be assumed.)

**2.8.** Write the coordinates of the vectors between all pairs of general equivalent positions in $P2_1/c$ with respect to the origin, and note that they are of two types. Remember that $-\frac{1}{2}$ and $+\frac{1}{2}$ in a coordinate are crystallographically equivalent, because we can always add or subtract 1 from a fractional coordinate without altering its crystallographic implication.

**2.9.** The orientation of the symmetry elements in the orthorhombic space group *Pban* may be written as follows:

> $\bar{1}$ at 0, 0, 0 (choice of origin)
> $b$-glide $\parallel (p, y, z)$
> $a$-glide $\parallel (x, q, z)$ (from the space-group symbol)
> $n$-glide $\parallel (x, y, r)$

Determine $p$, $q$, and $r$ from the following scheme, using the fact that $n.a.b \equiv \bar{1}$:

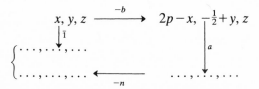

**\*2.10.** Construct a space-group diagram for *Pbam*, with the origin at the intersection of the three symmetry planes. List the coordinates of both the general equivalent positions and the centers of symmetry. Derive the standard coordinates for the general positions by transforming the origin to a center of symmetry.

**2.11.** Show that space groups *Pa*, *Pc*, and *Pn* represent the same pattern, but that *Ca* is different from *Cc* (*Cn*). What is the more usual symbol for space group *Ca*?

**2.12.** For each of the space groups $P2/c$, $Pca2_1$, $Cmcm$, $P\bar{4}2_1c$, $P6_322$, and $Pa3$:

 (a) Write down the parent point group and crystal system.
 (b) List the meaning conveyed by the symbol.
 (c) State the independent conditions limiting X-ray reflection.

**2.13.** Consider Figure 2.25. What would be the result of constructing this diagram with Z alone, and not using its mirror image?

# Preliminary Examination of Crystals by Optical and X-Ray Methods

## 3.1 Introduction

In this chapter we shall discuss the interaction between crystals and two different electromagnetic radiations, light and X-rays. Light, with its longer wavelength (5000–6000 Å), can reveal only limited information about crystal structures, whereas X-rays with wavelengths of less than about 2 Å can be used to determine the relative positions of atoms in crystals. A preliminary examination of a crystal aims to determine its space group and unit-cell dimensions, and may be carried out by a combination of optical and X-ray techniques. The optical methods described here are simple, but, nevertheless, often very effective; they should be regarded as a desirable prerequisite to an X-ray structure determination.

## 3.2 Polarized Light

An ordinary light source emits wave trains, or pulses of light, vibrating in all directions perpendicular to the direction of propagation (Figure 3.1); the light is said to be unpolarized. The vibrations of interest to us are those of the electric vector associated with the waves. Any one of these random vibrations can be resolved into two mutually perpendicular components, and the resultant vibration may, therefore, be considered as the sum of all components in these two perpendicular directions. In order to study the optical properties of crystals, we need to restrict the resultant vibration of the light source to one direction only by eliminating the component at right

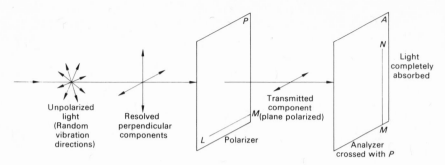

FIGURE 3.1. Production of plane-polarized light by passing unpolarized light through a sheet of Polaroid film (the polarizer, $P$). A second sheet of Polaroid (the analyzer, $A$), rotated through 90° with respect to $P$, completely absorbs all light transmitted by $P$. The lines $LM$ and $MN$ were parallel on the sheet from which $P$ and $A$ were cut.

angles to it. This is achieved with a special material known as Polaroid, which can be arranged to absorb the unwanted component.

Let us consider that a polarizer $(P)$, consisting of a sheet of Polaroid, transmits light vibrating in the horizontal direction $LM$ and absorbs all components vibrating in the direction perpendicular to $LM$. Thus, light passing through the polarizer vibrates in one plane only, and is said to be plane polarized. The plane contains the vibration direction, which is perpendicular to the direction of propagation, and the direction of propagation itself. A second Polaroid, the analyzer $(A)$, is placed after the polarizer and rotated so that its vibration transmission direction $(MN)$ is at 90° to that of the polarizer. It receives no component parallel to its transmission direction and, therefore, absorbs all the light transmitted by the polarizer. The two Polaroids are then said to be crossed. This effect may be demonstrated by cutting a Polaroid sheet marked with a straight line $LMN$ into two sections, $P$ and $A$ (Figure 3.1). When superimposed, the two halves will not transmit light if the reference lines $LM$ and $MN$ are exactly perpendicular. In intermediate positions, the intensity of light transmitted varies from a maximum, where they are parallel, to zero (crossed). The production and use of plane-polarized light by this method is used in the polarizing microscope.

## 3.3   Optical Classification of Crystals

Crystals may be grouped, optically, under two main headings, isotropic crystals and anisotropic (birefringent) crystals. All crystals belonging to the

TABLE 3.1.   Crystal Directions Readily Derivable from an Optical Study

| Optical classification | Crystal system | Information relating to crystal axes likely to be revealed |
|---|---|---|
| Isotropic | Cubic | Axes may be assigned from the crystal morphology |
| Anisotropic, uniaxial | Tetragonal | Direction of $Z$ axis |
| | Hexagonal | Direction of $Z$ axis |
| | Trigonal[a] | Direction of $Z$ axis |
| Anisotropic, biaxial | Orthorhombic | Direction of at least the $X$, $Y$, or $Z$ axis, possibly all three |
| | Monoclinic | Direction parallel to the $Y$ axis |
| | Triclinic | No special relationship between the crystal axes and vibration directions |

[a] Referred to hexagonal axes.

cubic system are optically isotropic; the refractive index of a cubic crystal is independent of direction, and its optical characteristics are similar to those of glass. Noncubic crystals exhibit a dependence on direction in their interaction with light.

Anisotropic crystals are divided into two groups, uniaxial crystals, which have one optically isotropic section and include the tetragonal, hexagonal, and trigonal crystal systems, and biaxial crystals, which have two optically isotropic sections and belong to the orthorhombic, monoclinic, and triclinic crystal systems.

A preliminary optical examination of a crystal will usually show whether it is isotropic, uniaxial, or biaxial. Distinction between the three biaxial crystal systems is often possible in practice and, depending on how well the crystals are developed, a similar differentiation may also be effected for the uniaxial crystals. Even if an unambiguous determination of the crystal system is not forthcoming, the examination should, at least, enable the principal symmetry directions to be identified; Table 3.1 summarizes this information.

## 3.3.1   Uniaxial Crystals

As an example of the use of the polarizing microscope, we shall consider a tetragonal crystal, such as potassium dihydrogen phosphate, lying on a

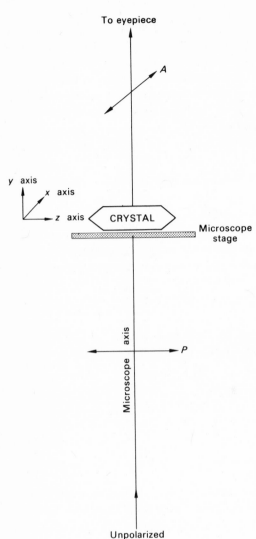

FIGURE 3.2. Schematic experimental arrangement for examining extinction directions. A tetragonal crystal is shown on the microscope stage, and the incident light is perpendicular to the $Z$ axis of the crystal.

microscope slide with its $Y$ axis parallel to the axis of the optical path through a microscope (Figure 3.2). The microscope is fitted with a polarizer $(P)$, and an analyzer $(A)$ which is crossed with respect to $P$ and may be removed from the optical path. The crystal can be rotated on the microscope stage between $P$ and $A$. With the Polaroids crossed and no crystal in between, the field of view is uniformly dark. However, with the crystal interposed, this situation will not necessarily be obtained.

The tetragonal crystal is lying with (010) on the microscope slide; both the $X$ and $Z$ axes are, therefore, perpendicular to the microscope axis. In general, some of the light passing through the crystal will be transmitted by the analyzer, even though $P$ and $A$ are crossed. The intensity of the transmitted light varies as the crystal is rotated on the microscope stage between the polarizer and the analyzer. During a complete revolution of the stage, the intensity of transmitted light passes through four maxima and four minima. At the minimum positions, the crystal is usually only just visible. These positions are called extinction positions, and they occur at exactly 90° intervals of rotation. Maximum intensity is observed with the crystal at 45° to these directions.

These changes would be observed if the crystal itself were replaced by a sheet of Polaroid. Extinction would occur when the vibrations of the "crystal Polaroid" were perpendicular to those of $P$ or $A$. A simple explanation of these effects is that the crystal behaves as a polarizer. Incident plane-polarized light from $P$ is resolved by the crystal into two perpendicular components (Figure 3.3). In our tetragonal crystal, the vibration directions

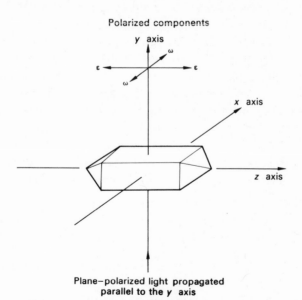

FIGURE 3.3.   Resolution of incident light into components vibrating parallel to the $X$ and $Z$ axes by a tetragonal crystal lying with its $Y$ axis parallel to the incident beam. $\omega$ and $\varepsilon$ are the refractive indices for light vibrating, respectively, perpendicular and parallel to $Z$.

associated with this polarizing effect are parallel to its $X$ and $Z$ axes. Rotating the crystal on the microscope stage will, therefore, produce extinction whenever $X$ and $Z$ are parallel to the vibration directions of $P$ and $A$. The $X$ and $Z$ axes of a tetragonal crystal correspond to its extinction directions. It should be noted that the $X$ and $Y$ directions are equivalent under the fourfold symmetry of the crystal.

### 3.3.2  Birefringence

The vibration components produced by the crystal are associated with different refractive indices. With reference to Figure 3.3, a tetragonal crystal with light vibrating parallel to the fourfold symmetry axis $(Z)$ has a refractive index $\varepsilon$, whereas light vibrating perpendicular to $Z$ has a different refractive index, $\omega$; the crystal is said to be birefringent, or optically anisotropic.

Figure 3.4 represents plane-polarized light incident in a general direction with respect to the crystallographic axes. It is resolved into two components, one with an associated refractive index $\omega$ and the other with an associated refractive index $\varepsilon'$, both vibrating perpendicular to each other and to the direction of incidence. In general, the value of $\varepsilon'$ lies between those of $\omega$ and $\varepsilon$. Two special cases arise: one, already discussed, where the incident light is perpendicular to $Z$, for which $\varepsilon' = \varepsilon$; the second arises where the incident light is parallel to $Z$, for which $\varepsilon' = \omega$. It follows that where the direction of incidence is parallel to the $Z$ axis, the refractive index is always $\omega$ for any vibration direction in the $XY$ plane. Plane-polarized incident light parallel to the $Z$ axis will pass through the crystal unmodified. In this particular direction, the crystal is optically isotropic, and if rotated on the microscope stage between crossed Polaroids, it remains in extinction. The $Z$ direction of a uniaxial crystal is called the optic axis, and there is only one such direction in the crystal.

### Identification of the $Z$ Axis of a Uniaxial Crystal

A polarizing microscope is usually fitted with eyepiece cross-wires arranged parallel and perpendicular to the vibration directions of the polarizer, and therefore we can relate the crystal vibration directions to its morphology. There are two important optical orientations for a tetragonal crystal, namely with the $Z$ axis either perpendicular or parallel to the axis of the microscope. These orientations are, in fact, important for all uniaxial crystals, and will be described in more detail.

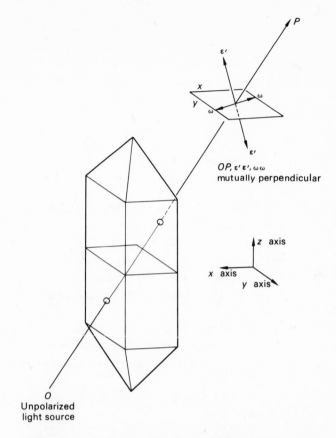

FIGURE 3.4. Uniaxial crystal showing a light ray *OP* resolved into two components. One component, with refractive index $\omega$, vibrates in the *XY* plane, the other, with refractive index $\varepsilon'$, vibrates parallel to both $\omega$ and the ray direction.

*Z Axis Perpendicular to the Microscope Axis.* In this position, a birefringent orientation is always presented to the incident light beam (Figure 3.5). Extinction will occur whenever the *Z* axis is parallel to the cross-wires, no matter how the crystal is rotated, or flipped over, *while keeping Z parallel to the microscope slide.* The success of this operation depends to a large extent on having a crystal with well-developed $(hk0)$ faces. The term straight extinction is used to indicate that the field of view is dark when a crystal edge is aligned with a cross-wire. A face of a uniaxial crystal for which one edge is parallel to *Z*, an $(hk0)$ face, or to its trace on a crystal face, for example, an $(h0l)$ face, will show straight extinction.

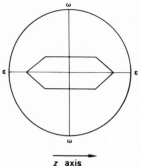

**z axis**

FIGURE 3.5. Extinction position for a tetragonal crystal lying with its $Z$ axis parallel to the microscope slide. Any $[UV0]$ direction may be parallel to the microscope axis; extinction will always be straight with respect to the $Z$ axis or its trace.

*Z Axis Parallel to the Microscope Axis.* The crystal now presents an isotropic section to the incident light beam, and will remain extinguished for all rotations of the crystal, *while keeping Z along the microscope axis.* A reasonably thin section of the crystal is required in order to observe this effect. Because of the needle-shaped habit (external development) of the crystal ($KH_2PO_4$), it would be necessary to cut the crystal carefully so as to obtain the desired specimen.

The section of a uniaxial crystal normal to the $Z$ axis, if well developed, may provide a clue to the crystal system. Tetragonal crystals often have edges at 90° to one another, whereas hexagonal and trigonal crystals often exhibit edges at 60° or 120° to one another. These angles are external manifestations of the internal symmetry; idealized uniaxial crystal sections are shown in Figure 3.6.

### 3.3.3   Biaxial Crystals

Biaxial crystals have two optic axes and, correspondingly, two isotropic directions. The reason for this effect lies in the low symmetry associated with

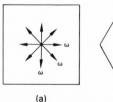
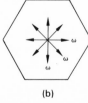
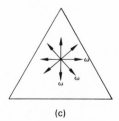

(a)                          (b)                          (c)

FIGURE 3.6. Idealized uniaxial crystals as seen along the $Z$ axis: (a) tetragonal, (b) hexagonal, (c) trigonal. The refractive index for light vibrating perpendicular to the $Z$ axis is always given the symbol $\omega$, and the crystal appears isotropic in this orientation.

the orthorhombic, monoclinic, and triclinic systems, which, in turn, results in less symmetric optical characteristics. Biaxial crystals have three principal refractive indices, $n_1$, $n_2$, and $n_3$ ($n_1 < n_2 < n_3$), associated with light vibrating parallel to three mutually perpendicular directions in the crystal. The optic axes that derive from this property are not directly related to the crystallographic axes. We shall not concern ourselves here with a detailed treatment of the optical properties of biaxial crystals, but will concentrate on relating the vibration, or extinction, directions to the crystal symmetry.

## Orthorhombic Crystals

In the orthorhombic system, the vibration directions associated with $n_1$, $n_2$, and $n_3$ are parallel to the crystallographic axes, but any combination of $X$, $Y$, and $Z$ with $n_1$, $n_2$, and $n_3$ may occur. Consequently, recognition of the extinction directions facilitates identification of the directions of the crystallographic axes. For a crystal with $X$, $Y$, or $Z$ perpendicular to the microscope axis, the extinction directions will be parallel (or perpendicular) to the axis in question, as shown in Figure 3.7. If the crystal is a well-developed orthorhombic prism, the three crystallographic axes may be identified by this optical method. A common alternative habit of orthorhombic crystals has one axis, $X$, for example, as a needle axis and the {011} form prominent. The appearance of such a crystal viewed along $X$ is illustrated in Figure 3.8, and is an example of symmetric extinction.

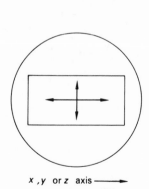

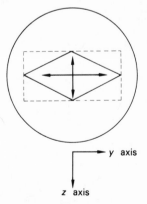

FIGURE 3.7. Extinction directions in an orthorhombic crystal viewed along the $X$, $Y$, or $Z$ axis.

FIGURE 3.8. Extinction directions as seen along the $X$ axis of an orthorhombic crystal with {011} development—an example of symmetric extinction.

## Monoclinic Crystals

The lower symmetry of monoclinic crystals results in a corresponding modification of the optical properties in this system. The symmetry axis $Y$ is chosen, conventionally, to be parallel to one of the vibration directions; $X$ and $Z$ are related arbitrarily to the other two vibration directions. Hence, two directions are of importance in monoclinic crystals, namely, perpendicular to and parallel to the $Y$ axis.

When viewed between crossed Polaroids, a monoclinic crystal lying with its $Y$ axis perpendicular to the microscope axis will always show straight extinction, with the cross-wires parallel (and perpendicular) to $Y$. Often, the $Y$ axis is a well-developed needle axis (Figure 3.9); rotation of the crystal about this axis while keeping it perpendicular to the microscope axis will not cause any change in the extinction positions.

If, on the other hand, a monoclinic crystal is arranged so that $Y$ is parallel to the microscope axis, the (010) plane will lie on the microscope slide. Extinction in this position will, in general, be oblique, as shown in Figure 3.10, thus giving further evidence for the position of the $Y$-axis direction. The appearance of extinction in a monoclinic crystal in this orientation may be somewhat similar to that of an orthorhombic crystal showing prominent $\{011\}$ development (compare Figures 3.8 and 3.10), and confusion may sometimes occur in practice.

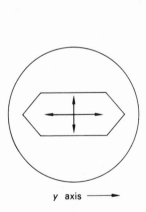

y axis ⟶

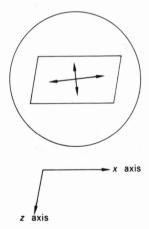

x axis

z axis

FIGURE 3.9. Extinction directions in a monoclinic crystal viewed perpendicular to the $Y$ axis—an example of straight extinction.

FIGURE 3.10. Extinction directions in a monoclinic crystal viewed along the $Y$ axis—an example of oblique extinction.

Triclinic Crystals

The mutually perpendicular vibration directions associated with $n_1, n_2$, and $n_3$ are arbitrarily related to the crystallographic axes, which are selected initially from morphological and X-ray studies.

Reference again to Table 3.1 should now enable the reader to consolidate the ideas presented in the discussion of extinction directions in the seven crystal systems. Although it gives only limited information* on the optical properties of crystals, a practical study of a crystal along these lines can often provide useful information about both its system and its axial directions.

### 3.3.4  Interference Figures

The effects which we have discussed so far may be observed when the crystal specimen is illuminated by a more or less parallel beam of plane-polarized light. There is another technique worthy of mention, in which the crystal is examined in a convergent beam of polarized light, which produces characteristic interference figures for uniaxial and biaxial crystals. This examination may be effected, at high magnification and between crossed Polaroids, either by removing the microscope eyepiece or by inserting a Bertrand lens† into the microscope system, below the eyepiece. Figure 3.11a shows an idealized interference figure from a section of a uniaxial crystal cut perpendicular to the optic axis, while Figures 3.11a and 3.11b are interference figures for a biaxial crystal section cut perpendicular to a bisector of the two optic axes.

If optical figures of good quality can be obtained, the distinction between uniaxial and biaxial specimens may be achieved with one orientation of the crystal. It may be confirmed by rotation of the crystal specimen about the microscope axis, which causes the dark brushes or isogyres in the biaxial figure to break up, as in Figure 3.11c, while those for the uniaxial interference figure remain intact.

## 3.4  Direction of Scattering of X-Rays by Crystals

The first experiments involving the scattering, or diffraction, of X-rays by crystals were initiated by von Laue in 1912. It is well known that similar

---

* For a fuller discussion, see Bibliography.
† A Bertrand lens is a normal accessory with a good polarizing microscope.

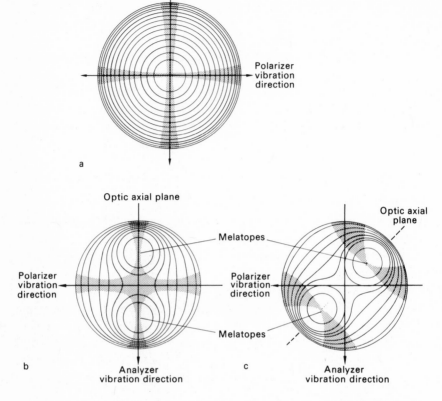

FIGURE 3.11. Interference figures: (a) uniaxial, along the optic axis; (b) biaxial, along a bisector of the optic axes and with the Polaroids crossed; (c) as in (b), but with the polarizer rotated by 45° (position of maximum transmitted intensity). [Reproduced from *An Introduction to Crystal Optics* by P. Gay, with the permission of Longmans Group Ltd., London.]

effects with visible light can be achieved with a ruled grating, provided that the rulings are spaced at about the same order of magnitude as the wavelength of the light. The analogy in a crystallographic experiment is that the X-rays must be of the same order of magnitude as the distance between the scattering units in the crystal. These scattering units are the electron clouds associated with the atoms in the structure, and their regularity is provided by the crystal lattice translations.

The wavelength range of X-rays used in crystallography is between about 0.7 and 2.0 Å. Crystals are, therefore, ideal materials for studying diffraction effects with an X-ray source, and, conversely, X-rays provide a

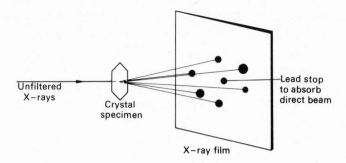

FIGURE 3.12.  Experimental arrangement for taking a Laue photograph on a flat-plat film.

powerful method for investigating crystal structure. The generation and
properties of X-rays are discussed in Appendix A.4.

Figure 3.12 shows, schematically, the experimental arrangement
required to produce a Laue X-ray photograph. The photograph is obtained
by irradiating a stationary single crystal with white X-radiation, which is
composed of a continuous range of wavelengths. Figure 3.13 is a Laue
photograph of $Al_2O_3$; it shows the symmetry of the point group $3m$. In
common with other X-ray photographs, the diagram shows two important

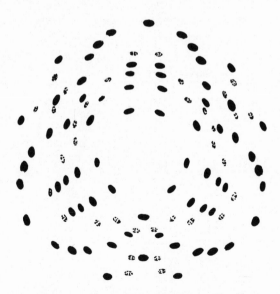

FIGURE 3.13.  Sketch of a Laue photograph of $\alpha$-$Al_2O_3$.

features: The spots occur in definite positions, which are determined by the wavelength of the X-rays and the size and orientation of the unit cell, and the intensity, or degree of blackening, varies from one spot to another. This second feature arises from both the geometry of the experiment and, most particularly, from the crystal structure itself. Thus, an X-ray photograph contains information about several aspects of the internal structure of a crystal, and is, therefore, a fingerprint of the particular specimen. X-ray structure analysis uses this information to deduce the positions of the atoms in the crystal.

This chapter is concerned with the interpretation of the positions of the spots on an X-ray photograph, that is, with the direction of scattering.

### 3.4.1 Laue Equations for X-Ray Scattering

Consider a row of scattering centers of regular spacing $b$ (Figure 3.14). X-rays are incident at an angle $\phi_2$ and are scattered at an angle $\psi_2$. The path difference between rays scattered by neighboring centers is given by

$$\delta_2 = AQ - BP \tag{3.1}$$

or

$$\delta_2 = b(\cos \psi_2 - \cos \phi_2) \tag{3.2}$$

For reinforcement of the scattered rays, the path difference must be an integral number of wavelengths, and we may write

$$b(\cos \psi_2 - \cos \phi_2) = k\lambda \tag{3.3}$$

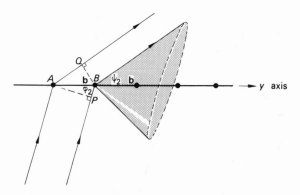

FIGURE 3.14. Diffraction from a row of scattering centers.

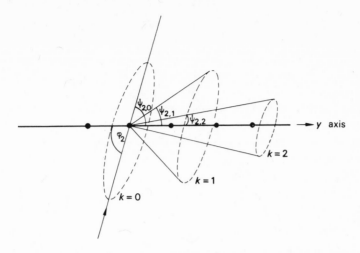

FIGURE 3.15. Several orders of diffraction from a row of scattering centers.

This equation is satisfied by the generators of a cone which is coaxial with the line of scattering centers and has a semi-vertical angle of $\psi_2$. For a series of values of $\phi_2$, there will be a number of such cones, each corresponding to an order of diffraction $k$ and a semi-vertical angle $\psi_{2,k}$ (Figure 3.15).

This discussion is readily extended to a net of scattering centers (Figure 3.16). For the rows parallel to the $X$ axis, we can write, by analogy with (3.3),

$$a(\cos\psi_1 - \cos\phi_1) = h\lambda \qquad (3.4)$$

When both (3.3) and (3.4) are satisfied simultaneously, as they are along the lines of intersection ($BR$ and $BS$) of the two cones, the entire net scatters in phase, producing $hk$ spectra. For the particular case that $BR$ and $BS$ coincide, the diffracted beam lies in the plane of the two-dimensional array of scattering centers.

Generalizing to three dimensions, we may write down the three Laue equations:

$$a(\cos\psi_1 - \cos\phi_1) = h\lambda$$
$$b(\cos\psi_2 - \cos\phi_2) = k\lambda \qquad (3.5)$$
$$c(\cos\psi_3 - \cos\phi_3) = l\lambda$$

Any of the three possible pairs of equations corresponds to scattering from the corresponding net. For the particular case that the three cones intersect

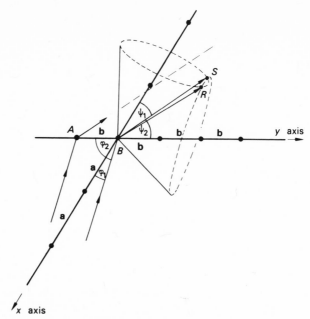

FIGURE 3.16. Diffraction from a net of scattering centers; for clarity only one row parallel to each axis has been drawn.

in a line, the entire three-dimensional array scatters in phase, producing the *hkl*th spectrum.

### 3.4.2   Bragg's Treatment of X-Ray Diffraction

The interaction of X-rays with a crystal is a complex process, often described as a diffraction phenomenon, although it is, strictly speaking, a combined scattering and interference effect. The Bragg treatment of X-ray diffraction, although an oversimplification of the complete process, gives a clear and accurate picture of the directional features of a diffraction pattern, and provides a valuable means for interpreting the positions of the spots on an X-ray photograph.

Any atom in a crystal structure is repeated by the symmetry operations of the space group. The simplest type of structure would consist of a single atom located at the lattice points associated with a primitive unit cell. Figure 3.17a is a representation of one unit cell of such a structure, based on an orthorhombic lattice. Figure 3.17b shows the *c*-axis projection of the same structure, but rotated so as to make the traces of (110) horizontal. Any other family of planes would be just as suitable in this discussion.

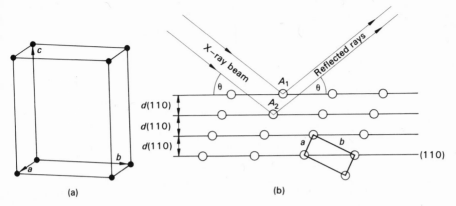

FIGURE 3.17. (a) One $P$ unit cell of an orthorhombic structure; (b) X-ray beam "reflected" by the (110) planes of the orthorhombic structure; two typical rays are shown.

Bragg considered that the crystal planes behaved as though they partially reflected the X-rays, like sheets of atomic mirrors. This approach was not *ad hoc*. The early experiments with X-ray diffraction showed that if a crystal was turned from one diffracting position to another through an angle $\alpha$, then the diffracted ray was rotated through an angle of $2\alpha$. The $\alpha, 2\alpha$ relationship is reminiscent of the reflection of visible light from a plane mirror. The analogy breaks down with X-rays because the Bragg equation (3.14) has to be satisfied, but the reflection treatment is useful, and we shall speak of Bragg reflection, or just reflection, of X-rays by crystals. In this description, the angles of incidence and reflection ($\theta$) are equal and are coplanar with the normal to the reflecting planes, but the angle of incidence used here is the complement of that employed in geometrical optics.

The part of the X-ray beam that is not reflected at a given level in the crystal passes on to be subjected to a similar process at the next level deeper into the crystal. In Figure 3.17b, Bragg reflection of two parallel rays is illustrated, and a special relationship between the interplanar spacing $d(hkl)$, the X-ray wavelength $\lambda$, and the Bragg angle $\theta$ exists when all planes in the $(hkl)$ family cooperate in the scattering process.

A more complete picture is given in Figure 3.18, which demonstrates also that all rays reflected from a given level remain in phase after reflection, since no path difference is introduced. The paths $AB$ and $CD$ are of equal length. However, two rays reflected from neighboring planes are, in general, out of phase because they travel different path lengths. The detailed geometry of this process is shown in Figure 3.19, where the typical path

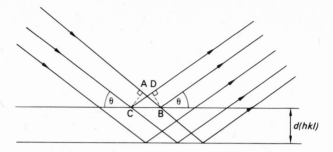

FIGURE 3.18. More complete picture of the Bragg reflection process. Two rays reflected from the same plane do not suffer any relative phase change or path difference ($AB = CD$).

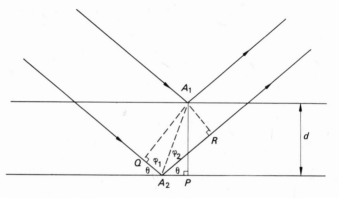

FIGURE 3.19. Detailed geometry of X-ray reflection. The path difference between the two typical rays reflected from successive planes is ($QA_2 + A_2R$), and is equal to $2d \sin \theta$.

difference $\delta$ between two rays is given by

$$\delta = QA_2 + A_2R = A_1A_2 \cos \phi_1 + A_1A_2 \cos \phi_2 \qquad (3.6)$$

or

$$\delta = A_1A_2(\cos \phi_1 + \cos \phi_2) = 2A_1A_2 \cos[(\phi_1 - \phi_2)/2] \cos[(\phi_1 + \phi_2)/2] \qquad (3.7)$$

The three components of (3.7) may be expressed as follows

$$A_1A_2 = d/\sin(\theta + \phi_2) \qquad (3.8)$$

$$\phi_1 + \phi_2 = 180° - 2\theta \qquad (3.9)$$

whence

$$\cos[(\phi_1 + \phi_2)/2] = \cos(90° - \theta) = \sin\theta \qquad (3.10)$$

$$\phi_1 - \phi_2 = 180° - 2(\theta + \phi_2) \qquad (3.11)$$

whence

$$\cos[(\phi_1 - \phi_2)/2] = \cos[90° - (\theta + \phi_2)] = \sin(\theta + \phi_2) \qquad (3.12)$$

Combining terms to give $\delta$, we obtain

$$\delta = 2d \sin\theta \qquad (3.13)$$

Since $\delta$ is independent of $\phi_1$ and $\phi_2$, this equation applies to all rays in the bundle reflected from two adjacent planes. By the usual rules applied to the combination of waves of the same wavelength, the rays reflected by these two planes will interfere with one another, the interference being at least partially destructive unless the path difference $\delta$ is equal to an integral number of wavelengths. Thus,

$$2d \sin\theta = n\lambda \qquad (3.14)$$

which is the Bragg equation, sometimes called Bragg's law. Reflection will be obtained when this equation is satisfied, which may be achieved in practice by varying one of the four quantities $\theta$, $d$, $\lambda$, or $n$.

In (3.14), $n$ is the order of the Bragg reflection. We can write this expression in another form if we recall from page 69 that

$$d(hkl)/n = d(nh, nk, nl) \qquad (3.15)$$

$d(nh, nk, nl)$ is usually replaced by $d(hkl)$, with $h$, $k$, and $l$ taking general values, with or without common factors. Hence, $n$ is included in the crystallographic definition of $d$, and the Bragg equation is now written as

$$2d(hkl) \sin\theta(hkl) = \lambda \qquad (3.16)$$

which means that we consider each Bragg reflection from a crystal as a first-order reflection from the family $(hkl)$, which is specified uniquely by its general Miller indices. To quantify this point further, the following example refers to reflections from planes parallel to (100) in a cube of unit cell side 12 Å.

| Original Bragg formulation | | | Current usage | |
| --- | --- | --- | --- | --- |
| Reflection | Order | $d$, Å | Reflection | $d$, Å |
| 100 | 1 | 12 | 100 | 12 |
| | 2 | 6 | 200 | 6 |
| | 3 | 4 | 300 | 4 |
| | 4 | 3 | 400 | 3 |

### 3.4.3  Equivalence of Laue and Bragg Treatments of X-Ray Diffraction

The treatments exemplified by (3.5) and (3.16) may be shown to be equivalent. Consider any three-dimensional array of scattering centers (Figure 3.20). From (3.3), we may write

$$p(\cos \psi - \cos \phi) = n\lambda \tag{3.17}$$

where $n$ is an integer. Expanding (3.17), we obtain

$$-2p \sin[(\psi + \phi)/2] \sin[(\psi - \phi)/2] = n\lambda \tag{3.18}$$

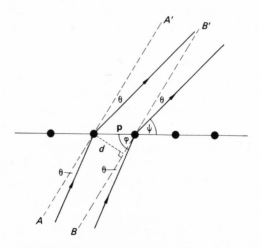

FIGURE 3.20. Equivalence of the Laue and
Bragg equations.

Now $AA'$ and $BB'$ must be the traces of part of the family of planes that make equal angles $\theta$ with the incident and diffracted beams. From the diagram,

$$\phi - \theta = \psi + \theta \tag{3.19}$$

or

$$\theta = (\phi - \psi)/2 \tag{3.20}$$

Furthermore,

$$d = p \, \sin(\phi - \theta) \tag{3.21}$$

which, from (3.20), becomes

$$d = p \, \sin[(\psi + \phi)/2] \tag{3.22}$$

Using (3.18), (3.20), and (3.22), we obtain

$$2d \, \sin \theta = n\lambda \tag{3.23}$$

and $n$ can be incorporated into $d$ in the manner described above. Hence, the two approaches to X-ray scattering by crystals are equivalent. We shall find both of them useful in our subsequent discussions.


## 3.5   X-Ray Techniques

The X-ray photographs in common use can be divided into two classes, single-crystal and powder photographs. If a detailed structure analysis is to be carried out, it is desirable to have well-formed single crystals of the given compound available, and the diffraction data are collected by one of the appropriate photographic methods, or with a single-crystal diffractometer (see Appendix A.5). If, on the other hand, it is required only to characterize a particular substance from its X-ray pattern, then it may be possible to effect identification from powder photographs, taken with a small amount of finely powdered material. Powder photographs are of minimal value in crystal structure analysis, and are not discussed in this book.*

* See Bibliography.

We now continue the preliminary examination of a single crystal by X-ray methods. X-ray photographs can be used to provide the information necessary to confirm the crystal system, to measure the unit-cell dimensions, to determine the number of chemical entities in the unit cell, and to establish, at least partially, the space group.

### 3.5.1   Laue Method

The three variables in the Bragg equation (3.16) provide a basis for the interpretation of X-ray crystallographic experiments. In the Laue method (Figure 3.12), the Bragg equation is satisfied by effectively varying $\lambda$, using a beam of continuous (white) radiation. Since the crystal is stationary with respect to the X-ray beam, it acts as a sort of filter, selecting the correct wavelengths for each reflection according to (3.16).

The spots on a Laue photograph lie on ellipses, all of which have one end of their major axis at the center of the photographic film (Figure 3.13). All spots on one ellipse arise through reflections from planes that lie in one and the same zone. In Figure 3.21, a zone axis for a given Bragg angle $\theta$ is represented by $ZZ'$. A reflected ray is labeled $R$, and we can simulate the effect of the zone by imagining the crystal to be rotated about $ZZ'$, taking the reflected beam with it. The rays, such as $R$, generate a cone, coaxial with $ZZ'$ and with a semivertical angle $\theta$. The lower limit, in the diagram, of $R$ is the direction $(XY)$ of the X-ray beam, and the general intersection of a circle with a plane is an ellipse. Hence, we can understand the general appearance of the Laue photograph. On each ellipse, discrete spots appear instead of continuous bands because only those orientations parallel to zone axes, such as $ZZ'$, that actually exist for crystal planes can give rise to X-ray reflections.

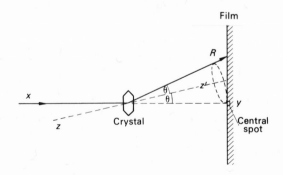

FIGURE 3.21.  Basic geometry of the Laue method.

*Symmetry in Laue Photographs.* One of the most useful features of Laue photographs is the symmetry observable in them. The crystal orientation with respect to the X-ray beam is selected by the experimenter from morphological and optical considerations. This orientation, together with the crystal point group, controls the symmetry on the Laue photograph.

In practice, a complication arises by the introduction of a center of symmetry into the diffraction pattern, in normal circumstances, whether or not the crystal is centrosymmetric. This situation is embodied in Friedel's law, the theoretical grounds for which are discussed in Chapter 4. As a result of this law, the diffraction pattern may not reveal the true point-group symmetry of a crystal. Table 1.6 shows the classification of the 32 crystallographic point groups according to Laue, or diffraction, symmetry.

It cannot be over-emphasized that the Laue group assigned to a crystal describes the symmetry of the *complete* X-ray diffraction pattern from that crystal. No single X-ray photograph can exhibit the complete diffraction symmetry, only that of a selected portion which is a projection, along the direction of the X-ray beam, of the symmetry information that would be encountered in that direction in a crystal having the Laue group of the given crystal.

It follows that in the triclinic system, no symmetry higher than 1 is ever observable in a Laue photograph. In other crystal systems, the Laue-projection symmetry depends on the orientation of the crystal with respect to the X-ray beam. Rotation axes of any order reveal their true symmetry when the X-ray beam is parallel to the symmetry axis. Even-order rotation axes, 2, 4, or 6, give rise to mirror diffraction symmetry in the plane normal to the rotation axis when the X-ray beam is normal to that axis. A mirror plane itself shows $m$ symmetry parallel to the mirror plane when the X-ray beam is contained by the plane. Various combinations of these effects may be observable, depending upon the Laue group in question.

The supplementary nature of the X-ray results to those obtained in the optical examination should be evident now. Uniaxial crystals can be allocated to their correct systems by a Laue photograph taken with the X-ray beam along the $Z$ axis. Figure 3.13 is an example of such a photograph.* Distinction between the monoclinic and orthorhombic systems, which is not always possible in an optical examination, is fairly straightforward with Laue photographs, as Table 1.6 shows. Cubic crystals can exhibit a variety of

---

* Laue projection symmetry $3m$.

symmetries, but with the X-ray beam along $\langle 100 \rangle$, differentiation between Laue groups $m3$ and $m3m$ is obvious.

It should be noted that, in practice, the symmetry pattern on a Laue photograph is very sensitive to precise orientation of the crystal.* Slight deviation from the ideal position will result in a distortion of the relative positions and intensities of the spots on the photographs.

### 3.5.2  Oscillation Method

The oscillation method is a somewhat more sophisticated technique for recording the X-ray diffraction patterns from single crystals. Reflections are produced, in accordance with the Bragg equation, by varying the angle $\theta$ for a given wavelength $\lambda$. The variation of $\theta$ is brought about by oscillating or rotating the crystal about a crystallographic axis, and $\lambda$ is "fixed" by the use of an appropriate filter (see Appendix A.4) placed in the path of the incident X-ray beam.

The basic arrangement used in the oscillation method is illustrated schematically in Figure 3.22. X-ray reflections produced by the moving crystal are recorded on a cylindrical film coaxial with the axis of oscillation of the crystal. The general appearance of an oscillation photograph is illustrated in Figure 3.23. For the moment, we shall concentrate on the periodicity of the crystal parallel to the oscillation axis. Thinking of the crystal as a row of scattering centers, and following the development of (3.1) to (3.3), we see that, for normal incidence ($\phi = 90°$), the diffracted beams will lie on cones that are coaxial with the oscillation axis and intersect the film in circular traces. When the film is flattened out for inspection, the spots are found to lie on parallel straight lines.

*Axial Spacings from Oscillation Photographs.* The equatorial layer line, or zero-layer line, $E$ passes through the origin $O$ where the direct X-ray beam intersects the film (Figures 3.22 and 3.23). The layer-line spacings, measured with respect to the zero layer, are denoted by $\nu(\pm n)$, where $\nu(+n) = \nu(-n)$, and are related to the repeat distance in the crystal parallel to the oscillation axis, as is shown in the following treatment.

Let the oscillation axis be $a$, and let $R$ be the radius of the film, measured in the same units as $\nu$. Consider the crystal to be acting as a one-dimensional diffraction grating with respect to the direction of the $a$ axis, and giving rise to spectra of order $\pm 1$, $\pm 2$, $\therefore$, $\pm n$; $\psi(n)$ is the scattering angle for the $n$th-order maximum, measured with respect to the

* See Bibliography.

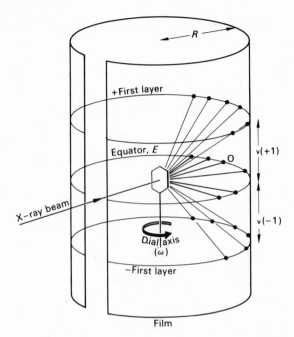

FIGURE 3.22. Basic geometry of the oscillation method, showing how diffraction spots are recorded on a cylindrical film placed around the crystal.

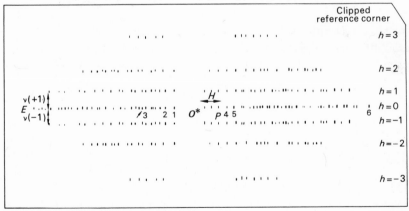

FIGURE 3.23. Sketch of a 15° oscillation photograph of an orthorhombic crystal mounted on the $a$ axis ($a = 6.167$ Å); the camera radius $R$ is 30.0 mm and $\lambda$ (Cu $K\alpha$) = 1.542 Å. The film is flattened out and the right-hand corner, looking toward the X-ray source, is clipped in order to provide a reference mark. $P$ represents any equatorial reflection at a distance $OP$ ($= H$ mm) from the center $O$. Reflections numbered 1–6 on the zero-level are indexed by the method given on page 131. Weissenberg and precession photographs for the same crystal are given in Figures 3.30 and 3.32, respectively. Linear scale of the diagram is $(1/1.78) \times$ true scale.

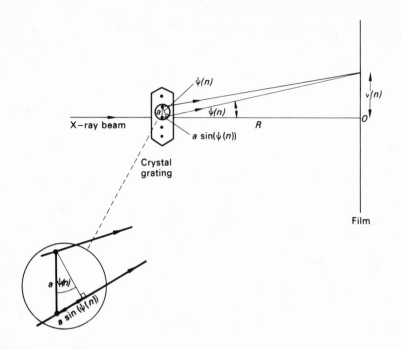

FIGURE 3.24. Diffraction grating analogy explaining the layer-line spacings on oscillation photographs. Monochromatic X-rays are incident normal to the $a$ axis (oscillation axis) of the crystal. The size of any spot at height such as $\nu(n)$ depends upon the experimental conditions.

direct beam. Normal beam diffraction uses the geometry of Figure 3.24. The path difference for rays scattered at an angle $\psi(n)$ by successive elements of the grating is $a \sin \psi(n)$, which, for maximum intensity, is equal to $n\lambda$. Hence, for layer $n$,

$$a \sin \psi(n) = n\lambda \tag{3.24}$$

where $\psi(n)$ is measured experimentally as

$$\tan \psi(n) = \nu(n)/R \tag{3.25}$$

Hence

$$a = \frac{n\lambda}{\sin\{\tan^{-1}[\nu(n)/R]\}} \tag{3.26}$$

For a known wavelength, this equation provides a convenient and reasonably accurate method for determining unit-cell spacings. In practice, we

TABLE 3.2.  Symmetry Indications from Oscillation Photographs

| Feature of photograph | Interpretation(s) |
|---|---|
| Horizontal $m$ line | Horizontal $m$ plane in the corresponding Laue group |
| Vertical $m$ line[a] | $m$ plane in Laue group of crystal, parallel to the plane defined by the oscillation axis and the beam |
| Twofold symmetry about the center of the photograph[a] | Twofold axis in the Laue group of the crystal, and parallel to the X-ray beam |
| Approximate $R$-fold symmetry around the central portion of the photograph[a] | $R$-fold axis in Laue group of crystal, and parallel to the X-ray beam |

[a] Symmetric oscillation photographs.

measure the double spacing between the $\pm n$th orders so as to enhance the precision of the result.

*Symmetry in Oscillation Photographs.* Oscillation photographs have several useful symmetry properties. A horizontal mirror line along the equator $E$ of a general oscillation photograph indicates a mirror plane perpendicular to the oscillation axis in the corresponding Laue group of the crystal (Table 3.2).

Further observations on the symmetry of the Laue group can be made by arranging for a particular crystal symmetry direction to be parallel to the X-ray beam at the midpoint of the oscillation range (symmetric oscillation photograph). The situations that can arise are also summarized in Table 3.2. Note that the highest symmetry observable by the oscillation method is $2mm$, obtained from a symmetric oscillation photograph of an orthorhombic crystal mounted on $a$, $b$, or $c$ and with one of these axes parallel to the X-ray beam at the center of the oscillation range. However, if an $R$-fold rotation axis $(R > 2)$ is parallel to the beam at the midpoint of the oscillation, then the central portion of the photograph will reveal an approximate $R$-fold symmetry pattern, particularly where the reciprocal unit cell is small. The true symmetry will not appear exactly, as it is degraded by the symmetry of the oscillation movement $(mmm)$; the exact symmetries are subgroups of the plane point group $2mm$.

Detection of threefold, fourfold, or sixfold rotational symmetry parallel to the oscillation axis may be effected by taking a series of photographs, the first of which is taken with the crystal oscillating about an arbitrary setting ($\omega_1$) of the dial axis (Figure 3.22). The identical appearance of succeeding photographs with the dial axis set at $\omega_1 + 60°$, $\omega_1 + 90°$, or $\omega_1 + 120°$ indicates sixfold, fourfold, or threefold (and sixfold) symmetry, respectively. A twofold axis cannot be detected by this method, because of Friedel's law.

*Indexing the Zero Level of a Crystal with an Orthogonal Lattice.* We discuss next the relatively small portion of the X-ray diffraction pattern produced in an oscillation photograph. It is of great importance in structure analysis to assign the correct indices *hkl* to each observed reflection. This process is known as indexing, and is reasonably straightforward. In this discussion, we shall consider the indexing of an orthogonal reciprocal lattice, using, as an example, an orthorhombic crystal mounted with *a* as the oscillation axis. It may be noted in passing that monoclinic *b*-axis and hexagonal and trigonal *c*-axis photographs can be indexed in a similar manner. Two prerequisites to indexing are the reciprocal unit-cell dimensions and the orientation of the reciprocal lattice axes perpendicular to the oscillation axis ($b^*$ and $c^*$ in the example) with respect to the incident X-ray beam. The reciprocal unit-cell dimensions may be derived from the corresponding direct space values through (2.17) and (2.18).

### 3.5.3 Ewald's Construction

The geometric interpretation of X-ray diffraction photographs is greatly facilitated by means of a device due to Ewald, and known as the Ewald sphere, or sphere of reflection. The sphere is centered on the crystal ($C$) and drawn with a radius of one reciprocal space unit (RU) on the X-ray beam ($AQ$) as diameter (Figure 3.25). The Bragg construction for reflection is superimposed, and a reflected beam *hkl* cuts the sphere in $P$. The points $A$, $P$, and $Q$ lie on a circular section of the sphere which passes through the center $C$.

From the construction

$$AQ = 2 \qquad \text{(by construction)} \qquad (3.27)$$

$$\widehat{APQ} = 90° \qquad \text{(angle in a semicircle)} \qquad (3.28)$$

Hence

$$QP = AQ \sin \theta(hkl) = 2 \sin \theta(hkl) \qquad (3.29)$$

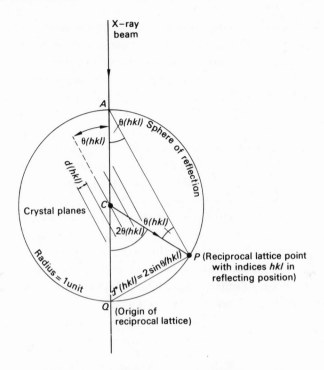

FIGURE 3.25. Ewald construction illustrating how an X-ray reflection may be considered to arise when a reciprocal lattice point $P$ passes through the sphere of reflection. $AP$ is parallel to the $(hkl)$ planes, and the reciprocal lattice vector $QP$ forms a right angle at $P$. (*Note*: If $a$ is the rotation axis and is normal to the circular section shown, this circle becomes the zero-layer circle and reflected beams such as $CP$ will all be denoted $0kl$.)

From Bragg's equation (3.16),

$$2 \sin \theta(hkl) = \lambda/d(hkl) \qquad (3.30)$$

and from the definition of the reciprocal lattice (page 68), we identify the point $P$ with the reciprocal lattice point $hkl$; hence

$$QP = d^*(hkl) \qquad (3.31)$$

with $K = \lambda$ [equation (2.11)], and

$$d^*(hkl) = 2 \sin \theta(hkl) \qquad (3.32)$$

We now have a mechanism for predicting the occurrence of X-ray reflections and their directions in terms of the sphere of reflection and the reciprocal lattice. The origin of the reciprocal lattice is taken at $Q$, and, although the crystal is at $C$, it may be helpful to imagine a conceptual crystal at $Q$ identical to the real crystal and moving about a parallel oscillation axis in a synchronous manner.

The condition that the crystal is in the correct orientation for a Bragg reflection $hkl$ to take place is that the corresponding reciprocal lattice point $P$ is on the sphere of reflection. As the crystal oscillates, an X-ray reflection flashes out each time a reciprocal lattice point cuts the sphere of reflection, and the direction of reflection is given by $CP$.

Ewald's construction provides an elegant illustration of the formation of layer lines on an oscillation photograph. Figure 3.26 shows an Ewald sphere and portions of several layers of an orthogonal reciprocal lattice. As the sphere and the X-ray beam oscillate about the $X$ axis, reciprocal lattice points cut the sphere of reflection in circles because the axis of the cylindrical film is arranged to be parallel to the oscillation axis $X$.

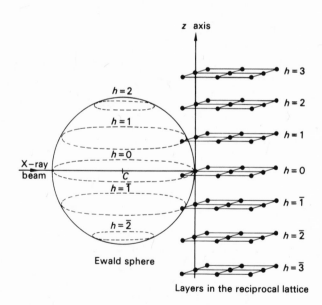

FIGURE 3.26. Formation of layer lines in terms of the Ewald construction. Layers with $|h| \geqslant 3$ lie outside the range of recording in this illustration. Note that except for $h = 0$, the circles labeled $\pm h$ $(h = 1, 2, 3, \ldots)$ cannot be identified with the circular section of Figure 3.25.

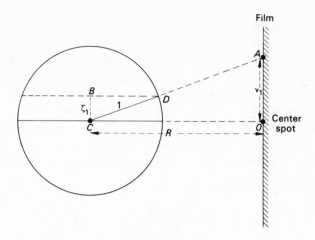

FIGURE 3.27 Relationship between reciprocal lattice spacing
and the corresponding layer-line spacing on a film.

The relationship between the reciprocal lattice spacing of the layers and the corresponding repeat distance $a$ in direct space is shown in Figure 3.27. From the similar triangles $AOC$ and $BCD$,

$$AO/OC = BC/BD \qquad (3.33)$$

or

$$\nu_1/R = \zeta_1/(1 - \zeta_1^2)^{1/2} \qquad (3.34)$$

where $\nu_1$ is the distance between the zero layer and the first layer line and $R$ is the radius of the film. If the lattice is orthogonal in the aspect illustrated ($X^*$ coinciding with $X$), then from (3.30) and (3.32), since $\zeta_1$ is equivalent to $d^*(100)$,

$$a = \lambda/\zeta_1 \qquad (3.35)$$

Although (3.35) holds generally, if the lattice is not orthogonal, $\zeta_1 \neq d^*(100)$ and the appropriate expressions are a little more complicated (see page 71), requiring a knowledge also of the interaxial angles.

*Indexing Procedure for an a-Axis Oscillation Photograph.* On the zero level of an $a$-axis oscillation photograph, reflections are of the type $0kl$.

The relevant portion of the reciprocal lattice is the $Y^*Z^*$ net which, for an orthorhombic crystal, is determined by $b^*\,(=\lambda/b)$, $c^*\,(=\lambda/c)$, and $\alpha^*\,(=90°)$. Without going into further detail, we note that the simplest method of determining $b^*$ and $c^*$ would be from the values of $b$ and $c$, through (3.26) in the appropriate forms.

A drawing of the reciprocal net is prepared carefully, using a convenient scale, for example, 1 RU = 50 mm. The values of $d^*(0kl)$ are obtained from measurements of $H(0kl)$ on the zero-layer line (Figure 3.23), noting also whether the spot lies to the left or the right of the center $O$. Since the angular deviation of the X-ray beam is $2\theta$ (Figure 3.25),

$$2\theta = H/R \qquad\qquad (3.36)$$

in radian measure. Hence,

$$d^*(0kl) = 2\,\sin\theta\,(hkl) = 2\,\sin[180H(0kl)/2\pi R]$$

in degree measure; $H$ and $R$ are, conveniently, measured in millimeters.

*Worked Example of Indexing.* The $a$-axis oscillation photograph of an orthorhombic crystal (Figure 3.23) was taken with $+b^*$ pointing toward the X-ray source at the start of a 15° anticlockwise oscillation. A sample of reflections recorded on the zero level of this photograph at $H$ mm from the center is listed in Table 3.3. For the X-ray wavelength used, $b^* = 0.1123$ and $c^* = 0.1127$ RU. A $Y^*Z^*$ reciprocal net was constructed and used to index these reflections.

The scheme for carrying out this indexing is illustrated in Figure 3.28, with the help of Table 3.4. Instead of thinking in terms of the crystal oscillating (first anticlockwise), we imagine that the sphere of reflection oscillates (first clockwise) about an axis through $Q$, normal to the X-ray beam, from $QA\,(0°)$ to $QA'(15°)$, taking the incident X-ray beam with it. The reciprocal lattice points that would intersect the sphere of reflection are

TABLE 3.3.  Measurements on the Zero-Layer $a$-Axis Photograph

| Reflection number | $H$, mm LHS of center | Reflection number | $H$, mm RHS of center |
|---|---|---|---|
| 1 | 10.75 | 4 | 14.50 |
| 2 | 15.30 | 5 | 17.50 |
| 3 | 27.30 | 6 | 82.50 |

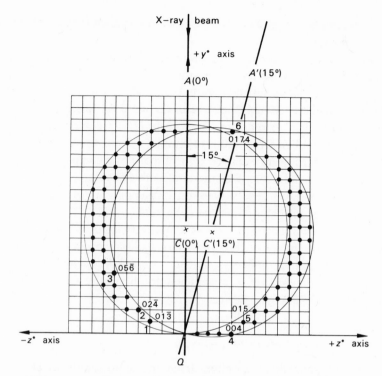

FIGURE 3.28. Indexing the 15° oscillation photograph in Figure 3.23: orthorhombic crystal, $a$ parallel to the rotation axis, $-Y^*$ along the direction of the incident X-ray beam at the start of the oscillation, scale $1\,RU = 50.0\,mm$. Possible reflections are shown as points within the lunes. Reflections 1–6 indexed in Table 3.3 are emphasized. Linear scale of the diagram is $(1/1.78) \times$ true scale.

those within the lunes swept out during this motion. Points lying within these lunes are shown in Figure 3.28, and they correspond to possible $0kl$ reflections for this oscillation movement. By measuring the $d^*$ values from $Q$ to reciprocal lattice points within the lunes on the left- or right-hand side,

TABLE 3.4.  Indexed Reflections for the $0kl$ Layer Line

| Reflection number | $d^*$ | $hkl$ | Reflection number | $d^*$ | $hkl$ |
|---|---|---|---|---|---|
| 1 | 0.356 | $01\bar{3}$ | 4 | 0.450 | 004 |
| 2 | 0.503 | $02\bar{4}$ | 5 | 0.574 | 015 |
| 3 | 0.879 | $05\bar{6}$ | 6 | 1.965 | 017,4 |

as appropriate, the required indices may be determined for the reflections that *do* occur for this crystal.

The construction of the lunes in the correct orientation greatly reduces the number of reciprocal lattice points to be considered. The indexed points are shown in both Table 3.4 and Figure 3.28. Use the diagram to determine which other reflections might have been recorded on this photograph. It should be clear that no reflection for which $d^*$ is greater than 2 can be observed; this number is the radius of another sphere, the limiting sphere, which is that sphere swept out in reciprocal space by a complete rotation of the Ewald sphere. Problem 4 at the end of this chapter is based on this example. It is sometimes necessary to consider a point just outside the lunes, depending on the probable experimental errors.

The Arndt–Wonnacott camera is a modern version of the oscillation camera, designed for use with crystals having axial spacings greater than about 100 Å (certain protein crystals, for example). Very small oscillations are used, and a complete recording of the diffraction pattern is facilitated by an automatic film-changing device.

### 3.5.4   Weissenberg Method

The oscillation method suffers from the main disadvantage that it presents three-dimensional information on a two-dimensional film. It is both a distorted and collapsed diagram of the reciprocal lattice. This situation leads, in turn, to overlapping spots, particularly if the reciprocal lattice dimensions are small, with attendant ambiguity in indexing. More advanced X-ray photographic methods permit rapid, unequivocal indexing without graphical construction.

Figure 3.29 is an extension of Figure 3.22. Each cone corresponds to possible reflections of constant $h$ index. In the Weissenberg method, all cones but one are excluded by means of adjustable metal screens, shown in the figure in the position which permits only the $0kl$ reflections to pass through and reach the film. If the film were kept stationary, the exposed record would still look like the zero-layer line in Figure 3.23. However, in the Weissenberg technique, the film is translated parallel to the oscillation axis, synchronously with the oscillatory motion of the crystal. The result of this procedure is a spreading of the spots over the surface of the film on characteristic straight lines and curves (Figure 3.30). Each spot has a particular position on the film, governed, in this example, by the values of $k$ and $l$. With practice, the film can be indexed by inspection.

If the layer-line screens are moved by the appropriate distance parallel

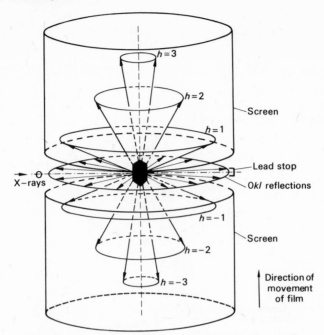

FIGURE 3.29. Cones representing possible directions of diffracted X-rays from a crystal rotating on its *a* axis. The screens are set to exclude all but the zero layer, 0*kl*.

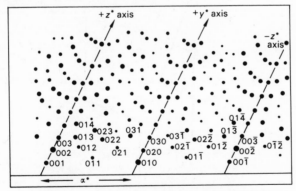

FIGURE 3.30. Sketch of one-half of a partly indexed zero-layer *a*-axis axis Weissenberg photograph of an orthorhombic crystal. The horizontal travel of the film is 1 mm per 2° rotation of the crystal, and $\alpha^* = 90°$, which is correct in the orthorhombic system. Some of the reflections on this diagram may be seen on the zero layer of the 15° oscillation photograph (Figure 3.23). The spots form a distorted 0*kl* reciprocal net on the Weissenberg photograph. A precession photograph for the 1.542 Å; $R = 57.3$ mm). Linear scale of the diagram is $(1/1.78) \times$ true scale.

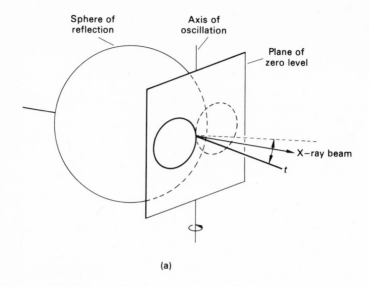

(a)

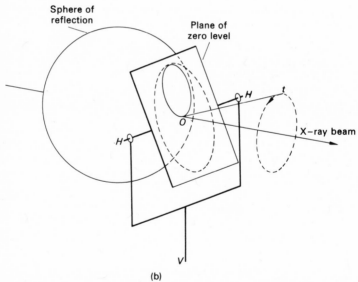

(b)

FIGURE 3.31. Oscillation and precession geometry compared: (a) reciprocal lattice zero-level plane whose normal $t$ is oscillating to equal limits on
each side of the X-ray beam. Maximum symmetry information about the
direction $t$ is $2mm$, the symmetry of an oscillation movement; (b) reciprocal
lattice zero-level plane whose normal $t$ is precessing about the X-ray beam. $H$
and $V$ are horizontal and vertical axes. [Reproduced from *The Precession
Method* by M. J. Buerger, with the permission of John Wiley and Sons Inc.,
New York.]

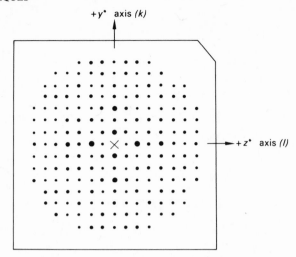

FIGURE 3.32.  Sketch of a precession photograph of an orthorhombic crystal precessing about $a$; an undistorted $0kl$ layer is obtained, permitting values of $b^*$, $c^*$, and $\alpha^*$ to be read directly from the film ($b^*$ and $c^*$ are magnified by a factor equal to the crystal-to-film distance $G$). Oscillation and Weissenberg photographs for the same crystals are given in Figures 3.23 and 3.30, respectively. (Cu $K\alpha$ radiation, $\lambda = 1.542$ Å; $G = 50.0$ mm.) Linear scale of the diagram is $(1/1.78) \times$ true scale.

to the oscillation axis, the first layer ($h = 1$) can be recorded on the film. Generally, nonzero layers are recorded by the equi-inclination technique,* which ensures that their interpretation is very similar to that for the corresponding zero layer.

The Weissenberg photograph is clearly much simpler to interpret than the oscillation photograph, but the apparatus required is more complex. Although the photograph of the reciprocal lattice is not collapsed, it is clear from Figure 3.30 that it is still distorted; the orthorhombic reciprocal net does not appear as rectangular.

### 3.5.5  Precession Method

The precession method produces an undistorted picture of the reciprocal lattice. This is achieved, in principle, by ensuring that a crystal axis $t$ precesses about the X-ray beam and that the film follows the precession motion in such a way that the film is always perpendicular to the crystal axis (Figure 3.31). This is much easier to say than to carry out, and an apparatus of appreciable mechanical complexity is required.* However, the precession photograph is symmetry-true and readily indexed, as shown by Figure 3.32. It is a sort of "contact print" of a layer of the reciprocal lattice.

* See Bibliography.

# Bibliography

### Crystal Optics

GAY, P., *An Introduction to Crystal Optics*, London, Longmans.

HARTSHORNE, N. H., and STUART, A., *Crystals and the Polarising Micro-scope*, London, Arnold.

### X-Ray Scattering and Reciprocal Lattice

BUERGER, M. J., *X-Ray Crystallography*, New York, Wiley.

JEFFERY, J. W., *Methods in X-Ray Crystallography*, London, Academic Press.

WOOLFSON, M. M., *An Introduction to X-Ray Crystallography*, Cambridge, University Press.

### Interpretation of X-Ray Diffraction Photographs

HENRY, N. F. M., LIPSON, H., and WOOSTER, W. A., *The Interpretation of X-Ray Diffraction Photographs*, London, Macmillan.

JEFFERY, J. W., *Methods in X-Ray Crystallography*, London, Academic Press.

### Powder Methods

AZAROFF, L. V., and BUERGER, M. J., *The Powder Method in X-Ray Crystallography*, New York, McGraw-Hill.

### Precession Method

BUERGER, M. J., *The Precession Method*, New York, Wiley.

# Problems

**3.1.** Crystals of $KH_2PO_4$ are needle shaped and show straight extinction parallel to the needle axis. A Laue photograph taken with the X-rays parallel to the needle axis shows symmetry $4mm$.

(a) What is the crystal system, and how is the optic axis oriented?

(b) Describe and explain the appearance between crossed Polaroids of a section cut perpendicular to the needle axis.

(c) What minimum symmetry would be observed on both general and symmetric oscillation photographs taken with the crystal mounted on the needle axis?

**3.2.** Crystals of acetanilide ($C_8H_9NO$) are brick-shaped parallelepipeda, showing straight extinction for sections cut normal to each of the three edges of the "brick."

(a) What system would you assign to the crystals?

(b) Allocate suitable crystallographic axes.

(c) What minimum symmetry would be shown by general oscillation photographs taken, in turn, about each of the three crystallographic axes?

(d) What symmetry would an oscillation photograph exhibit where the crystal is oscillating about the $a$ axis such that $b$ is parallel to the X-ray beam at the center of the oscillation range?

**3.3.** Crystals of sucrose show the extinction directions indicated on the crystal drawing of Figure 3.P1; the arrows indicate the directions of the cross-wires at extinction.

(a) To what crystal system does sucrose belong?

(b) How are the morphological directions, $p$, $q$, and $r$ related to the crystallographic axes?

(c) How would you mount the crystal in order to test your conclusions with (i) Laue photographs, in a single mounting of the crystal, and (ii) oscillation photographs? In each case, indicate the symmetry you

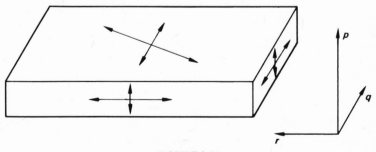

FIGURE 3.P1.

would expect the photographs to exhibit in the orientations you have chosen.

**3.4.** General oscillation photographs of an orthorhombic crystal mounted, in turn, about its $a$, $b$, and $c$ axes had layer-line spacings, measured between the zero and first levels, of 5.07, 7.75, and 9.43 mm, respectively. If $\lambda = 1.50$ Å and $R = 30.0$ mm, calculate $a$, $b$, and $c$ and $a^*$, $b^*$, and $c^*$ for the given wavelength.

Explain why the 146 reflection for this crystal could not be recorded with X-rays of the given wavelength. What symmetry would be observed on the above oscillation photographs? Is this evidence alone conclusive that the crystal is orthorhombic?

**3.5.** (a) A tetragonal crystal is oscillated through 15° (i) about [110] and (ii) about the $c$ axis. The layer-line spacings, measured between the +2 and −2 layers in each case, were 3.5 and 2.7 cm, respectively. If $\lambda = 1.54$ Å and $R = 3.00$ cm, calculate the $a$ and $c$ dimensions of the unit cell.

(b) How many layers could be recorded in position (ii) if the overall film dimension parallel to the oscillation axis is 12.4 cm?

(c) What is the minimum symmetry obtainable on films such as (i) and (ii)?

(d) How would a third photograph taken after rotating the crystal in (ii) through 90° compare with photograph (ii)?

**3.6.** Euphenyl iodoacetate is monoclinic, with $a = 7.260$, $b = 11.55$, and $c = 19.22$ Å. Figure 3.P2 is the Weissenberg zero level ($h0l$). If the film translation constant is 1 mm per 2° rotation, determine $\beta^*$ and $\beta$.

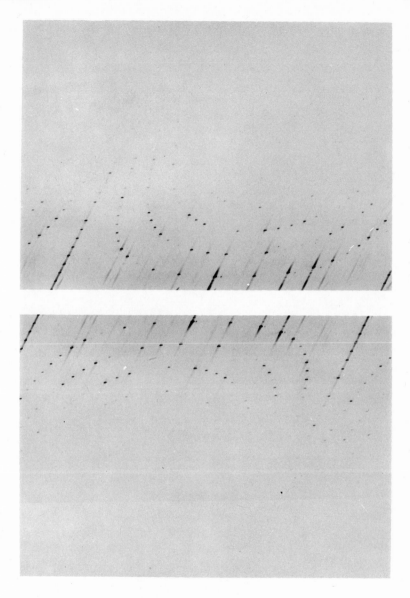

FIGURE 3.P2. Weissenberg photograph of the $h0l$ layer of euphenyl iodoacetate.

# 4

# Intensity of Scattering of X-Rays by Crystals

## 4.1 Introduction

In the previous chapter, we showed how Bragg's equation is used to interpret the geometric features of X-ray photographs. An understanding of the variation in intensity from one diffraction spot to another requires further development of the underlying theory.

The Bragg equation was derived by considering a simple structure in which the atoms were situated at lattice points. Such structures are not unknown, but they are not of a sufficiently general nature for present purposes. Figure 4.1 shows again the reflection of X-rays from any $(hkl)$ family of planes. The plane $O$ passes through the origin of the unit cell, and may be called the zero $(hkl)$ plane. The plane $O'$ is the first $(hkl)$ from the origin and, therefore, is at a perpendicular distance $d(hkl)$ from $O$. The path

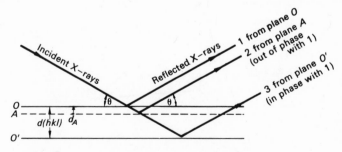

FIGURE 4.1. Construction for Bragg reflection from $(hkl)$ planes, taking into account the variation of atomic distribution in the unit cell. Planes $O$ and $O'$ contain a different sample of atoms from that contained by the parallel conceptual plane $A$. The ray 2 reflected from plane $A$ is out of phase with those reflected (1 and 3) from planes $O$ and $O'$ in the same $(hkl)$ family.

difference between X-rays reflected from the plane $O$ and those reflected from the adjacent plane $O'$ is, from (3.16), $2d(hkl)\sin\theta(hkl)$.

Consider an atom lying between the planes $O$ and $O'$; we can imagine a plane $A$, parallel to $(hkl)$, passing through this atom. For an incident X-ray beam in the correct reflecting position, the planes $O$ and $O'$ will reflect in phase with each other, but, in general, out of phase with the reflection from plane $A$. We require to determine the path difference between waves reflected by planes $O$ and $A$ in any general situation.

## 4.2   Path Difference

In Figure 4.2, the plane $LMN$, at a perpendicular distance $p$ from the origin O, makes intercepts $a_p$, $b_p$, and $c_p$ with the general triclinic axes $X$, $Y$, and $Z$, respectively. The perpendicular $OP$ makes angles $\chi$, $\psi$, and $\omega$ with the same axes. The intercept form of the equation of this plane is, from (1.10),

$$X/a_p + Y/b_p + Z/c_p = 1 \qquad (4.1)$$

where

$$a_p = p/\cos\chi, \qquad b_p = p/\cos\psi, \qquad c_p = p/\cos\omega \qquad (4.2)$$

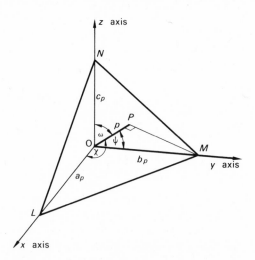

FIGURE 4.2. Plane $LMN$ referred to general (triclinic) axes.

Equation (4.1) may be rewritten as

$$X \cos \chi + Y \cos \psi + Z \cos \omega = p \qquad (4.3)$$

This equation represents a plane parallel to $LMN$, in terms of the perpendicular distance from the origin, which may be regarded as a variable. Two planes of interest, $O$ and $A$ (Figure 4.1), can be expressed in this way. If plane $LMN$ is parallel to $(hkl)$, then from (4.2)

$$\cos \chi = \frac{d(hkl)}{a/h}, \qquad \cos \psi = \frac{d(hkl)}{b/k}, \qquad \cos \omega = \frac{d(hkl)}{c/l} \qquad (4.4)$$

Since plane $A$ is parallel to $(hkl)$ and at a distance $d_A$ from the origin, its equation is

$$X \cos \chi + Y \cos \psi + Z \cos \omega = d_A \qquad (4.5)$$

Substituting for the cosines from (4.4), we have

$$[(hX/a) + (kY/b) + (lZ/c)]d(hkl) = d_A \qquad (4.6)$$

Let atom $A$ have fractional coordinates $x_A = X_A/a$, $y_A = Y_A/b$, and $z_A = Z_A/c$. Then

$$d_A = (hx_A + ky_A + lz_A)d(hkl) \qquad (4.7)$$

By analogy with the Bragg equation (3.16), the path difference $\delta_A$ between X-rays reflected from planes $O$ and $A$ is given by

$$\delta_A = 2d_A \sin \theta(hkl) \qquad (4.8)$$

or

$$\delta_A = 2d(hkl)[\sin \theta(hkl)](hx_A + ky_A + lz_A) \qquad (4.9)$$

which, from Bragg's equation, gives

$$\delta_A = \lambda(hx_A + ky_A + lz_A) \qquad (4.10)$$

This relationship is important; it provides a quantitative measure of the path difference between rays reflected by different conceptual planes of atoms, all parallel to any given $(hkl)$ plane, with respect to the parallel plane through the origin. It must be realized that in a real situation there need be no atom lying on the $(hkl)$ planes; the electron density which is associated with an atom pervades the whole unit cell space.

## 4.3  Combination of Two Waves

A given $hkl$ reflection consists of the combined scattering by all atoms in the structure. One unit cell is a sufficient representative portion of the structure: Waves scattered by neighboring unit cells will be in phase if Bragg's equation is satisfied. We shall assume that the crystal consists of a geometrically perfect stacking of unit cells; this assumption is considered in Appendix A.5.

Scattered X-rays of interest in crystal structure analysis are those scattered without change in wavelength, or frequency, and so we consider next the combination of two waves (scattered by two atoms) of the same frequency, but having, in general, different amplitudes and phases.

Figure 4.3 illustrates two such waves $W_1$ and $W_2$ of angular frequency $\omega$ and having amplitudes $f_1$ and $f_2$ and corresponding phases $\phi_1$ and $\phi_2$. These waves may be represented analytically as

$$W_1 = f_1 \cos(\omega t - \phi_1) = f_1[\cos \omega t \cos \phi_1 + \sin \omega t \sin \phi_1] \qquad (4.11)$$

$$W_2 = f_2 \cos(\omega t - \phi_2) = f_2[\cos \omega t \cos \phi_2 + \sin \omega t \sin \phi_2] \qquad (4.12)$$

and their sum $W_1 + W_2$ is

$$W = (\cos \omega t)(f_1 \cos \phi_1 + f_2 \cos \phi_2) + (\sin \omega t)(f_1 \sin \phi_1 + f_2 \sin \phi_2) \qquad (4.13)$$

Let $W$ be of the form $F \cos(\omega t - \phi)$. Then

$$W = F[\cos \omega t \cos \phi + \sin \omega t \sin \phi] \qquad (4.14)$$

By comparing coefficients in (4.13) and (4.14), we obtain

$$F \cos \phi = f_1 \cos \phi_1 + f_2 \cos \phi_2 \qquad (4.15)$$

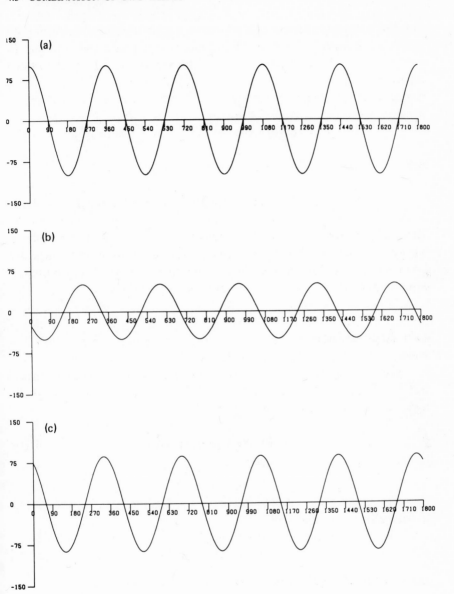

FIGURE 4.3. Combination of two waves of different amplitudes $f_1$ and $f_2$ and different phases $\phi_1$ and $\phi_2$, respectively. The resultant wave has an amplitude $|F|$ and a phase $\phi$. No change of wavelength (or frequency) takes place: (a) $f_1 = 100$, $\phi_1 = 0°$, (b) $f_2 = 50$, $\phi_2 = 240°$; (c) $|F| = [(100 + 50 \cos 240°)^2 + (50 \sin 240°)^2]^{1/2} = 86.6$, $\phi = \tan^{-1}[(50 \sin 240°)/(100 + 50 \cos 240°)] = 330°$.

and

$$F \sin \phi = f_1 \sin \phi_1 + f_2 \sin \phi_2 \qquad (4.16)$$

Hence

$$F = [(f_1 \cos \phi_1 + f_2 \cos \phi_2)^2 + (f_1 \sin \phi_1 + f_2 \sin \phi_2)^2]^{1/2} \qquad (4.17)$$

and

$$\tan \phi = (f_1 \sin \phi_1 + f_2 \sin \phi_2)/(f_1 \cos \phi_1 + f_2 \cos \phi_2) \qquad (4.18)$$

This analysis shows that the combination of two waves of the same frequency but with different amplitudes and phases results in a third wave of the same frequency but with its own amplitude and phase (Figure 4.3). Continued combination of several such waves gives further similar results.

## 4.4  Argand Diagram

The combination of waves may be represented in a clear and concise manner by an Argand diagram. The waves are represented as vectors with real and imaginary components. Thus,

$$\mathbf{f}_1 = f_1 \cos \phi_1 + if_1 \sin \phi_1 \qquad (4.19)$$

$$\mathbf{f}_2 = f_2 \cos \phi_2 + if_2 \sin \phi_2 \qquad (4.20)$$

These equations are illustrated in Figure 4.4; $\mathbf{F}$ is the resultant vector. De Moivre's theorem states that

$$e^{\pm i\phi} = \cos \phi \pm i \sin \phi \qquad (4.21)$$

and provides an even more convenient summary of these expressions:

$$\mathbf{f}_1 = f_1 e^{i\phi_1}, \qquad \mathbf{f}_2 = f_2 e^{i\phi_2} \qquad (4.22)$$

Hence,

$$\mathbf{F} = f_1 e^{i\phi_1} + f_2 e^{i\phi_2} \qquad (4.23)$$

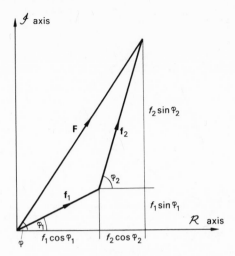

FIGURE  4.4.  Combination of two waves as
vectors on an Argand diagram.

$\mathbf{F}$, $\mathbf{f}_1$, and $\mathbf{f}_2$ are all vectors,* having both magnitude and direction on the
Argand diagram; $e^{i\phi}$ may be regarded as a mathematical operator which
rotates a vector anticlockwise in the complex plane through an angle $\phi$
measured from the real axis.

## 4.5  Combination of $N$ Waves

Extension of the foregoing analysis enables us to combine any number
of waves. The resultant of $N$ waves is, from (4.23),

$$\mathbf{F}=f_1e^{i\phi_1}+f_2e^{i\phi_2}+f_3e^{i\phi_3}+\cdots+f_je^{i\phi_j}+\cdots+f_Ne^{i\phi_N} \qquad (4.24)$$

or

$$\mathbf{F}=\sum_{j=1}^{N} f_je^{i\phi_j} \qquad (4.25)$$

* In many textbooks, $\mathbf{F}$ is not written in bold (vector) type in relation to its representation
through the Argand diagram. Our nomenclature has been chosen in order to draw attention to
the amplitude and phase characteristics of $\mathbf{F}$.

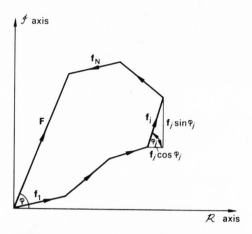

FIGURE 4.5. Combination of $N$ waves $(N = 6)$
on an Argand diagram.

On an Argand diagram, (4.25) expresses a polygon of vectors (Figure 4.5);
the resultant **F** is given by

$$\mathbf{F} = |F| e^{i\phi} \tag{4.26}$$

The amplitude $|F|$ is given by

$$|F|^2 = \mathbf{F}\mathbf{F}^* \tag{4.27}$$

where $\mathbf{F}^*$ is the complex conjugate of $\mathbf{F}$:

$$\mathbf{F}^* = |F| e^{-i\phi} \tag{4.28}$$

By analogy with (4.17),

$$|F| = (A'^2 + B'^2)^{1/2} \tag{4.29}$$

where

$$A' = \sum_{j=1}^{N} f_j \cos \phi_j \tag{4.30}$$

and

$$B' = \sum_{j=1}^{N} f_j \sin \phi_j \tag{4.31}$$

$A'$ and $B'$ are, respectively, the real and imaginary components of $\mathbf{F}$, and the phase angle $\phi$ is given by

$$\tan \phi = B'/A' \qquad (4.32)$$

## 4.6 Combined Scattering of X-Rays from the Contents of the Unit Cell

We may now employ the formulae just deduced in order to obtain an equation for the resultant scattering by all the atoms in a unit cell. We shall consider a general structure consisting of $N$ atoms, not necessarily of the same species, occupying fractional coordinates $x_j, y_j, z_j$ $(j = 1, 2, \ldots, N)$ in a unit cell.

### 4.6.1 Phase Difference

The path difference associated with waves scattered by an atom $j$ whose position relative to the origin is specified by the coordinates $x_j, y_j, z_j$ is given, by analogy with (4.10), as

$$\delta_j = \lambda (hx_j + ky_j + lz_j) \qquad (4.33)$$

The corresponding phase difference (angular measure) is given by

$$\phi_j = (2\pi/\lambda)\delta_j \qquad (4.34)$$

or

$$\phi_j = 2\pi(hx_j + ky_j + lz_j) \qquad (4.35)$$

### 4.6.2 Scattering by Atoms

In order to evaluate the combined scattering from the atoms in the unit cell, we need also the amplitudes of the waves scattered by the atoms, the atomic scattering factors. Although well known in practice, the evaluation of atomic scattering factors is a complicated calculation; they are tabulated* as functions of $(\sin \theta)/\lambda$, and denoted as $f_{j,\theta}$, or just $f_j$. The atomic scattering

---

* See Bibliography.

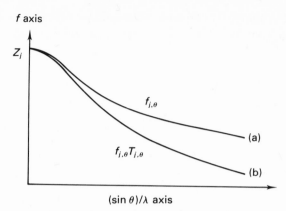

FIGURE 4.6  Atomic scattering factors: (a) stationary atom, $f_{j,\theta}$, (b) atom corrected for thermal vibration, $f_{j,\theta}T_{j,\theta}$.

factor depends upon the nature of the atom, the direction of scattering, the wavelength of X-rays used, and the thermal vibrations of the atom; it is a measure of the efficiency with which an atom scatters with respect to a single electron.

In the first place, $f$ depends upon the number of extranuclear electrons in the atom; its maximum value for a given atom $j$ is $Z_j$, the atomic number of the $j$th atomic species. Along the direction of the incident beam $[\sin \theta(hkl) = 0]$, $f$ has its maximum value:

$$f_{j,\theta(\theta=0)} = Z_j \tag{4.36}$$

$f$ is measured in electrons.

Figure 4.6a shows the general form of the variation of $f_j$ with $(\sin \theta)/\lambda$ for an atom at rest. Since the electrons are distributed over a finite volume in an atom, interference takes place within the atom, and the overall effect of $Z_j$ electrons is diminished. In the forward direction $(\sin \theta = 0)$ there is no interference [equation (4.36)]. Interference increases with increasing $\sin \theta$ and decreasing $\lambda$, since, in both cases, the path differences within the atom become relatively larger. The decrease of $f$, being a function of $(\sin \theta)/\lambda$, is also, for a given crystal, a function of $hkl$.

## Thermal Vibration and Temperature Factor

The fourth factor influencing the amplitude of scattering from an atom is the thermal vibration of the particular atom in a given crystal. Each atom in a structure vibrates, in general, in an anisotropic manner, and an exact description of this motion involves several parameters which are dependent upon direction. For simplicity, we shall assume isotropic vibration, in which case the temperature factor correction for the $j$th atom is

$$T_{j,\theta} = \exp[-B_j(\sin^2\theta)/\lambda^2] \tag{4.37}$$

$B_j$ is the temperature factor of atom $j$. It is given by

$$B_j = 8\pi^2\overline{U_j^2} \tag{4.38}$$

where $\overline{U_j^2}$ is the mean square amplitude of vibration of the $j$th atom from its equilibrium position in a direction normal to the reflecting plane, and is a function of temperature. The factor $T$, like $f$, is a function of $(\sin\theta)/\lambda$ and, hence, of $hkl$. We may write for the temperature-corrected atomic scattering factor

$$g_j = f_{j,\theta}T_{j,\theta} \tag{4.39}$$

Thermal vibrations increase the effective volume of the atom, and interference within the atom becomes more noticeable. Consequently, $f$ falls off with increasing $\sin\theta$ more rapidly the higher the temperature (Figure 4.6b).

## 4.7 Structure Factor

The structure factor $\mathbf{F}(hkl)$ expresses the combined scattering of all atoms in the unit cell compared to that of a single electron; its amplitude $|F(hkl)|$ is measured in electrons. The components required for the combined scattered wave from the $(hkl)$ planes are $g_{j,\theta}$ and $\phi_j(hkl)$, which, *for brevity*, will often be referred to as $g_j$ and $\phi_j$, respectively. The individual scattered waves from the various atoms $j$ are vectors of the form given by the general term in (4.24). The resultant wave for the unit cell is therefore

$$\mathbf{F}(hkl) = \sum_{j=1}^{N} g_j \exp(i\phi_j) = \sum_{j=1}^{N} g_j \exp[i2\pi(hx_j + ky_j + lz_j)] \tag{4.40}$$

From Figure 4.7, it may be seen that

$$\mathbf{F}(hkl) = A'(hkl) + iB'(hkl) \tag{4.41}$$

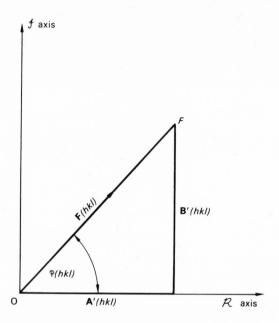

FIGURE 4.7. Structure factor $\mathbf{F}(hkl)$ plotted on an Argand diagram; $\phi(hkl)$ is the resultant phase, and the amplitude $|F(hkl)|$ is represented by $OF$.

where

$$A'(hkl) = \sum_{j=1}^{N} g_j \cos 2\pi(hx_j + ky_j + lz_j) \qquad (4.42)$$

and

$$B'(hkl) = \sum_{j=1}^{N} g_j \sin 2\pi(hx_j + ky_j + lz_j) \qquad (4.43)$$

By comparison with (4.26), we may write

$$\mathbf{F}(hkl) = |F(hkl)|e^{i\phi(hkl)} \qquad (4.44)$$

where the amplitude is given by

$$|F(hkl)| = [A'^2(hkl) + B'^2(hkl)]^{1/2} \qquad (4.45)$$

and the phase by

$$\tan \phi(hkl) = B'(hkl)/A'(hkl) \tag{4.46}$$

From Figure 4.7, it may be seen that

$$A'(hkl) = |F(hkl)| \cos \phi(hkl) \tag{4.47}$$

and

$$B'(hkl) = |F(hkl)| \sin \phi(hkl) \tag{4.48}$$

## 4.8  Intensity Expressions

The energy associated with a cosine wave is proportional to the square of the amplitude of the wave. In an X-ray diffraction experiment, it is expressed in terms of the intensity of the scattered wave from the unit cell, $I_o(hkl)$; the subscript $o$ signifies an experimentally observed quantity. Since the amplitude of the combined wave is, from (4.45), $|F(hkl)|$, we use the symbol $I(hkl)$ to represent $|F(hkl)|^2$, sometimes called the ideal intensity. Hence,

$$I_o(hkl) \propto |F_o(hkl)|^2 \tag{4.49}$$

or

$$I_o(hkl) = K^2 C(hkl)|F_o(hkl)|^2 \tag{4.50}$$

where $C(hkl)$ combines several geometric and physical factors which depend upon both $hkl$ and the experimental conditions, notably the Lorentz, polarization, and absorption corrections (see Appendix A.5). The factor $K$ is a scale factor associated with $|F_o(hkl)|$.

Equation (4.50) forms the basis of X-ray structure analysis: The experimental quantities $I_o(hkl)$ are directly related to the structure through $|F_o(hkl)|$ [vide (4.40) and (4.44)], and represent the information from which the crystallographer begins to unravel a structure in terms of the atomic coordinates.

## 4.9   Phase Problem in Structure Analysis

The determination of a crystal structure cannot proceed directly from the observed intensity data. Structure analysis is hampered fundamentally by the inability to determine, in an X-ray diffraction experiment, the complete vectorial structure factor. The modulus $|F(hkl)|$ can be obtained from the intensity data, but the corresponding phase $\phi(hkl)$ is not directly measurable. Determination of the structure must, however, be carried out in terms of both amplitude and phase.

Image formation in a microscope is obtained mechanically by focusing the lens system. Scattered X-rays cannot be focused by any known experimental procedure. Values for the phase angles have to be obtained, and this process calls for a solution of the phase problem—the central problem in X-ray structure analysis—and is the main subject of Chapter 6.

## 4.10   Applications of the Structure Factor Equation

The remainder of this chapter will be concerned with applications of the structure factor equation to preliminary structure analysis, and the development of relationships which have a bearing on the more detailed techniques to be discussed later.

### 4.10.1   Friedel's Law

In normal circumstances, the X-ray diffraction pattern from a crystal is centrosymmetric, whatever the crystal class (see pages 45 and 123). A diffraction pattern may be thought of as a reciprocal lattice with each point weighted by the corresponding value of $I(hkl)$. Friedel's law states the centrosymmetric property of the diffraction pattern as

$$I(hkl) = I(\bar{h}\bar{k}\bar{l}) \tag{4.51}$$

which may be derived as follows.

Since the atomic scattering factor is a function of $(\sin\theta)/\lambda$, $g_j$ will be the same for both the $hkl$ and the $\bar{h}\bar{k}\bar{l}$ reflections. Thus,

$$g_{j,\theta} = g_{j,-\theta} \tag{4.52}$$

because reflection from opposite sides of any plane will occur at the same

Bragg angle $\theta$. This equality may be shown also through (2.16) and (3.32); however, it should be noted that it depends upon the spherically symmetric model of an atom, which is generally used in the calculation of $f$ values.

From (4.40),

$$\mathbf{F}(hkl) = \sum_{j=1}^{N} g_{j,\theta} \exp[i2\pi(hx_j + ky_j + lz_j)] \tag{4.53}$$

and

$$\mathbf{F}(\bar{h}\bar{k}\bar{l}) = \sum_{j=1}^{N} g_{j,-\theta} \exp[-i2\pi(hx_j + ky_j + lz_j)] \tag{4.54}$$

From (4.41),

$$\mathbf{F}(\bar{h}\bar{k}\bar{l}) = A'(\bar{h}\bar{k}\bar{l}) + iB'(\bar{h}\bar{k}\bar{l}) \tag{4.55}$$

where $A'(\bar{h}\bar{k}\bar{l})$ and $B'(\bar{h}\bar{k}\bar{l})$ are given by (4.42) and (4.43), respectively.

From (4.21), and (4.52)–(4.55), we can now derive

$$\mathbf{F}(\bar{h}\bar{k}\bar{l}) = A'(hkl) - iB'(hkl) \tag{4.56}$$

The vectorial representations of $\mathbf{F}(hkl)$ and $\mathbf{F}(\bar{h}\bar{k}\bar{l})$ are shown on an Argand diagram in Figure 4.8. Several important relationships now follow:

$$\phi(\bar{h}\bar{k}\bar{l}) = -\phi(hkl) \tag{4.57}$$

$$|F(hkl)| = |F(\bar{h}\bar{k}\bar{l})| = [A'^2(hkl) + B'^2(hkl)]^{1/2} \tag{4.58}$$

Since $I(hkl) \propto |F(hkl)|^2$, from (4.49),

$$I(hkl) = I(\bar{h}\bar{k}\bar{l}) \tag{4.59}$$

which is Friedel's law. Apart from minor variations connected with anomalous dispersion (page 293), this equality is observable within the limits of experimental error.

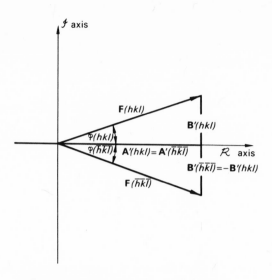

FIGURE 4.8. Relationship between $\mathbf{F}(hkl)$ and $\mathbf{F}(\bar{h}\bar{k}\bar{l})$ leading to Friedel's law, from which $|F(hkl)| = |F(\bar{h}\bar{k}\bar{l})|$.

## 4.10.2  Structure Factor for a Centrosymmetric Crystal

One of the questions arising at the outset of a crystal structure analysis is whether the space group is centrosymmetric or not. In the following discussion of the determination of space groups from systematic absences, it turns out that often an ambiguity arises which is connected with centrosymmetry.

Consider a centrosymmetric structure with $N$ atoms per unit cell, the origin of coordinates being at a center of symmetry in the unit cell. There will be $N/2$ atoms in the unit cell with positions independent of the center of symmetry, assuming that no atoms lie on a center of symmetry. Thus, for any atom at $x_j$, $y_j$, $z_j$, there is a centrosymmetrically related atom of the same type at $\bar{x}_j$, $\bar{y}_j$, $\bar{z}_j$, and the real and imaginary components of the structure factor (4.42) and (4.43) become

$$A'(hkl) = \sum_{j=1}^{N/2} g_j[\cos 2\pi(hx_j + ky_j + lz_j)$$

$$+ \cos 2\pi(-hx_j - ky_j - lz_j)] \qquad (4.60)$$

and

$$B'(hkl) = \sum_{j=1}^{N/2} g_j[\sin 2\pi(hx_j + ky_j + lz_j)$$

$$+ \sin 2\pi(-hx_j - ky_j - lz_j)] \qquad (4.61)$$

Since, for any angle $\phi$, $\cos(-\phi) = \cos \phi$ and $\sin(-\phi) = -\sin \phi$.

$$A'(hkl) = 2 \sum_{j=1}^{N/2} g_j \cos 2\pi(hx_j + ky_j + lz_j) \qquad (4.62)$$

$$B'(hkl) = 0 \qquad (4.63)$$

The summation in (4.62) is taken over $N/2$ noncentrosymmetrically related atoms, and $A'(hkl)$ is now equivalent to $F(hkl)$.

This important result has the further consequence that the phase angle has only two possible values. From (4.46), if $A'(hkl)$ is positive,

$$\phi(hkl) = 0 \qquad (4.64)$$

whereas if $A'(hkl)$ is negative,

$$\phi(hkl) = \pi \qquad (4.65)$$

From (4.47) and (4.48), we see that the phase angle then attaches itself to $|F(hkl)|$ as a positive or negative sign, and we often speak of the signs of the reflections in centrosymmetric crystals instead of the phases. If we use the symbol $s(hkl)$ as the sign of a centrosymmetric reflection, then

$$F(hkl) = s(hkl)|F(hkl)| \qquad (4.66)$$

These results are illustrated in Figure 4.9. By taking the origin at a center of symmetry, the value of the phase angle is restricted to 0 or $\pi$. The phase problem is much simplified, and centrosymmetric crystals usually present fewer difficulties to the structure analyst than do noncentrosymmetric crystals, in which the phase angles range between 0 and $2\pi$.

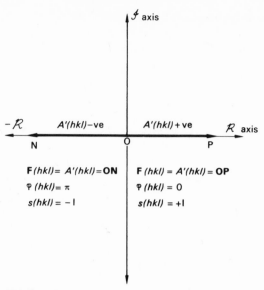

FIGURE 4.9. Structure factor for a centrosymmetric crystal (origin on $\bar{1}$); $\mathbf{F}(hkl) = A'(hkl)$ and can have one of two possible phases, 0 (right-hand side) or $\pi$ (left-hand side).

## 4.10.3  Limiting Conditions and Systematic Absences

The X-ray diffraction pattern can always be used to determine the unit-cell type corresponding to the chosen axial system. From the form of (4.40), it would be an unexpected coincidence for many intensities to be zero. With primitive unit cells, the intensity of reflection is not, in general, zero for any particular combination of $h$, $k$, and $l$. Very weak intensities may arise for certain reflections because of the particular structure under investigation; we call these unobservable reflections "accidental absences."

### Body-Centered Unit Cell

In any body-centered unit cell an atom at $x_j$, $y_j$, $z_j$ is related by translation to another atom of the same type at $\frac{1}{2}+x_j$, $\frac{1}{2}+y_j$, $\frac{1}{2}+z_j$. From (4.40), assuming $N$ atoms in the unit cell, we can write a summation for the $N/2$ atoms *not* related by translation:

$$\mathbf{F}(hkl) = \sum_{j=1}^{N/2} g_j \{\exp[\mathrm{i}2\pi(hx_j + ky_j + lz_j)]\cdot$$

$$+\exp[\mathrm{i}2\pi(hx_j + ky_j + lz_j + \tfrac{1}{2}(h + k + l))]\} \qquad (4.67)$$

The term in curly brackets may be expressed as

$$\{\exp[i2\pi(hx_j + ky_j + lz_j)]\}\{1 + \exp[i\pi(h + k + l)]\} \qquad (4.68)$$

From (4.21), since $(h + k + l)$ is an integer, say $m$,

$$1 + e^{i\pi(h+k+l)} = 1 + \cos(m\pi) \qquad (4.69)$$

and

$$1 + \cos(m\pi) = 2\cos^2(m\pi/2) \qquad (4.70)$$

or

$$2\cos^2(m\pi/2) = 2\cos^2[2\pi(h + k + l)/4] = G(hkl), \quad \text{say} \qquad (4.71)$$

From (4.67)–(4.71) we can state some useful results. The reduced structure factor equation for a body-centered unit cell may be written as

$$\mathbf{F}(hkl) = 2\cos^2[2\pi(h + k + l)/4] \sum_{j=1}^{N/2} g_j \exp[i2\pi(hx_j + ky_j + lz_j)] \qquad (4.72)$$

and this equation can be broken down into its components $A'(hkl)$ and $B'(hkl)$ in the usual way. Further simplification is possible since $G(hkl)$, or $2\cos^2[2\pi(h + k + l)/4]$, can take only two values: If $h + k + l$ is an even number,

$$G(hkl) = 2 \qquad (4.73)$$

whereas if it is odd,

$$G(hkl) = 0 \qquad (4.74)$$

Hence, we may write the limiting conditions for a body-centered unit cell, the conditions under which reflection can occur, as

$$hkl: \quad h + k + l = 2n \qquad (n = 0, \pm 1, \pm 2, \ldots) \qquad (4.75)$$

The same situation expressed in terms of systematic absences, the conditions under which reflection cannot occur, is

$$hkl: \quad h+k+l=2n+1 \tag{4.76}$$

Both terms are in common use, and the reader should distinguish between them carefully.

### 4.10.4  Determination of Unit-Cell Type

Expressions analogous to (4.72) may be derived for any unit-cell type. The reader should attempt the derivations for $C$ and $F$, and check the results against Table 4.1.

We may summarize these results by saying that if a reflection arises from a centered unit cell, the structure factor has the same form as that for the primitive unit cell, but multiplied by a factor $G(hkl)$, which is the number of lattice points associated with the particular unit-cell type. The summation in the reduced structure factor equation is taken over the number of atoms in the unit cell *not* related by the unit-cell translational symmetry.

In practice, the diffraction pattern is recorded and indexed, and a scrutiny of the reflections present leads to a deduction of the unit-cell type, following Table 4.1.

TABLE 4.1.  Limiting Conditions for Unit-Cell Type

| Unit-cell type | Limiting conditions | Translations associated with the unit-cell type | $G(hkl)$ |
|---|---|---|---|
| $P$ | None | None | 1 |
| $A$ | $hkl$: $k+l=2n$ | $b/2+c/2$ | 2 |
| $B$ | $hkl$: $h+l=2n$ | $a/2+c/2$ | 2 |
| $C$ | $hkl$: $h+k=2n$ | $a/2+b/2$ | 2 |
| $I$ | $hkl$: $h+k+l=2n$ | $a/2+b/2+c/2$ | 2 |
| $F$ | $\begin{cases} hkl: \quad h+k=2n \\ hkl: \quad k+l=2n \\ hkl: \quad (h+l=2n)^a \end{cases}$ | $\begin{cases} a/2+b/2 \\ b/2+c/2 \\ a/2+c/2 \end{cases}$ | 4 |
| $R_{\text{hex}}{}^b$ | $hkl$: $\quad -h+k+l=3n$ (obv) <br> or <br> $hkl$: $\quad h-k+l=3n$ (rev) | $\begin{cases} a/3+2b/3+2c/3 \\ 2a/3+b/3+c/3 \end{cases}$ <br><br> $\begin{cases} a/3+2b/3+c/3 \\ 2a/3+b/3+2c/3 \end{cases}$ | 3 <br><br> 3 |

$^a$ This condition is not independent of the other two, as may be shown easily.
$^b$ See page 64 and Table 2.3.

### 4.10.5   Structure Factors and Symmetry Elements

In the structure factor equation (4.40), some of the $N$ atoms in the unit cell may be equivalent under symmetry operations other than unit-cell centering translations. Let $N_A$ be the number of atoms in the asymmetric unit (not related by symmetry) and $Z$ the number of asymmetric units in one unit cell. Thus,

$$N = N_A Z \qquad (4.77)$$

Symbolically, we may write

$$\sum_{j=1}^{N} \equiv \sum_{r=1}^{N_A} \times \sum_{s=1}^{Z} \qquad (4.78)$$

where the sum over $r$ refers to the symmetry-independent atoms, and that over $s$ to the symmetry-related atoms. The structure factor equation thus contains two parts, which may be considered separately.

The summation over $Z$ symmetry-related atoms of any one type is expressed in terms of the coordinates of the general equivalent positions. Following (4.42) and (4.43),

$$A(hkl) = \sum_{s=1}^{Z} \cos 2\pi (hx_s + ky_s + lz_s) \qquad (4.79)$$

and

$$B(hkl) = \sum_{s=1}^{Z} \sin 2\pi (hx_s + ky_z + lz_s) \qquad (4.80)$$

Extending to the $N_A$ atoms in the asymmetric unit, with one such term for each atom $r$,

$$A'(hkl) = \sum_{r=1}^{N_A} g_r A_r(hkl) \qquad (4.81)$$

and

$$B'(hkl) = \sum_{r=1}^{N_A} g_r B_r(hkl) \qquad (4.82)$$

The terms $A(hkl)$ and $B(hkl)$ are independent of the structural arrangement of atoms in the asymmetric unit; they are a property of the space group and are called geometric structure factors. Much can be learnt about the diffraction pattern of a crystal from its geometric structure factors. We shall consider several examples, starting in the monoclinic system. Generally, for convenience, the subscript $s$ in (4.79) and (4.80) will not be retained, as all $Z$ positions are related to $x$, $y$, $z$ by symmetry. It may be noted, in passing, that these and similar results assume that the atoms in the unit cell are vibrating isotropically. However, where anisotropic vibration is postulated, similar relationships can be derived, but the treatment is correspondingly complex and too detailed for inclusion here.

Space Group $P2_1$

General equivalent positions: $x$, $y$, $z$; $\bar{x}$, $\frac{1}{2}+y$, $\bar{z}$ (see Figure 2.30).
Geometric structure factors:

$$A(hkl) = \cos 2\pi(hx + ky + lz) + \cos 2\pi(-hx + ky - lz + k/2) \tag{4.83}$$

Using

$$\cos P + \cos Q = 2 \cos[(P+Q)/2] \cos[(P-Q)/2] \tag{4.84}$$

we obtain

$$A(hkl) = 2 \cos 2\pi(hx + lz - k/4) \cos 2\pi(ky + k/4) \tag{4.85}$$

$$B(hkl) = \sin 2\pi(hx + ky + lz) + \sin 2\pi(-hx + ky - lz + k/2) \tag{4.86}$$

Using

$$\sin P + \sin Q = 2 \sin[(P+Q)/2] \cos[(P-Q)/2] \tag{4.87}$$

we find

$$B(hkl) = 2 \cos 2\pi(hx + lz - k/4) \sin 2\pi(ky + k/4) \tag{4.88}$$

The reader should have no difficulty in evaluating the corresponding expressions for space group $P2$.

*Systematic Absences in* $P2_1$. Geometric structure factors enable one to determine limiting conditions, or to predict which reflections will be systematically absent in the X-ray diffraction pattern. If we can show, for given values of $h, k,$ and $l$, that both $A(hkl)$ and $B(hkl)$ are zero, then $I(hkl)$ is zero, regardless of the structure. For $P2_1$, we can cast (4.85) and (4.88) in the following forms, according to the parity (oddness or evenness) of $k$. Expanding (4.85), using

$$\cos(P \pm Q) = \cos P \cos Q \mp \sin P \sin Q \qquad (4.89)$$

we may write

$$A(hkl)/2 = [\cos 2\pi(hx + lz) \cos 2\pi(k/4) + \sin 2\pi(hx + lz) \sin 2\pi(k/4)]$$
$$\times [\cos 2\pi ky \cos 2\pi(k/4) - \sin 2\pi ky \sin 2\pi(k/4)] \qquad (4.90)$$

In multiplying the right-hand side of (4.90), terms such as

$$\cos 2\pi(hx + lz) \cos 2\pi(k/4) \sin 2\pi ky \sin 2\pi(k/4) \qquad (4.91)$$

occur. Using the identity

$$\sin 2P = 2 \sin P \cos P \qquad (4.92)$$

we find that (4.91) becomes

$$\tfrac{1}{2} \cos 2\pi(hx + lz) \sin 2\pi ky \sin 4\pi(k/4) \qquad (4.93)$$

which is zero, since $k$ is an integer. Hence, (4.90) becomes

$$A(hkl)/2 = \cos 2\pi(hx + lz) \cos 2\pi ky \cos^2 2\pi(k/4)$$
$$- \sin 2\pi(hx + lz) \sin 2\pi ky \sin^2 2\pi(k/4) \qquad (4.94)$$

In a similar manner, and using

$$\sin(P + Q) = \sin P \cos Q + \cos P \sin Q \qquad (4.95)$$

we can derive

$$B(hkl)/2 = \cos 2\pi(hx + lz)\sin 2\pi ky \cos^2 2\pi(k/4)$$
$$+\sin 2\pi(hx + lz)\cos 2\pi ky \sin^2 2\pi(k/4) \qquad (4.96)$$

Separating for $k$ even and odd, we obtain

$$k = 2n: \qquad A(hkl) = 2\cos 2\pi(hx + lz)\cos 2\pi ky \qquad (4.97)$$

$$B(hkl) = 2\cos 2\pi(hx + lz)\sin 2\pi ky \qquad (4.98)$$

$$k = 2n + 1: \qquad A(hkl) = -2\sin 2\pi(hx + lz)\sin 2\pi ky \qquad (4.99)$$

$$B(hkl) = 2\sin 2\pi(hx + lz)\cos 2\pi ky \qquad (4.100)$$

Only one systematic condition can be extracted from these equations: If both $h$ and $l$ are zero, then from (4.99) and (4.100)

$$A(hkl) = B(hkl) = 0 \qquad (4.101)$$

In other words, the limiting condition, associated with the $2_1$ axis, is

$$0k0: \quad k = 2n$$

### 4.10.6    Limiting Conditions from Screw-Axis Symmetry

The example of the $2_1$ axis has been treated in detail; it shows again the ability of the diffraction pattern to reveal translational symmetry elements (cf. the $I$ unit cell on page 160). We can show how the limiting conditions for a $2_1$ axis arise from a consideration of the Bragg construction. Figure 4.10 is a schematic illustration of a $2_1$ symmetry pattern; the motif ♥ represents a structure at a height $z$, and ○ the structure at a height $\bar{z}$ after operating on it with the $2_1$ axis. $MM'$ represents a family of $(0k0)$ planes and $NN'$ a family of $(02k,0)$ planes.

Reflections of the type $0k0$ from $MM'$ planes are cancelled by the reflections from the $NN'$ planes because their phase change relative to $MM'$ is $180°$. Clearly, this result is not obtained with the $02k,0$ reflections. Although the figure illustrates the situation for $k = 1$, the same argument can be applied with any value of $k$.

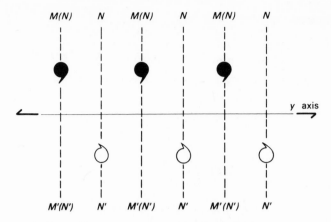

FIGURE 4.10. Pattern of a structure containing a $2_1$ axis: $d(NN') = d(MM')/2$; the $MM'$ planes are "halved" by the $NN'$ family.

Limiting conditions for other screw axes, and in other orientations, can be written down by analogy. Try to decide what the conditions for the following screw axes will be, and then check your results from Table 4.2:

$2_1$ parallel to $a$
$3_2$ parallel to $c$
$4_3$ parallel to $b$
$6_5$ parallel to $c$

Notice that pure rotation axes $R$, as in $P2$, do not introduce any limiting conditions.

TABLE 4.2. Limiting Conditions for Screw Axes

| Screw axis | Orientation | Limiting condition | | Translation component |
|---|---|---|---|---|
| $2_1$ | $\parallel a$ | $h00$: | $h = 2n$ | $a/2$ |
| $2_1$ | $\parallel b$ | $0k0$: | $k = 2n$ | $b/2$ |
| $2_1$ | $\parallel c$ | $00l$: | $l = 2n$ | $c/2$ |
| $3_1$ or $3_2$ | $\parallel c$ | $000l$: | $l = 3n$ | $c/3, 2c/3$ |
| $4_1$ or $4_3$ | $\parallel c$ | $00l$: | $l = 4n$ | $c/4$ |
| $4_2$ | $\parallel c$ | $00l$: | $l = 2n$ | $2c/4$ $(c/2)$ |
| $6_1$ or $6_5$ | $\parallel c$ | $000l$: | $l = 6n$ | $c/6, 5c/6$ |
| $6_2$ or $6_4$ | $\parallel c$ | $000l$: | $l = 3n$ | $2c/6, 4c/6$ $(c/3, 2c/3)$ |
| $6_3$ | $\parallel c$ | $000l$: | $l = 2n$ | $3c/6$ $(c/2)$ |

## 4.10.7    Centrosymmetric Zones

In space group $P2_1$ and other space groups of crystal class 2, the $h0l$ reflections are of special interest. Among equations (4.97)–(4.100), only (4.97) is relevant, because zero is an even number, and $\sin(2\pi 0 y) = 0$. Hence,

$$A(h0l) = 2 \cos 2\pi(hx + lz) \qquad (4.102)$$

and

$$B(h0l) = 0 \qquad (4.103)$$

From (4.46), $\phi(h0l)$ is either 0 or $\pi$; in other words, the $[h0l]$ zone is centrosymmetric, which is of importance in a structure analysis in this space group. Centrosymmetric zones occur in noncentrosymmetric space groups which have 2 as a subgroup of their class (page 45).

### 4.10.7.1   Space Group $Pc$

General equivalent positions: $x, y, z$; $x, \bar{y}, \frac{1}{2} + z$ (Figure 4.11).
Geometric structure factors: proceeding as before, we obtain

$$A(hkl) = 2 \cos 2\pi(hx + lz + l/4) \cos 2\pi(ky - l/4) \qquad (4.104)$$

$$B(hkl) = 2 \sin 2\pi(hx + lz + l/4) \cos 2\pi(ky - l/4) \qquad (4.105)$$

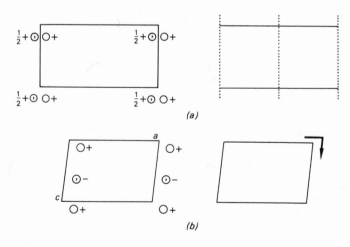

FIGURE 4.11. Space group $Pc$: (a) viewed along $c$, (b) viewed along $b$.

If we separate these equations for $l$ even and odd, we find systematic absences occur only for the $h0l$ reflections:

$$h0l: \quad l = 2n + 1$$

Again, there is a relationship between the index ($l$) involved in the condition and the symmetry translation ($c/2$).

### 4.10.8 Limiting Conditions from Glide-Plane Symmetry

Table 4.3 indicates the conditions relating to glide planes of interest to us. They may be deduced, or written down by analogy with $Pc$.

Space Group $P2_1/c$

This space group, which is often encountered in practice, exhibits a combination of the two translational symmetry elements already considered in detail, a $2_1$ axis along $b$ and a $c$-glide plane perpendicular to $b$ (Figure 2.32 and Problem 2.7a).

$P2_1/c$ is a centrosymmetric space group, and the general equivalent positions may be summarized as

$$\pm\{x, y, z; \ x, \tfrac{1}{2} - y, \tfrac{1}{2} + z\}$$

Geometric structure factors: From the discussion on pages 158–159, we may write down

$$A(hkl) = 2\{\cos 2\pi(hx + ky + lz) + \cos 2\pi[hx - ky + lz + \tfrac{1}{2}(k + l)]\} \quad (4.106)$$

$$B(hkl) = 0 \quad (4.107)$$

TABLE 4.3. Limiting Conditions for Glide Planes

| Glide plane | Orientation | Limiting condition | | Translation component |
|:---:|:---:|:---:|:---:|:---:|
| $a$ | $\perp b$ | $h0l:$ | $h = 2n$ | $a/2$ |
| $a$ | $\perp c$ | $hk0:$ | $h = 2n$ | $a/2$ |
| $b$ | $\perp a$ | $0kl:$ | $k = 2n$ | $b/2$ |
| $b$ | $\perp c$ | $hk0:$ | $k = 2n$ | $b/2$ |
| $c$ | $\perp a$ | $0kl:$ | $l = 2n$ | $c/2$ |
| $c$ | $\perp b$ | $h0l:$ | $l = 2n$ | $c/2$ |
| $n$ | $\perp a$ | $0kl:$ | $k + l = 2n$ | $b/2 + c/2$ |
| $n$ | $\perp b$ | $h0l:$ | $h + l = 2n$ | $a/2 + c/2$ |
| $n$ | $\perp c$ | $hk0:$ | $h + k = 2n$ | $a/2 + b/2$ |

Carrying out the analysis as before leads to

$$A(hkl) = 4 \cos 2\pi[hx + lz + \tfrac{1}{4}(k+l)] \cos 2\pi[ky - \tfrac{1}{4}(k+l)] \quad (4.108)$$

Separating for $k+l$ even and odd, we obtain

$$k+l = 2n: \qquad A(hkl) = 4 \cos 2\pi(hx + lz) \cos 2\pi ky \quad (4.109)$$

$$k+l = 2n+1: \qquad A(hkl) = -4 \sin 2\pi(hx + lz) \sin 2\pi ky \quad (4.110)$$

We can now *deduce* the limiting conditions:

$hkl$:   None   —   $P$ unit cell
$h0l$:   $l = 2n$   —   $c$-glide plane $\perp b$
$0k0$:   $k = 2n$   —   $2_1$ axis $\| b$

These regions are important in monoclinic reciprocal space, because only here can we find *any* systematic absences. Despite Friedel's law, the diffraction symmetry determines the true space group in this case: The

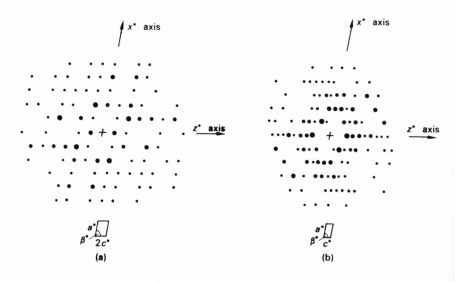

FIGURE 4.12. $X^*Z^*$ reciprocal nets for space group $Pc$, $P2/c$, or $P2_1/c$: (a) $k = 0$, (b) $k > 0$. The $c$-glide plane, which is perpendicular to $b$, causes a halving of the rows parallel to $X^*$ when $k = 0$, so that only the rows with $l = 2n$ are present. Hence, the true $c^*$ spacing is not observed on the level $k = 0$, but can be detected on higher reciprocal lattice levels. The symmetry on both levels is 2, in keeping with the diffraction symmetry $2/m$, so that $|F(hkl)|^2 = |F(\bar{h}k\bar{l})|^2$. The reciprocal lattice points in the diagram are weighted according to $|F(hkl)|^2$.

center of symmetry is revealed through the interaction of the $c$ and $2_1$ symmetry elements. Figure 4.12 illustrates weighted reciprocal lattice levels for a possible monoclinic crystal with space group $Pc$, $P2/c$, or $P2_1/c$.

We shall complete this symmetry study with two examples from the orthorhombic system.

## Space Group *Pma2*

From the data in Figure 4.13, we can write down the expression for the geometric structure factors:

$$A(hkl) = \cos 2\pi(hx + ky + lz) + \cos 2\pi(-hx - ky + lz)$$

$$+ \cos 2\pi(-hx + ky + lz + h/2) + \cos 2\pi(hx - ky + lz + h/2) \tag{4.111}$$

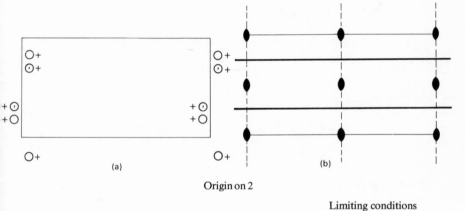

Origin on 2

| | | | | | Limiting conditions | |
|---|---|---|---|---|---|---|
| 4 | $d$ | 1 | $x, y, z;$ | $\bar{x}, \bar{y}, z;$   $\frac{1}{2}-x, y, z;$   $\frac{1}{2}+x, \bar{y}, z.$ | $hkl$: | None |
| | | | | | $0kl$: | None |
| | | | | | $h0l$: | $h = 2n$ |
| | | | | | $hk0$: | None |
| | | | | | $h00$: | $(h = 2n)$ |
| | | | | | $0k0$: | None |
| | | | | | $00l$: | None |
| 2 | $c$ | $m$ | $\frac{1}{4}, y, z;$ | $\frac{3}{4}, \bar{y}, z.$ | As above | |
| 2 | $b$ | 2 | $0, \frac{1}{2}, z;$ | $\frac{1}{2}, \frac{1}{2}, z.$ | As above + | |
| 2 | $a$ | 2 | $0, 0, z;$ | $\frac{1}{2}, 0, z.$ | $hkl$:   $h = 2n$ | |

Symmetry of special projections

(001) $p2mg$      (100) $pm1(p1m1)$      (010) $p1m$ $(p11m)a' = a/2$

FIGURE 4.13. Space group *Pma2*.

Combining the first and third, and second and fourth terms,

$$A(hkl) = 2 \cos 2\pi(ky + lz + h/4) \cos 2\pi(hx - h/4)$$
$$+ 2 \cos 2\pi(-ky + lz + h/4) \cos 2\pi(hx + h/4) \quad (4.112)$$

Further simplification of this expression requires the separate parts to contain a common factor. We return to (4.111) and make a minor alteration to the term $\cos 2\pi(hx - ky + lz + h/2)$. Since $h$ is an integer, we may write this term as $\cos(\phi + h\pi)$, where $\phi = 2\pi(hx + ky + lz)$. But $\cos(\phi + h\pi) = \cos(\phi - h\pi)$; hence, we change the term $\cos 2\pi(hx - ky + lz + h/2)$ to $\cos 2\pi(hx - ky + lz - h/2)$ in (4.111). Another way of looking at this process is that the fourth general equivalent position in the list has been converted to $-\frac{1}{2} + x, \bar{y}, z$, or moved through one unit-cell repeat $a$ in the negative $x$ direction to a crystallographically equivalent position, a perfectly valid and generally applicable tactic.

Returning to $Pma2$, (4.112) now becomes

$$A(hkl) = 2 \cos 2\pi(ky + lz + h/4) \cos 2\pi(hx - h/4)$$
$$+ 2 \cos 2\pi(-ky + lz - h/4) \cos 2\pi(hx - h/4) \quad (4.113)$$

which simplifies to

$$A(hkl) = 2[\cos 2\pi(hx - h/4)][\cos 2\pi(ky + lz + h/4)$$
$$+ \cos 2\pi(-ky + lz - h/4)] \quad (4.114)$$

Combining again,

$$A(hkl) = 4[\cos 2\pi(hx - h/4)] \cos 2\pi(ky + h/4) \cos 2\pi lz \quad (4.115)$$

Similarly,

$$B(hkl) = 4[\cos 2\pi(hx - h/4)] \cos 2\pi(ky + h/4) \sin 2\pi lz \quad (4.116)$$

In the orthorhombic system, the regions of reciprocal space of particular interest are listed under Figure 4.13. Separating (4.115) and (4.116) for even and odd $h$, we obtain

$$h = 2n: \qquad A(hkl) = 4 \cos 2\pi hx \cos 2\pi ky \cos 2\pi lz \quad (4.117)$$
$$B(hkl) = 4 \cos 2\pi hx \cos 2\pi ky \sin 2\pi lz \quad (4.118)$$

$$h = 2n + 1: \qquad A(hkl) = -4 \sin 2\pi hx \sin 2\pi ky \cos 2\pi lz \qquad (4.119)$$

$$B(hkl) = -4 \sin 2\pi hx \sin 2\pi ky \sin 2\pi lz \qquad (4.120)$$

from which we find the limiting condition, listed in Figure 4.13:

$$h0l: \quad h = 2n$$

The condition

$$h00: \quad (h = 2n)$$

should be considered carefully. One might be excused for thinking that it implies the existence of a $2_1$ axis parallel to $a$, but for the knowledge that there are no symmetry axes parallel to $a$ in class $mm2$. This condition is dependent upon the previous one. We must emphasize the danger of considering limiting conditions out of the following hierarchal order:

| | |
|---|---|
| $hkl$ | for the unit cell type |
| $\left.\begin{array}{l} 0kl \\ h0l \\ hk0 \end{array}\right\}$ | for glide planes |
| $\left.\begin{array}{l} h00 \\ 0k0 \\ 00l \end{array}\right\}$ | for screw axes |

order of inspection

One should proceed to a lower level in this list only after considering the full implications of a condition arising from a higher level. Conditions such as that for $h00$ in $Pma2$ are called redundant, or dependent, and are placed in parentheses. Reflections involved in such conditions are certainly absent from the diffraction record, but do not, necessarily, contribute to a determination of the space-group symmetry.

Space Group *Pman*

This space group may be derived from $Pma2$ by the inclusion of an $n$-glide plane perpendicular to the $c$ axis [with a translational component of $(a + b)/2$]. We have now seen on several occasions that it is advantageous to take the origin at $\bar{1}$ wherever possible; Figure 4.14 shows *Pman* drawn in this orientation. It is left to the reader to show that the geometric structure

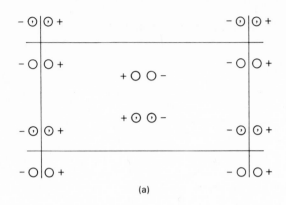

(a)

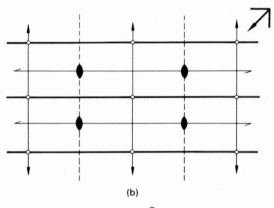

(b)

Origin at $\bar{1}$

Limiting conditions

$$\pm\{x, y, z;\quad \bar{x}, y, z;\quad \tfrac{1}{2}+x, \tfrac{1}{2}-y, z;\quad \tfrac{1}{2}-x, \tfrac{1}{2}-y, z\}$$

| | |
|---|---|
| $hkl$: | None |
| $0kl$: | None |
| $h0l$: | $h = 2n$ |
| $hk0$: | $h + k = 2n$ |
| $h00$: | $(h = 2n)$ |
| $0k0$: | $(k = 2n)$ |
| $00l$: | None |

FIGURE 4.14.  Space group *Pman*: (a) general equivalent positions, (b) symmetry elements, coordinates of general equivalent positions and limiting conditions.

factors are

$$A(hkl) = 8 \cos 2\pi h x \, \cos 2\pi[ky - (h + k)/4] \cos 2\pi[lz + (h + k)/4]$$

$$(4.121)$$

$$B(hkl) = 0 \qquad\qquad\qquad\qquad\qquad\qquad\qquad\qquad (4.122)$$

TABLE 4.4  Preliminary Stages in a Structure Analysis

| Stage | Procedures |
|-------|------------|
| 1 | Optical examination and selection of crystals |
| 2 | Determination of crystal system and unit-cell dimensions |
| 3 | Measurement of crystal density |
| 4 | Calculation of unit-cell contents and enumeration of atom types present |
| 5 | Recording of $I(hkl)$ for the practicable ranges of $h$, $k$, and $l$ |
| 6 | Determination of space group |

and, thence, to derive the limiting conditions. Notice that redundant conditions sometimes do and sometimes do not relate to the presence of the corresponding symmetry element.

# 4.11  Preliminary Structure Analysis

We may now make a brief survey of the various aspects of structure analysis which we have introduced so far. Table 4.4 summarizes the steps which are usually taken in a preliminary examination of a crystal.

Stages 1 and 2 have been discussed in Chapter 3. Stages 3 and 4 would normally be undertaken before 5 and 6, but for the moment we will by-pass 3 and 4, and assume that stage 5 has been taken far enough for the determination of the space group of the crystal.

## 4.11.1  Practical Determination of Space Groups

Crystal structure analysis is always aided by a knowledge of the space group at the outset of the detailed investigation. We shall examine samples of several diffraction patterns of monoclinic and orthorhombic crystals in order to discuss the process of space-group determination.

It is important to bear in mind that X-ray techniques can reveal directly the presence of four types of translational symmetry:

1. Translations relating to the unit cell ($a$, $b$, and $c$).
2. Translations relating to the centering of the unit cell (integer fractions of $a$, $b$, and $c$).
3. Translations relating to glide planes.
4. Translations relating to screw axes.

Categories 2–4 are concerned with systematic absences.

TABLE 4.5.  Some Reflection Data for
Monoclinic Crystal (I)

| hkl | hkl | hkl | hkl |
| --- | --- | --- | --- |
| 200 | 401 | 112 | 510 |
| 201 | 402 | 113 | 020 |
| 202 | 600 | 114 | 040 |
| 203 | 110 | 310 | 060 |
| 400 | 111 | 311 | 080 |

## Monoclinic Space Groups

Single-crystal X-ray photographs taken with a monoclinic crystal showed typically the reflections listed in Table 4.5. From the important reflection types, $hkl$, $h0l$, and $0k0$, we can deduce the conditions

$hkl$:  $h + k = 2n$
$h0l$:  $(h = 2n)$
$0k0$:  $(k = 2n)$

from which we must conclude, using Table 4.7, that the space group could be $C2$, $Cm$, or $C2/m$. At this stage in the structure analysis of this crystal, there is no way of distinguishing among these three possibilities.

Table 4.6 provides the next list of reflections for inspection. Now there is no limiting condition on $hkl$, but $h0l$ are restricted by $l$ being even, and $0k0$ are restricted by $k$ being even. This space group is determined uniquely as $P2_1/c$. The ambiguity which arose for monoclinic crystal (I) is not unusual, and we shall consider in Chapter 6 how it might be resolved.

The limiting conditions for the 13 monoclinic space groups are listed in Table 4.7 in their standard orientations. In practice, it is possible, by an inadvertent choice of axes, to find oneself working with a nonstandard

TABLE 4.6.  Some Reflection Data for
Monoclinic Crystal (II)

| hkl | hkl | hkl | hkl |
| --- | --- | --- | --- |
| 100 | 204 | 111 | 322 |
| 200 | 402 | 122 | 020 |
| 300 | 502 | 113 | 040 |
| 400 | 110 | 311 | 060 |
| 202 | 310 | 322 | 080 |

TABLE 4.7. Limiting Conditions for Monoclinic Space Groups

| Conditions limiting possible X-ray reflections | | Possible space groups |
|---|---|---|
| $hkl$: | none | |
| $h0l$: | none | $P2, Pm, P2/m$ |
| $0k0$: | none | |
| $hkl$: | none | |
| $h0l$: | none | $P2_1, P2_1/m$ |
| $0k0$: | $k = 2n$ | |
| $hkl$: | none | |
| $h0l$: | $l = 2n$ | $Pc, P2/c$ |
| $0k0$: | none | |
| $hkl$: | none | |
| $h0l$: | $l = 2n$ | $P2_1/c$ |
| $0k0$: | $k = 2n$ | |
| $hkl$: | $h + k = 2n$ | |
| $h0l$: | none | $C2, Cm, C2/m$ |
| $0k0$: | none | |
| $hkl$: | $h + k = 2n$ | |
| $h0l$: | $l = 2n, (h = 2n)$ | $Cc, C2/c$ |
| $0k0$: | $(k = 2n)$ | |

space-group symbol. Generally, a fairly straightforward transformation of axes will provide the standard setting (see Problems 2.11 and 4.6b and c).

## Orthorhombic Space Groups

We begin with sample data in Table 4.8.

TABLE 4.8. Some Reflection Data for Orthorhombic Crystal (I)

| $hkl$ | $hkl$ | $hkl$ | $hkl$ |
|---|---|---|---|
| 111 | 011 | 110 | 020 |
| 112 | 021 | 120 | 040 |
| 212 | 012 | 310 | 060 |
| 312 | 101 | 200 | 002 |
| 322 | 203 | 400 | 004 |
| 332 | 303 | 600 | 006 |

From these data, we deduce the following conditions.

$hkl$:  none    $h00$:  $h = 2n$
$0kl$:  none    $0k0$:  $k = 2n$
$h0l$:  none    $00l$:  $l = 2n$
$hk0$:  none

All of these reflection classes must be examined in the orthorhombic system, in the correct hierarchy. We deduce $2_1$ axes parallel to $a$, $b$, and $c$: The space group is uniquely determined as $P2_12_12_1$ (see pages 94 and 95).

In the next two examples we shall state only the conclusions obtained from the inspection of reflection data. Consider the following list:

$hkl$:  none       $h00$:  none
$0kl$:  $k = 2n$   $0k0$:  $(k = 2n)$
$h0l$:  $l = 2n$   $00l$:  $(l = 2n)$
$hk0$:  none

These conditions apply to space groups $Pbc2_1$ and $Pbcm$; the difference between them involves a center of symmetry. In the list

$hkl$:  none       $h00$:  $(h = 2n)$
$0kl$:  $k = 2n$   $0k0$:  $(k = 2n)$
$h0l$:  $l = 2n$   $00l$:  $(l = 2n)$
$hk0$:  $h = 2n$

space group $Pbca$ is uniquely determined.

These results seem quite reasonable and straightforward, but, nevertheless, the beginner might be tempted to question their validity. For example, in orthorhombic crystal (I), is there a space group in class $mmm$ which would give the same systematic absences as those in Table 4.8? Experience tells us that there is not. Since no glide planes are indicated by the systematic absences, the three symmetry planes would have to be mirror planes. Three mirror planes could not be involved with three screw axes except in a centered space group, such as $Immm$, and so our original conclusion was correct.

The practicing crystallographer is assisted by the space group information which is tabulated fully in Volume I of the *International Tables for X-Ray Crystallography*.* Combined with a working knowledge of sym-

* See Bibliography.

metry, these tables enable most situations arising in the course of a structure analysis to be treated correctly.

## Bibliography

### Structure Factor and Intensity

STOUT, G. H., and JENSEN, L. H., *X-Ray Structure Determination—A Practical Guide*, New York, Macmillan.

WOOLFSON, M. M., *An Introduction to X-Ray Crystallography*, Cambridge, University Press.

### Determination of Space Groups

HENRY, N. F. M., and LONSDALE, K. (Editors), *International Tables for X-Ray Crystallography*, Vol. I, Birmingham, Kynoch Press.

STOUT, G. H., and JENSEN, L. H., *X-Ray Structure Determination—A Practical Guide*, New York, Macmillan.

### Atomic Scattering Factor Data

IBERS, JAMES A., and HAMILTON, W. C. (Editors), *International Tables for X-Ray Crystallography*, Vol. IV, Birmingham, Kynoch Press.

## Problems

**4.1.** A two-dimensional structure has four atoms per unit cell, two of type $P$ and two of type $Q$, with the following fractional coordinates:

|        | $x$  | $y$  |
|--------|------|------|
| $P_1$  | 0.1  | 0.2  |
| $P_2$  | 0.9  | 0.8  |
| $Q_1$  | 0.2  | 0.7  |
| $Q_2$  | 0.8  | 0.3  |

Calculate $\mathbf{F}(hkl)$ for the reflections 5, 0; 0, 5; 5, 5; and 5, 10 in terms of the scattering factors for the two species, $g_P$ and $g_Q$. If $g_P$ is equal to $2g_Q$, what are the phase angles for these reflections?

**4.2.** $\alpha$-Uranium (U) crystallizes in the orthorhombic system with four U in the special positions

$$\pm\{0, y, \tfrac{1}{4}; \ \tfrac{1}{2}, \tfrac{1}{2}+y, \tfrac{1}{4}\}$$

Use the data below to decide whether $y$ is best chosen as 0.10 or 0.15.

| hkl | $|F_o(hkl)|$ | $g_U(hkl)$ |
|-----|--------------|------------|
| 020 | 88.5 | 70.0 |
| 110 | 268.9 | 80.0 |

**4.3.** The unit-cell dimensions of $\alpha$-U are $a = 2.85$, $b = 5.87$, and $c = 5.00$ Å. Use the value of $y_U$ from problem 4.2 to determine the shortest U—U distance in the structure. It may be helpful to plot the U atom positions in a few neighboring unit cells.

**4.4.** In the examples listed below, conditions limiting possible X-ray reflections are given for crystals in the monoclinic system. In each case, write down the symbols of the possible space groups corresponding to this information.

(a)  $hkl$:  no conditions
   $h0l$:  no conditions
   $0k0$:  $k = 2n$.

(b)  $hkl$:  no conditions
   $h0l$:  $h = 2n$
   $0k0$:  no conditions.

(c)  $hkl$:  $h + k = 2n$
   $h0l$:  $l = 2n$; $(h = 2n)$
   $0k0$:  $(k = 2n)$.

(d)  $hkl$:  no conditions
   $h0l$:  no conditions
   $0k0$:  no conditions.

**4.5.** Repeat question 4.4 for the conditions given below relating to orthorhombic space groups.

(a)  $hkl$:
   $0kl$:  no conditions
   $h0l$:
   $hk0$:

   $h00$:  $h = 2n$
   $0k0$:  $k = 2n$
   $00l$:  no conditions

(b)  $hkl$:  no conditions       $h00$:  no conditions

$0kl$:  $k = 2n$              $0k0$:  $k = 2n$

$h0l$:  no conditions        $00l$:  no conditions

$hk0$:  no conditions

(c)  $hkl$:  $h + k + l = 2n$     $h00$:  $h = 2n$

$0kl$:  $k = 2n, l = 2n$      $0k0$:  $k = 2n$

$h0l$:  $h + l = 2n$          $00l$:  $l = 2n$

$hk0$:  $h + k = 2n$

Rewrite (c) with parentheses to indicate the redundant conditions.

**4.6.** (a) Write the conditions limiting possible X-ray reflections for the following space groups: (i) $P2_1/a$; (ii) $Pc$; (iii) $C2$; (iv) $P2_122$; (v) $Pcc2$; (vi) $Imam$. In each case write the symbols of the space groups, if any, in the same crystal system with the same conditions.

(b) Write the conditions limiting possible reflections in the monoclinic space group $P2_1/n$ (nonstandard symbol).

(c) Give the conventional symbols for space groups $A2/a$ and $B2_122_1$.

# Methods in X-Ray Structure Analysis. I

## 5.1 Introduction

We have reached the stage where we begin to consider the contents of the unit cell and attempt to assign possible types of locations to atoms or molecules in the asymmetric unit.

## 5.2 Analysis of the Unit-Cell Contents

The density $D_m$ of the crystals under examination may be measured by suspending them in a liquid or liquid mixture.* The composition of the liquid is altered until the crystals neither rise nor fall; then the density of the liquid, equal to $D_m$, is measured. The flotation procedure is best carried out in a thermostat. It may still happen, however, that the demarcation between sinking and floating is a little ill defined. Generally, inclusion of air or solvent in the crystal will lead to a smaller apparent density, and a position corresponding to maximum measured density should be appropriate.

If the crystal contains $Z$ chemical species per unit cell, then the following relationship holds:

$$D_m = ZMm/V_c \qquad (5.1)$$

where $M$ is the formula weight, $m$ is the atomic mass unit ($1.660 \times 10^{-24}$ g), and $V_c$ is the unit-cell volume, given by (2.19). In practice, the volume is expressed in $\text{Å}^3$, in which case $m$ is replaced by 1.660. In terms of the known quantities,

$$Z = D_m V_c / 1.660M \qquad (5.2)$$

* If a large quantity of crystals is available, a simple displacement method may be used.

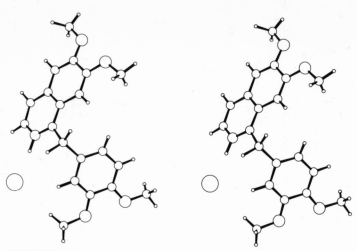

FIGURE 5.1. Molecular conformation of papaverine hydrochloride; the circles, in decreasing order of size, represent Cl, N, C, and H.

and, depending on the accuracy of the measurements involved, the value for $Z$ should be an integer, or nearly so. Knowing $Z$ and the formula of the species, we can determine the number of each atom type in the unit cell, and then try to decide whether they are located in general or special positions. A few examples will serve to illustrate some of the situations which arise in practice.

### 5.2.1 Papaverine Hydrochloride, $C_{20}H_{21}NO_4 \cdot HCl$

Crystal Data

> System: monoclinic.
> Unit-cell dimensions: $a = 13.059, b = 15.620, c = 9.130$ Å, $\beta = 92.13°$.
> $V_c$: 1861.0 Å$^3$.
> $D_m$: 1.33 g cm$^{-3}$.
> $M$: 375.8.
> $Z$: 3.98, or 4 to the nearest integer.
> Unit-cell contents: 80 C, 88 H, 4 N, 16 O, 4 Cl atoms.
> Absent spectra: $h0l$: $l$ odd; $0k0$: $k$ odd.
> Space group: $P2_1/c$ (Figure 2.32 and page 91).

All atoms are in general equivalent positions. The molecular conformation, obtained by a complete structural analysis,* is shown in Figure 5.1.

* C. D. Reynolds et al., Journal of Crystal and Molecular Structure 4, 213 (1974).

## 5.2.2 Naphthalene, $C_{10}H_8$

Crystal Data

System: monoclinic.
Unit-cell dimensions: $a = 8.658$, $b = 6.003$, $c = 8.235$ Å, $\beta = 122.92°$.
$V_c$: 359.2 Å$^3$.
$D_m$: 1.152 g cm$^{-3}$.
$M$: 128.2.
$Z$: 1.94, or 2 to the nearest integer.
Unit-cell contents: 20 C, 16 H atoms.
Absent spectra: $h0l$: $l = 2n + 1$; $0k0$: $l = 2n + 1$.
Space group: $P2_1/c$.

### 5.2.3 Molecular Symmetry

In papaverine hydrochloride, the four molecules in the unit cell are, as complete entities, able to satisfy the requisite general positions in the space group. The molecules are said to be in general positions, and each atom at coordinates $x_j$, $y_j$, $z_j$ ($j = 1, 2, \ldots, 48$) is repeated by the space-group symmetry mechanism to build up the crystal structure. There are, therefore, 48 atoms, including hydrogen, in the asymmetric unit to be located by the structure analysis.

Naphthalene is not quite so straightforward. With two molecules per unit cell, there are 20 carbon atoms and 16 hydrogen atoms, which may be distributed in four equivalent-position sets of five and four atoms, respectively, in the unit cell. This means that in order to solve the structure, we have to locate five carbon atoms and four hydrogen atoms. This number is only half that expected: since $Z$ is 2, each atom is related by one of the symmetry elements of the space group to a second atom of the same type in the same molecule, so as to generate $C_{10}H_8$ from $C_5H_4$. There are three different symmetry elements to consider: the $2_1$ axis, the $c$-glide plane, and the center of symmetry. The screw axis and glide plane are eliminated because they involve translational symmetry, which would generate an infinite molecule with translational repeats. We must, therefore, conclude that the atom pairs are related by a center of symmetry, which means that the molecule of naphthalene is centrosymmetric.

The symmetry analysis for naphthalene has served two very useful purposes: It has halved the work of the subsequent structure analysis, and shown that the molecules in the crystal exhibit a certain minimum symmetry ($\bar{1}$). This result is, of course, in agreement with chemical knowledge, which

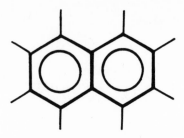

FIGURE 5.2. Naphthalene molecule.

ordinarily we are quite entitled to use. The conventional notion that naphthalene should have *mmm* symmetry (Figure 5.2) is not supported directly, although the crystal structure analysis shows that this symmetry holds within experimental error.

### 5.2.4 Special Positions

The molecules of naphthalene are said to lie on special positions in $P2_1/c$ (Figure 5.3). Special position sites correspond in symmetry to one of the 32 crystallographic point groups, and in subsequent examples, we shall see that both atoms and molecules can occupy special positions.

Glide planes and screw axes do not usually accommodate atoms or molecules; an atom lying exactly on a translational symmetry element would introduce a pseudo-half-axial translation, thus creating special conditions which, depending on the atomic number, may be observable among the X-ray reflections (see Problem 5.1 at the end of this chapter).

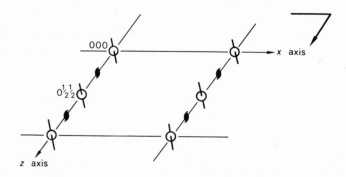

FIGURE 5.3. Grouping of one of the special position sets in $P2_1/c$; the arrangement of molecules (symmetry $\bar{1}$) with their centers at $0, 0, 0$ and $0, \frac{1}{2}, \frac{1}{2}$ is shown.

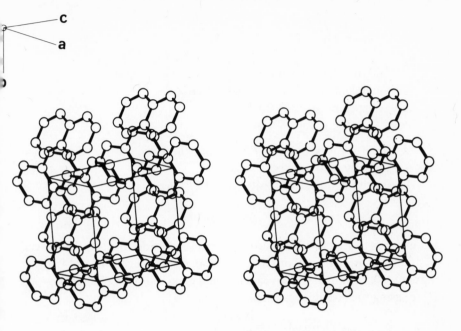

FIGURE 5.4. Stereoviews of the naphthalene structure (H atoms are not shown).

Although they are in special positions, the molecules of naphthalene are subject to the space-group mechanism inherent in the general positions: If one molecule is located at 0, 0, 0, then the second molecule is at $0, \frac{1}{2}, \frac{1}{2}$. This set may be determined by substituting $x = y = z = 0$ into the set of general positions. The structure of naphthalene* is shown in Figure 5.4. The reader may like to consider the three other possible sets of special positions that could be used to represent this structure, and then show from the structure factor equation that $|F(hkl)|$ is invariant with respect to each set of special positions.

### 5.2.5 Nickel Tungstate, NiWO₄

Crystal Data

System: monoclinic.
Unit-cell dimensions: $a = 4.60$, $b = 5.66$, $c = 4.91$ Å, $\beta = 90.1°$.

---

* D. W. J. Cruickshank, *Acta Crystallographica* **10**, 504 (1957), who used the nonstandard space group $P2_1/a$, equivalent to $P2_1/c$, with the $a$ and $c$ axes quoted here interchanged.

$V_c$: 127.8 Å$^3$.

$D_m$: 7.964 g cm$^{-3}$.

$M$: 306.5.

$Z$: 2.00, or 2 to the nearest integer.

Unit-cell contents: 2 Ni, 2 W, 8 O atoms.

Absent spectra: $h0l$: $l = 2n + 1$.

Possible space groups: $Pc$ or $P2/c$.

We shall use space group $P2/c$, since the structure was determined successfully only with this space group.*

The general equivalent positions in $P2/c$ are

$$\pm\{x, y, z;\ x, \bar{y}, \tfrac{1}{2} + z\}$$

but in order to study NiWO$_4$ further, we must consider the possible special positions for this space group; they are located on either the twofold axes or the centers of symmetry. The reader should make a drawing for space group $P2/c$, using the coordinates listed above.

## Special Positions on Twofold Axes

The twofold axes lie along the lines $[0, y, \tfrac{1}{4}]$, $[\tfrac{1}{2}, y, \tfrac{1}{4}]$, $[0, y, \tfrac{3}{4}]$ and $[\tfrac{1}{2}, y, \tfrac{3}{4}]$. The equivalent positions generated by the space-group symmetry show that the special position sets are

$$\pm\{0, y, \tfrac{1}{4}\}\quad \text{or}\quad \pm\{\tfrac{1}{2}, y, \tfrac{1}{4}\}$$

and each set satisfies $P2/c$ symmetry by accommodating two structural entities with symmetry 2 in the unit cell.

## Special Positions on Centers of Symmetry

If we repeat the above analysis for the eight centers of symmetry in the space group, we will develop four special position sets:

$$0, 0, 0;\quad 0, 0, \tfrac{1}{2}$$

$$\tfrac{1}{2}, 0, 0;\quad \tfrac{1}{2}, 0, \tfrac{1}{2}$$

$$0, \tfrac{1}{2}, 0;\quad 0, \tfrac{1}{2}, \tfrac{1}{2}$$

$$\tfrac{1}{2}, \tfrac{1}{2}, 0;\quad \tfrac{1}{2}, \tfrac{1}{2}, \tfrac{1}{2}$$

* R. O. Keeling, *Acta Crystallographica* **10**, 209 (1957).

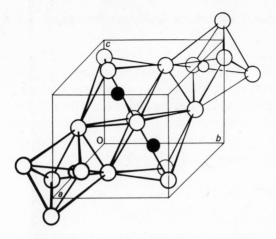

FIGURE 5.5. $WO_6$ and $NiO_6$ octahedra in $NiWO_4$: large open circles O, small open circles W, small black circles Ni.

The Ni and W atoms must lie on special positions, with either 2 or $\bar{1}$ symmetry. Nothing can be said about the position of the oxygen atoms, and without further detailed analysis we cannot define this structure further. However, to complete the picture, we list the atomic parameters for this structure, and illustrate it in Figure 5.5:

$$2\,Ni \quad \pm\{\tfrac{1}{2}, 0.653, \tfrac{1}{4}\}$$

$$2\,W \quad \pm\{0, 0.180, \tfrac{1}{4}\}$$

$$4\,O \quad \pm\{0.22, 0.11, 0.96; 0.22, 0.89, 0.46\}$$

$$4\,O' \quad \pm\{0.26, 0.38, 0.39; 0.26, 0.62, 0.89\}$$

The heavy atoms were found to occupy the four twofold axes in pairs. This conclusion, although not uniquely derivable from the symmetry analysis alone, was at least partially indicated by it. Once again, a pencil and paper operation saved considerable effort in the subsequent detailed structure analysis by pointing to the proper course of action.

In these few examples, we have shown the value of the symmetry analysis in the early stages of a structure determination. This procedure may be regarded as a routine to be carried out before the more detailed calculations required in the elucidation of the atomic parameters.

## 5.3   Two Early Structure Analyses Revisited

The structures of NaCl and $FeS_2$ (pyrite) are largely of historical interest. Both structures were solved by Sir Lawrence Bragg, and although we are not considering entirely the methods which he used in 1913, we are reminded through these structures that he laid the firm foundations for the subject of this book. We shall show that they can provide both further examples of procedures already discussed and introduce new features in our study of crystal structure analysis.

### 5.3.1   Sodium Chloride, NaCl

Crystal Data

> System: cubic.
> Unit-cell dimensions: $a = 5.638$ Å.
> $V_c$: 179.2 Å$^3$.
> $D_m$: 2.165 g cm$^{-3}$.
> $M$: 58.44.
> $Z$: 4.00, or 4 to the nearest integer.
> Unit-cell contents: 4 Na and 4 Cl atoms.
> Conditions limiting possible X-ray reflections: $hkl$: $h + k = 2n$, $k + l = 2n$, $(l + h = 2n)$.
> Possible space groups: $F23$, $Fm3$, $F432$, $F\bar{4}3m$, $Fm3m$.

Symmetry Analysis

Since $Z = 4$, we can define an origin in the unit cell by placing one atom, say Na, at $0, 0, 0$. An atom can be used conveniently to define an origin in this manner if it occurs $N_L$ times per unit cell, where $N_L$ is the number of lattice points per unit cell.

The symmetry-equivalent Na atoms occupy the positions

$$0, 0, 0; \quad 0, \tfrac{1}{2}, \tfrac{1}{2}; \quad \tfrac{1}{2}, 0, \tfrac{1}{2}; \quad \tfrac{1}{2}, \tfrac{1}{2}, 0$$

A survey of cubic $F$ space groups* shows that there are only two possible situations for the Cl atoms:

(a) $\tfrac{1}{2}, 0, 0; \quad 0, \tfrac{1}{2}, 0; \quad 0, 0, \tfrac{1}{2}; \quad \tfrac{1}{2}, \tfrac{1}{2}, \tfrac{1}{2}$

(b) $\tfrac{1}{4}, \tfrac{1}{4}, \tfrac{1}{4}; \quad \tfrac{3}{4}, \tfrac{3}{4}, \tfrac{1}{4}; \quad \tfrac{1}{4}, \tfrac{3}{4}, \tfrac{3}{4}; \quad \tfrac{3}{4}, \tfrac{1}{4}, \tfrac{3}{4}$

* See Bibliography.

TABLE 5.1. Observed and Calculated Structure Factor Amplitudes
for NaCl

| $hkl$ | $|F_o|$ | $|F_c|$ | $K|F_o|$ | $\Delta F(=K|F_o|-|F_c|)$ |
|---|---|---|---|---|
| 200 | 209.0 | 81.6 | 86.5 | 4.9 |
| 400 | 115.6 | 45.6 | 47.9 | 2.3 |
| 600 | 53.2 | 25.3 | 22.0 | −3.3 |
| 800 | 26.7 | 12.7 | 11.1 | −1.6 |
| 220 | 162.9 | 64.3 | 67.4 | 3.1 |
| 440 | 61.5 | 28.1 | 25.5 | −2.6 |
| 660 | 24.5 | 10.6 | 10.1 | −0.5 |
| 880 | 4.2 | 3.1 | 1.7 | −1.4 |
| 111 | 49.7 | 19.0 | 20.6 | 1.6 |
| 222 | 127.7 | 53.3 | 52.9 | −0.4 |
| 333 | 19.6 | 7.3 | 8.1 | 0.8 |
| 444 | 41.7 | 18.6 | 17.3 | −1.3 |
| 555 | 4.9 | 3.7 | 2.0 | −1.7 |

$\sum|F_o|=901.2$  $\sum|F_c|=373.2$  $\sum K|F_o|=373.1$  $\sum|\Delta F|=25.5$

Although there are five possible space groups, there are no positional parameters to determine, and it is of no consequence to know the true space group at this stage.*

## Structure Factor Calculations

Distinction between structural arrangements (a) and (b) can be effected by comparing the *observed* structure factor amplitudes $|F_o(hkl)|$ listed in Table 5.1 with the corresponding calculated values $|F_c(hkl)|$ based on each of the models in turn.

In any $F$ unit cell, we may write, from (4.42), (4.43), and Table 4.1, for $hkl$ all odd or all even,

$$A'(hkl) = 4 \sum_{j=1}^{N/4} g_j \cos 2\pi(hx_j + ky_j + lz_j) \qquad (5.3)$$

$$B'(hkl) = 4 \sum_{j=1}^{N/4} g_j \sin 2\pi(hx_j + ky_j + lz_j) \qquad (5.4)$$

For all other $hkl$, $A'(hkl) = B'(hkl) = 0$.

---

* This information is necessary in order to effect true refinement in the final stages of the analysis. Alternative (b) corresponds only to $F23$ and $Fm3$. The Laue symmetry (Table 1.6) $m3m$ provides a partial answer, because $F23$ and $Fm3$ correspond to Laue group $m3$, whereas $F432$, $F\bar{4}3m$, and $Fm3m$ belong to Laue group $m3m$.

Structure (a) is centrosymmetric, with the origin on $\bar{1}$. Hence, from (5.3) and (5.4),

$$A'(hkl) = 4\{g_{Na^+} + g_{Cl^-} \cos 2\pi[(h + k + l)/2]\} \qquad (5.5)$$

$$B'(hkl) = 0 \qquad (5.6)$$

A similar analysis for structure (b), keeping the same positions for Na, leads to the following equations:

$$A'(hkl) = 4\{g_{Na^+} + g_{Cl^-} \cos 2\pi[(h + k + l)/4]\} \qquad (5.7)$$

$$B'(hkl) = 4g_{Cl^-} \sin 2\pi[(h + k + l)/4] \qquad (5.8)$$

Determination of the Structure

It is not difficult to decide at this stage which structure model, (a) or (b), is correct. Table 5.2 lists expressions for $|F_c(hkl)|$ for a few selected reflections for comparison with Table 5.1, and it is easy to see that only model (a) will produce the correct *pattern* of calculated structure factor amplitudes ($Z_{Na^+} = 10$, $Z_{Cl^-} = 18$). Table 5.1 lists also the results of a quantitative comparison of $|F_o|$ and $|F_c|$ over all the experimental data. Values of $g_{Na^+}$ and $g_{Cl^-}$ have been derived by correcting the calculated values of $f_{Na^+}$ and $f_{Cl^-}$ for isotropic thermal vibration, taking $\overline{U_{Na^+}^2} = \overline{U_{Cl^-}^2} = 0.025$ Å$^2$. The $|F_o|$ data have been scaled by a factor $K$ of 0.414, since the experimental data are obtained on a true relative but arbitrary scale. The crystal structure is illustrated in Figure 1.1.

Scale Factor

Assuming that all the atoms in the unit cell are included in the model, the $|F_o|$ values may be scaled by making their sum over all values of $hkl$ equal

TABLE 5.2.  Structure Factors Calculated for Models (a) and (b)

| $hkl$ | $|F_c(hkl)|$, model (a) | $|F_c(hkl)|$, model (b) |
|-------|------------------------|--------------------------|
| 200 | $4(g_{Na^+} + g_{Cl^-})$ | $4(g_{Na^+} - g_{Cl^-})$ |
| 220 | $4(g_{Na^+} + g_{Cl^-})$ | $4(g_{Na^+} + g_{Cl^-})$ |
| 111 | $4(g_{Na^+} - g_{Cl^-})$ | $4(g_{Na^+}^2 + g_{Cl^-}^2)^{1/2}$ |
| 222 | $4(g_{Na^+} + g_{Cl^-})$ | $4(g_{Na^+} - g_{Cl^-})$ |

to the same sum with respect to $|F_c|$. The scale factor $K$ is thus given by

$$K = \sum_{hkl} |F_c(hkl)| \Big/ \sum_{hkl} |F_o(hkl)| \qquad (5.9)$$

where the sum extends over all symmetry-independent reflections. If only a fraction of the atoms in the unit cell are present in a trial structure, a realistic value of $K$ is given by

$$K \times \sum_{j=1}^{T} Z_j \Big/ \sum_{j=1}^{N} Z_j \qquad (5.10)$$

where $N$ is the number of atoms in the unit cell, $T$ is the number of atoms in the trial structure, and $Z_j$ is the atomic number of the $j$th atom.

Reliability Factor

The differences between the scaled-observed and the calculated structure-factor amplitudes are a measure of the quality of the trial structure. Large differences correspond to poor reliability, and vice versa. An overall reliability factor ($R$-factor) is defined as

$$R = \sum_{hkl} |K|F_o| - |F_c|| \Big/ \sum_{hkl} K|F_o| \qquad (5.11)$$

For a well-refined structure model, the value of $R$ approaches a small value, corresponding to the errors in both the experimental data and the model. In the early stages of the analysis, however, it may be between 0.4 and 0.5. For the NaCl data in Table 5.1, the bottom line shows the components of $R$ (0.068) and $K$ (0.414). In completing the structure analysis of NaCl, we have introduced the first criterion of correctness, namely, good agreement between $|F_o|$ and $|F_c|$, expressed through the $R$-factor. It should be noted that trial structures with an $R$-factor of more than 50% have been known to be capable of refinement—it is only a rough guide at that stage of the analysis. A better basis for judgment is a comparison of the *pattern* of $|F_o|$ and $|F_c|$, which requires care and experience.

### 5.3.2 Pyrite, FeS$_2$

This problem is a little more complicated than that of NaCl, because a positional parameter has to be determined to specify the crystal structure.

Crystal Data

System: cubic.
Unit-cell dimensions: $a = 5.407$ Å.
$V_c$: 158.1 Å$^3$.
$D_m$: 4.87 g cm$^{-3}$.
$M$: 120.0.
$Z$: 3.87, or 4 to the nearest integer.
Unit-cell contents: 4 Fe and 8 S atoms.
Absent spectra: $0kl$: $k = 2n + 1$.
Space group: $Pa3$.

Table 5.3 lists $|F_o|^2$ values for pyrite, relative to $|F_o(200)|^2 = 100.0$.

Structure Analysis

A symmetry analysis of space group $Pa3$ shows that the Fe and S atoms must lie in special equivalent positions:

4Fe:     $0, 0, 0;$   $0, \frac{1}{2}, \frac{1}{2};$   $\frac{1}{2}, 0, \frac{1}{2};$   $\frac{1}{2}, \frac{1}{2}, 0$

8 S:     $\pm\{x, x, x;$   $\frac{1}{2}+x, \frac{1}{2}-x, \bar{x};$   $\bar{x}, \frac{1}{2}+x, \frac{1}{2}-x;$   $\frac{1}{2}-x, \bar{x}, \frac{1}{2}+x\}$

The iron atoms occupy sites which correspond to an $F$ unit cell. These atoms dominate the X-ray diffraction pattern, since the atomic numbers of Fe and S are 26 and 16, respectively. For this reason, only reflections with $h + k$, $k + l$, and $(l + h)$ even are listed (see Table 4.1), the others being comparatively weak.

TABLE 5.3. $|F_o(hkl)|^2$ Data for Pyrite, Relative to $|F_o(200)|^2 = 100.0$

| hkl | $|F_o|^2$ | hkl | $|F_o|^2$ | hkl | $|F_o|^2$ |
|---|---|---|---|---|---|
| 200 | 100.0 | 220 | 55.0 | 111 | 44.0 |
| 400 | 0.4 | 440 | 37.0 | 222 | 36.0 |
| 600 | 3.2 | 660 | 4.0 | 333 | 19.0 |
| 800 | 14.4 | 880 | 4.0 | 444 | 1.0 |
| 10,00 | 4.0 | | | 555 | 10.0 |

Using methods developed already, we can show that the structure factor equations assume the following forms* for pyrite:

$$A'(hkl) = 4\{g_{Fe} + 2g_S \cos 2\pi[hx_S + (h+k)/4]$$
$$\times \cos 2\pi[kx_S - (k+l)/4] \cos 2\pi[lx_S - (h-l)/4]\}$$

$$(5.12)$$

$$B'(hkl) = 0 \qquad (5.13)$$

We can solve this structure from the $h00$ and $hhh$ reflections, for which cases (5.12) becomes

$$A'(h00) = 4(g_{Fe} + 2g_S \cos 2\pi hx_S) \qquad (5.14)$$

with $h$ even, and with $h+k$ and $k+l$ both even

$$A'(hhh) = 4(g_{Fe} + 2g_S \cos^3 2\pi hx_S) \qquad (5.15)$$

From an inspection of Table 5.3, we see that among the $h00$ reflections, the intensities of 400 and 600 are comparatively weak, whereas that of 800, bearing in mind that it is a high-order reflection, for which the temperature-corrected $f$ factors $(g)$ will be fairly small, is quite strong. From (5.14), it is to be expected that intensities will tend to be weakest where the Fe and S contributions are in opposition ($\cos 2\pi hx_S \approx -1$).

The implications of these observations are summarized in Table 5.4, which also contains information for two $hhh$ reflections. The most consistent values of $x_S$ are those marked with asterisks under "Possible values," which, in fact, represent a unique value for $x_S$ since they are, in each case, related by the center of symmetry at the origin. Selecting the fourth column under "Possible values" and finding the mean value, we have $x_S = 0.39$. A few further structure factor calculations confirm this result. The comparison of $|F_o|$ and $|F_c|$ and the calculation of the $R$-factor are left as an exercise for the reader. The crystal structure is illustrated in Figure 5.6.

---

This is easily proved, particularly if the last S position is changed to $\frac{1}{2}+x$, $x$, $-\frac{1}{2}-x$.

TABLE 5.4. Solution of the $x$ Parameter of the S atom in $FeS_2$

| $hkl$ | $\lvert F_o(hkl)\rvert^2$ | Interpretation | Solution in terms of possible values for $x_S$ in the unit cell | Possible values of $x_S$ in 120ths[a] |
|---|---|---|---|---|
| 400 | 0.4 | $A'(400) = 4(f_{Fe} + 2f_S \cos 8\pi x_S)$<br>If $\lvert A'\rvert \to$ min, $\cos 8\pi x_S \to -1$ | $(2n+1)/8 = \frac{1}{8}, \frac{3}{8}, \frac{5}{8}, \frac{7}{8}$ | — 15 — 45* — 75* — 105 |
| 600 | 3.2 | $A'(600) = 4(f_{Fe} + 2f_S \cos 12\pi x_S)$<br>If $\lvert A'\rvert \to$ min, $\cos 12\pi x_S \to -1$ | $(2n+1)/12 = \frac{1}{12}, \frac{3}{12}, \frac{5}{12},$ $\frac{7}{12}, \frac{9}{12}, \frac{11}{12}$ | — 10 30 50* — 70* 90 110 |
| 800 | 14.4 | $A'(800) = 4(f_{Fe} + 2f_S \cos 16\pi x_S)$<br>If $\lvert A'\rvert \to$ max, $\cos 16\pi x_S \to +1$ | $n/8 = 0, \frac{1}{8}, \frac{2}{8}, \frac{3}{8},$ $\frac{4}{8}, \frac{5}{8}, \frac{6}{8}, \frac{7}{8}$ | 0 15 30 45* 60 75* 90 105 |
| 444 | 1.0 | $A'(444) = 4(f_{Fe} + 2f_S \cos^3 8\pi x_S)$<br>If $\lvert A'\rvert \to$ min, $\cos^3 8\pi x_S \to -1$ | $(2n+1)/8 = \frac{1}{8}, \frac{3}{8}, \frac{5}{8}, \frac{7}{8}$ | — 15 — 45* — 75* — 105 |
| 555 | 10.0 | $A'(555) = 4(f_{Fe} + 2f_S \cos^3 10\pi x_S)$<br>If $\lvert A'\rvert \to$ max, $\cos^3 10\pi x_S \to +1$ | $n/5 = 0, \frac{1}{5}, \frac{2}{5}, \frac{3}{5}, \frac{4}{5}$ | 0 — 24 48* — 72* 96 — |

[a] The most consistent values of $x$ are marked with asterisks. Other agreements among the first four rows show the inability to differentiate between $x_S$ and $\frac{1}{2} - x_S$ when $h$ is an even number.

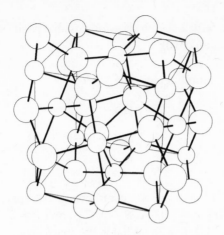

FIGURE 5.6. Unit cell of the pyrite structure: large circles S, small circles Fe.

# Bibliography

## Introductory Structure Analysis

BRAGG, W. L., *A General Survey* (*The Crystalline State*, Vol. I), London, Bell.

HENRY, N. F. M., and LONSDALE, K. (Editors), *International Tables for X-Ray Crystallography*, Vol. I, Birmingham, Kynoch Press.

# Problems

**5.1.** A structure with the apparent space group $P2_1/c$ consists of atoms at $0.2, \frac{1}{4}, 0.1$ and the symmetry-related positions; the center of symmetry is at the origin. Evaluate the geometric structure factor for the four given positions in the unit cell, and, hence, determine the systematic absences among the $hkl$ reflections. What are the consequences of these absences as far as the true structure is concerned? Sketch the structure in projection along $b$. What is the true space group?

**5.2.** $Rh_2B$ crystallizes in space group $Pnma$ with $a = 5.42$, $b = 3.98$, $c = 7.44$ Å, and $Z = 4$. Consider Figure 2.36. Show that if no two Rh atoms may approach within 2.5 Å of each other, they cannot lie in general positions. Where might the Rh atoms be placed? Illustrate your answer with a sketch showing possible positions for these atoms in projection on (010).

**5.3.** Trimethylammonium chloride,

$$
\left[\begin{array}{c}
H_3C \\
\searrow \\
H_3C-N-H \\
\nearrow \\
H_3C
\end{array}\right]^+ Cl^-,
$$

crystallizes in a monoclinic, centrosymmetric space group, with $a = 6.09$, $b = 7.03$, $c = 7.03$ Å, $\beta = 95.73°$, and $Z = 2$. The only limiting condition is $0k0$: $k = 2n$. What is the space group? Comment on the probable positions of (a) Cl, (b) C, (c) N, and (d) H atoms.

**5.4.** Potassium hexachloroplatinate(IV), $K_2[PtCl_6]$, is cubic, with $a = 9.755$ Å. The atomic positions are as follows $(Z = 4)$:

$$
(0,0,0; \quad 0,\tfrac{1}{2},\tfrac{1}{2}; \quad \tfrac{1}{2},0,\tfrac{1}{2}; \quad \tfrac{1}{2},\tfrac{1}{2},0)+
$$

4   Pt:   0, 0, 0

8   K:   $\tfrac{1}{4},\tfrac{1}{4},\tfrac{1}{4}; \quad \tfrac{3}{4},\tfrac{3}{4},\tfrac{3}{4}$

24   Cl:   $\pm\{x, 0, 0; \quad 0, x, 0; \quad 0, 0, x\}$

Show that $F_c(hhh) = A'(hhh)$, where

$$
A'(hhh) = 4g_{Pt} + 8g_K \cos(3\pi h/2) + 24g_{Cl} \cos 2\pi h x_{Cl}
$$

Calculate $|F_c(hhh)|$ for the values of $h$ tabulated below, with $x_{Cl} = 0.23$ and 0.24. Obtain $R$-factors for the scaled $|F_o|$ data for the two values of $x_{Cl}$, and indicate which value of $x_{Cl}$ is the more acceptable. Calculate the Pt—Cl distance, and sketch the $[PtCl_6]^{2-}$ ion. What is the point group of this species?

| $hkl$ | 111 | 222 | 333 |
|-------|-----|-----|-----|
| $|F_o|$ | 491 | 223 | 281 |

Atomic scattering factor curves are given in Figure 5.P1, and may be taken to be temperature-corrected values.

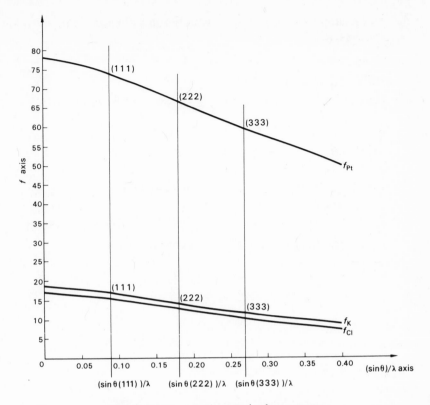

FIGURE 5.P1. Atomic scattering factor curves.

**5.5.** USi crystallizes in space group *Pbnm*, with $a = 5.65$, $b = 7.65$, $c = 3.90$ Å, and $Z = 4$. The U atoms lie at the positions

$$\pm\{x, y, \tfrac{1}{4};\quad \tfrac{1}{2} - x, \tfrac{1}{2} + y, \tfrac{1}{4}\}$$

Obtain a reduced expression for the geometric structure factor ($\bar{1}$ at 0, 0, 0) for the U atoms. From the data below, determine approximate values for $x_U$ and $y_U$; the Si contributions may be neglected.

| hkl | 200 | 111 | 210 | 231 | 040 | 101 | 021 | 310 |
|-----|-----|-----|-----|-----|-----|-----|-----|-----|
| $I_o(hkl)$ | 0 | 236 | 251 | 200 | 0 | 170 | 177 | 0 |

Proceed by using 200 to find a probable value for $x_U$. Then find $y_U$ from 111, 231, and 040.

**5.6.**   Methylamine forms a complex with boron trifluoride of composition $CH_3NH_2BF_3$.

*Crystal data*

System: monoclinic.
Unit-cell dimensions: $a = 5.06$, $b = 7.28$, $c = 5.81$ Å, $\beta = 101.5°$.
$V_c$: 209.7 Å$^3$.
$D_m$: 1.54 g cm$^{-3}$.
$M$: 98.86.
$Z$: 1.97, or 2 to the nearest integer.
Unit-cell contents: 2 C, 10 H, 2 N, 2 B, and 6 F atoms.
Absent spectra: $0k0$: $k = 2n + 1$.
Possible space groups: $P2_1$ or $P2_1/m$ ($P2_1/m$ may be assumed).

Determine what you can about the crystal structure.

# Methods in X-Ray Structure Analysis. II

## 6.1 Introduction

In this chapter, we shall introduce Fourier series and show how they are used in structure analysis. This discussion leads naturally to the Patterson function and the very important heavy-atom and isomorphous replacement techniques for solving the phase problem.

## 6.2 Fourier Series

According to Fourier's theorem, a continuous, single-valued, periodic function can be represented by a series composed of sine and cosine terms. A typical periodic function $\psi(X)$ is shown in Figure 6.1, in which the repeat period along the $X$ axis is $a$. The corresponding Fourier series may be written as

$$\psi(X) = \frac{1}{a} \sum_{h=-\infty}^{\infty} \left[ C(h) \cos 2\pi \frac{hX}{a} + S(h) \sin 2\pi \frac{hX}{a} \right] \qquad (6.1)$$

The index $h$ of the $h$th term in this equation represents its frequency or wavenumber, which is the number of times its own wavelength fits into the

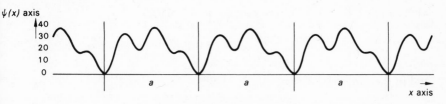

FIGURE 6.1. Periodic function.

201

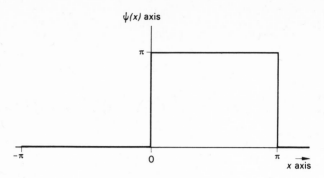

FIGURE 6.2. Square wave.

repeat period. The coefficients $C(h)$ and $S(h)$ may be expressed by the equations

$$C(h) = \int_{-a/2}^{a/2} \psi(X) \cos(2\pi hX/a)\, dx \qquad (6.2)$$

and

$$S(h) = \int_{-a/2}^{a/2} \psi(X) \sin(2\pi hX/a)\, dx \qquad (6.3)$$

If the form of the function $\psi(X)$ is known, $C(h)$ and $S(h)$ can be evaluated for values of $h$. We shall carry out this process for the square wave in Figure 6.2.

### 6.2.1   Computation of $\psi(X)$ for a Square Wave

This function is defined in the range $-\pi$ to $\pi$, with a repeat of $2\pi$. Hence,

$$\psi(X) = \psi(X \pm m2\pi) \qquad (6.4)$$

where $m = 0, 1, \ldots, \infty$. In the range $-\pi \leq X \leq 0$,

$$\psi(X) = 0 \qquad (6.5)$$

and in the range $0 \leq X \leq \pi$,

$$\psi(X) = \pi \qquad (6.6)$$

From (6.2), with $a/2$ replaced by $\pi$ and using (6.5) and (6.6),

$$C(h) = \int_{-\pi}^{0} 0 \cos hX \, dx + \int_{0}^{\pi} \pi \cos hX \, dx \tag{6.7}$$

Integrating with respect to $X$, and since the first term on the right-hand side is zero,

$$C(h) = \frac{\pi \sin hX}{h} \Big]_{0}^{\pi} \tag{6.8}$$

The limit $h = 0$ must be considered separately from (6.7), since $h$ occurs in both the numerator and the denominator. Thus,

$$C(0) = \int_{0}^{\pi} \pi \, dx \tag{6.9}$$

Integrating,

$$C(0) = \pi^2 \tag{6.10}$$

Similarly for $S(h)$, from (6.3),

$$S(h) = \int_{-\pi}^{0} 0 \sin hX \, dx + \int_{0}^{\pi} \pi \sin hX \, dx \tag{6.11}$$

Integrating as before,

$$S(h) = -\frac{\pi}{h} \cos hX \Big]_{0}^{\pi} = \frac{\pi}{h}(1 - \cos h\pi) \tag{6.12}$$

From (6.11), it is clear that

$$S(0) = 0 \tag{6.13}$$

Substituting for $C(h)$ and $S(h)$ in (6.1),

$$\psi(X) = \frac{1}{2\pi} \left[ \pi^2 + \sum_{\substack{h=-\infty \\ h \neq 0}}^{\infty} \frac{\pi}{h}(1 - \cos h\pi) \sin hX \right] \tag{6.14}$$

The term corresponding to $h = 0$ is taken outside the summation. From (6.12),

$$S(-h) = -S(h) \tag{6.15}$$

Now,

$$(1/h) \sin hX = (1/-h) \sin(-hX) \tag{6.16}$$

and since $1 - \cos h\pi$ is equal to $1 - (-1)^h$, $\psi(X)$ is defined only for *odd values* of $h$. Thus,

$$\psi(X) = \frac{\pi}{2} + 2 \sum_{h=1}^{\infty} \frac{1}{h} \sin hX, \qquad (h = 2n + 1) \tag{6.17}$$

Range of $X$

The variable $X$ defines a sampling point in any repeat interval of the function. We can describe the function $\psi(X)$ by choosing values of $X$ from any point $X_0$ to $X_0 + 2\pi$. For convenience, we shall choose $X_0$ as 0, and define a sampling interval of $2\pi/50$. In this example, we shall calculate $\psi(X)$ at $0, 2\pi/50, 4\pi/50, \ldots, 98\pi/50, 100\pi/50$, but we shall note from the results that we could, and, in general, would, have quite properly made use of the reflection symmetry of the function at $X = (2m + 1)\pi/2$, $m = 0, 1, 2, \ldots, \infty$.

Range of $h$

The summations in Fourier series extend, theoretically, from $-\infty$ to $\infty$. In practice, however, this range is $h_{min} \leq h \leq h_{max}$, where $h_{min}$ and $h_{max}$ are some preset limits of the frequency variable. In this example, $h_{min}$ is unity, and the effect of two different values for $h_{max}$ is illustrated in Figures 6.3a and 6.3b, drawn from the results in Table 6.1. Even at this level, the value of $h_{max}$ has a dramatic effect on the series. By increasing the number of terms, the series (6.17) approaches more closely a representation of the square-wave function. In general, the more independent terms that can be included in a Fourier series, the better it represents the periodic function under investigation, from which the terms are derived.

The process of determining the coefficients of a Fourier series is called Fourier analysis, and the process of reconstructing the function by summa-

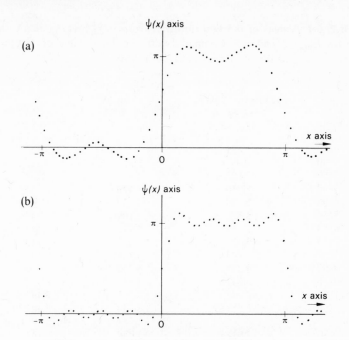

FIGURE 6.3. Calculated square waves: (a) $h_{max} = 3$, (b) $h_{max} = 5$. The positive and negative fluctuations across $\psi(X) = 0$ and $\pi$ are known as series termination errors; they arise because there are insufficient terms to provide good convergence of the series.

tion of a series such as (6.17) is called Fourier synthesis. A microscope, in forming an image of an object, effectively performs its own Fourier synthesis of the light scattered by the object (see also page 156).

### 6.2.2 Exponential Form of Fourier Expressions

Using de Moivre's theorem, (4.21), we obtain

$$\cos(2\pi hX/a) = (e^{i2\pi hX/a} + e^{-i2\pi hX/a})/2 \qquad (6.18)$$

$$\sin(2\pi hX/a) = (e^{i2\pi hX/a} - e^{-i2\pi hX/a})/2i \qquad (6.19)$$

and substituting (6.18) and (6.19) in (6.1),

$$\psi(X) = \frac{1}{2a} \sum_{h=-\infty}^{\infty} [C(h)(e^{i2\pi hX/a} + e^{-i2\pi hX/a}) - iS(h)(e^{i2\pi hX/a} - e^{-i2\pi hX/a})]$$

$$(6.20)$$

TABLE 6.1.  Values of the Function $\psi(X) = \frac{1}{2}\pi + 2\sum_{h=1}^{h_{max}} (1/h) \sin hX$ ($h$ odd) for $h_{max} = 3$ and $= 5$, Compared with the True Value of $\psi(X) = \pi$ ($0 < X < \pi$) or $\psi(X) = 0$ ($\pi < X < 2\pi$)[a]

| $X/2\pi$ | $\psi(X) \approx \frac{1}{2}\pi + 2 \sin X$ $+ \frac{2}{3} \sin 3X$ $h_{max} = 3$ | $\psi(X) \approx \frac{1}{2}\pi + 2 \sin X$ $+ \frac{2}{3} \sin 3X$ $+ \frac{2}{5} \sin 5X$ $h_{max} = 5$ | $\psi(X)$ |
|---|---|---|---|
| 0/50 | 1.571 | 1.571 | 3.142 |
| 1/50 | 2.067 | 2.522 | 3.142 |
| 2/50 | 2.525 | 3.186 | 3.142 |
| 3/50 | 2.910 | 3.428 | 3.142 |
| 4/50 | 3.200 | 3.330 | 3.142 |
| 5/50 | 3.380 | 3.109 | 3.142 |
| 6/50 | 3.454 | 2.977 | 3.142 |
| 7/50 | 3.433 | 3.017 | 3.142 |
| 8/50 | 3.343 | 3.158 | 3.142 |
| 9/50 | 3.215 | 3.265 | 3.142 |
| 10/50 | 3.081 | 3.249 | 3.142 |
| 11/50 | 2.972 | 3.137 | 3.142 |
| 12/50 | 2.912 | 3.034 | 3.142 |
| _m_———   |  |  | ———_m_ |
| 13/50 | 2.912 | 3.034 | 3.142 |
| 14/50 | 2.972 | 3.137 | 3.142 |
| 15/50 | 3.081 | 3.249 | 3.142 |
| 16/50 | 3.215 | 3.265 | 3.142 |
| 17/50 | 3.343 | 3.158 | 3.142 |
| 18/50 | 3.433 | 3.017 | 3.142 |
| 19/50 | 3.454 | 2.977 | 3.142 |
| 20/50 | 3.380 | 3.109 | 3.142 |
| 21/50 | 3.200 | 3.330 | 3.142 |
| 22/50 | 2.910 | 3.428 | 3.142 |
| 23/50 | 2.525 | 3.186 | 3.142 |
| 24/50 | 2.067 | 2.522 | 3.142 |
| 25/50 | 1.571 | 1.571 | 3.142 |
| 26/50 | 1.075 | 0.619 | 0 |
| 27/50 | 0.617 | −0.044 | 0 |
| 28/50 | 0.231 | −0.287 | 0 |
| 29/50 | −0.058 | −0.188 | 0 |
| 30/50 | −0.239 | 0.033 | 0 |
| 31/50 | −0.312 | 0.164 | 0 |
| 32/50 | −0.291 | 0.125 | 0 |
| 33/50 | −0.201 | −0.017 | 0 |
| 34/50 | −0.073 | −0.123 | 0 |
| 35/50 | 0.061 | −0.107 | 0 |
| 36/50 | 0.169 | 0.005 | 0 |
| 37/50 | 0.230 | 0.108 | 0 |
| _m_———   |  |  | ———_m_ |
| 38/50 | 0.230 | 0.108 | 0 |

TABLE 6.1—*cont.*

| $X/2\pi$ | $\psi(X) \approx \frac{1}{2}\pi + 2\sin X$ $+\frac{2}{3}\sin 3X$ $h_{max} = 3$ | $\psi(X) \approx \frac{1}{2}\pi + 2\sin X$ $+\frac{2}{3}\sin 3X$ $+\frac{2}{5}\sin 5X$ $h_{max} = 5$ | $\psi(X)$ |
|---|---|---|---|
| 39/50 | 0.169 | 0.005 | 0 |
| 40/50 | 0.061 | −0.107 | 0 |
| 41/50 | −0.073 | −0.123 | 0 |
| 42/50 | −0.201 | −0.017 | 0 |
| 43/50 | −0.291 | 0.125 | 0 |
| 44/50 | −0.312 | 0.164 | 0 |
| 45/50 | −0.239 | 0.033 | 0 |
| 46/50 | −0.058 | −0.188 | 0 |
| 47/50 | 0.231 | −0.287 | 0 |
| 48/50 | 0.617 | −0.044 | 0 |
| 49/50 | 1.075 | 0.619 | 0 |
| 50/50 | 1.571 | 1.571 | 3.142 |

[a] The corresponding curves are shown in Figure 6.3.

Collecting terms, and following the notation of (4.41),

$$\psi(X) = \frac{1}{2a} \sum_{h=-\infty}^{\infty} [\mathbf{G}(h)e^{-i2\pi hX/a} + \mathbf{G}(-h)e^{i2\pi hX/a}] \qquad (6.21)$$

where

$$\mathbf{G}(h) = C(h) + iS(h) \qquad (6.22)$$

and

$$\mathbf{G}(-h) = C(h) - iS(h) \qquad (6.23)$$

Since $h$ ranges from $-\infty$ to $\infty$, both expressions under the summation in (6.21) take a sequence of identical values within the range of the variable $h$. Hence, (6.21) may be written

$$\psi(X) = \frac{1}{a} \sum_{h=-\infty}^{\infty} \mathbf{G}(h)e^{-i2\pi hX/a} \qquad (6.24)$$

From (6.2), (6.3), and (6.22),

$$G(h) = \int_{-a/2}^{a/2} \psi(X)[\cos(2\pi hX/a) + i \sin(2\pi hX/a)] \, dX \qquad (6.25)$$

and, using (4.21),

$$G(h) = \int_{-a/2}^{a/2} \psi(X) e^{i2\pi hX/a} \, dX \qquad (6.26)$$

Although not of great practical use, the complex forms of (6.24) and (6.26) can provide a useful starting point for further manipulation of Fourier equations. The functions $\psi(X)$ and $G(h)$, as defined in these equations, are said to be Fourier transforms of each other. If $G(h)$ is known for all values of $h$, we can calculate $\psi(X)$. Similarly, if $\psi(X)$ is known over the periodic range $-a/2$ to $a/2$, we can calculate $G(h)$. Equation (6.24) represents the Fourier synthesis and (6.26) the Fourier analysis of the function $\psi(X)$.

## 6.3   Representation of Crystal Structures by Fourier Series

Because of the underlying lattice structure of crystals, the contents of the unit cell are repeated periodically by the translations $a$, $b$, and $c$. We may consider an analogy between the square wave and a conceptual one-dimensional crystal, or a projection of a crystal structure onto an axis. The first applications of Fourier series in crystallography used this restricted geometric form, and for good reason. The mathematical treatment for three-dimensional periodicity is correspondingly more complicated, but it follows the form of the one-dimensional example, and we shall obtain the necessary equations by analogy with (6.24) and (6.26). It is important, however, to grasp the significance of the functions in crystallography that correspond to $\psi(X)$ and $G(h)$.

### 6.3.1   Electron Density and Structure Factors

We have shown in Chapter 4 how X-rays are scattered by the electrons associated with the atoms in a crystal. Atoms with high atomic numbers provide a greater concentration of electrons than do atoms of low atomic numbers. This concentration of electrons and its distribution around the atom is called the electron density $\rho$, and is usually measured in electrons per $\text{Å}^3$. Since it is, in general, a function of position, we specify the electron density at the point $X$, $Y$, $Z$ as $\rho(X, Y, Z)$. The periodicities of the lattice are impressed on the electron density in the crystal, and, therefore, we become concerned with a three-dimensional, periodic electron density function: We may identify $\rho(X)$ with $\psi(X)$ of the previous example [see (6.24)], but we must determine the meaning of $\mathbf{G}(h)$.

In Chapter 4, we considered the electrons in an atom as though they were concentrated at a point, and their distribution was specified by a shape factor, the atomic scattering factor $f$. The exponential term modifying $f$ represented the phase at the atomic position with respect to the origin of the unit cell.

Consider the one-dimensional electron density function in Figure 6.4. In a small interval $dX$ along the $X$ axis, the electron density $\rho(X)$ may be regarded as constant. The electron count in this strip is, therefore, $\rho(X)\,dX$. Following (4.25) and (4.35), its contribution to the $h$th structure factor is given by

$$\rho(X)e^{i2\pi hX/a}\,dX \tag{6.27}$$

Hence, $\mathbf{F}(h)$ is given by the integration of this expression over the unit cell:

$$\mathbf{F}(h) = \int_0^a \rho(X)e^{i2\pi hX/a}\,dX \tag{6.28}$$

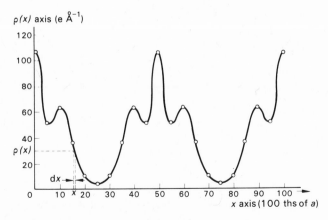

FIGURE 6.4. One-dimensional electron density projection for pyrite $(FeS_2)$.

Equation (6.28) is a generalized form for the structure factor; similar representations for two- and three-dimensional structure factors can be written down by analogy with (6.28). We have already identified $\psi(X)$ with $\rho(X)$, and substituting (6.24) in (6.28) gives

$$\mathbf{F}(h) = \int_0^a \frac{1}{a} \sum_{h'=-\infty}^{\infty} \mathbf{G}(h') e^{-i2\pi h'X/a} e^{i2\pi hX/a} \, dX \qquad (6.29)$$

where $h'$ indicates a range of values of $h$ under the summation sign. Since the integral of the sum in (6.29) may be thought of as a sum of the integrals of the separate terms, we write

$$\mathbf{F}(h) = \frac{1}{a} \sum_{h'=-\infty}^{\infty} \mathbf{G}(h') \int_0^a e^{i2\pi X(h-h')/a} \, dX \qquad (6.30)$$

Performing the integration in (6.30) leads to

$$\frac{\exp[i2\pi X(h-h')/a]}{i2\pi(h-h')/a} \Bigg]_0^a \qquad (6.31)$$

which, on substituting the limits, becomes

$$\frac{\exp[i2\pi(h-h')]-1}{i2\pi(h-h')/a} \qquad (6.32)$$

Since $h$ and $h'$ are both integers, the numerator of (6.32) is zero, from (4.21), unless $h = h'$. In this special case (6.32) is indeterminate, but from (6.30) we see that the integral is now

$$\int_0^a dX \qquad (6.33)$$

which has the value $a$. Since only a *single* value of $h'$ (namely $h$) leads to a nonzero value for the integral in (6.30), this equation becomes

$$\mathbf{F}(h) = \mathbf{G}(h) \qquad (6.34)$$

and from (6.24)

$$\rho(X) = \frac{1}{a} \sum_{h=-\infty}^{\infty} \mathbf{F}(h)e^{-i2\pi hX/a} \qquad (6.35)$$

Since $\rho(X)$ is periodic, we could use the limits $-a/2$ to $a/2$ instead of 0 to $a$, and, by analogy with (6.24) and (6.26), we see that (6.28) is the Fourier transform of the electron density (6.35). In other words, the structure factors provide the coefficients for the Fourier synthesis of the electron density, which returns us to the phase problem (page 156). We have, in fact, considered already simple examples of the solution of the phase problem in studying NaCl and $FeS_2$ (pages 190–197). Once the atomic positions are known, even approximately, phases can be calculated for each value of $|F_o(hkl)|$ and a Fourier synthesis performed. This has been done for $FeS_2$ in Figure 6.4, from which we can deduce that $x_S$ is 0.11 (compare page 195).

### 6.3.2  Electron Density Equations

The general electron density function is expressed as a three-dimensional Fourier series. By analogy with (6.24), and introducing fractional coordinates ($x = X/a$, etc.), we may write

$$\rho(x, y, z) = \frac{1}{V_c} \sum_h \sum_k \sum_l^{\infty} \mathbf{F}(hkl)e^{-i2\pi(hx+ky+lz)} \qquad (6.36)$$

where $V_c$ is the unit-cell volume, (2.19). Distinguish carefully between the points $x, y, z$ at which the electron density is calculated, and the points $x_j, y_j, z_j$ (in the structure factor equation), which represent *actual* atomic positions; both sets of points cover the same field, the unit cell. The summations, in practice, extend over a finite set of $hkl$ values, which are limited by both the nature of the crystal and the particular experimental arrangement.

The exponential form of (6.36) implies that the electron density is a complex function, whereas, in fact, it is real throughout the unit cell. The derivations of practicable expressions depend upon the use of (4.21), (4.41), (4.44), and (4.56), leading to

$$\rho(x, y, z) = \frac{1}{V_c} \sum_h \sum_k \sum_l^{\infty} |F(hkl)|e^{i\phi(hkl)}e^{-i2\pi(hx+ky+lz)} \qquad (6.37)$$

or

$$\rho(x, y, z) = \frac{1}{V_c} \sum_h \sum_{k-\infty}^{\infty} \sum_l [A'(hkl) \cos 2\pi(hx + ky + lz)$$

$$+ B'(hkl) \sin 2\pi(hx + ky + lz)] \qquad (6.38)$$

where the summations are taken over the appropriate practical values of $h$, $k$, and $l$. Using (4.47) and (4.48), (6.38) becomes

$$\rho(x, y, z) = \frac{1}{V_c} \sum_h \sum_{k-\infty}^{\infty} \sum_l |F(hkl)| \cos[2\pi(hx + ky + lz) - \phi(hkl)] \quad (6.39)$$

Further simplification of (6.38) and (6.39) depends on the use of Friedel's law (4.51). Hence, (6.38) becomes

$$\rho(x, y, z) = \frac{2}{V_c} \sum_{h=0} \sum_{k-\infty}^{\infty} \sum_l [A'(hkl) \cos 2\pi(hx + ky + lz)$$

$$+ B'(hkl) \sin 2\pi(hx + ky + lz)] \qquad (6.40)$$

and (6.39),

$$\rho(x, y, z) = \frac{2}{V_c} \sum_{h=0} \sum_{k-\infty}^{\infty} \sum_l |F(hkl)| \cos[2\pi(hx + ky + lz) - \phi(hkl)] \quad (6.41)$$

since the maximum necessary summations now take place over a hemisphere in reciprocal space. In these and similar equations, the term $F(000)$, which represents the total number of electrons in the unit cell, is multiplied by $1/V_c$ and, in practice, the term $F(000)/V_c$ is added separately to the result of the summations. It does not matter which of $h$, $k$, or $l$ has zero as its lower summation limit.

Equation (6.40) is the most useful for electron density calculations, but (6.41) shows clearly the dependence of $\rho(x, y, z)$ on $\phi(hkl)$. The one-dimensional function in Figure 6.1 is shown again, now called $\rho(x)$, in Figure 6.5, together with its four components corresponding to $h = 0, 1, 2,$ and 3. This function may be written, following (6.41), as

$$\rho(x) = \frac{20}{a} + \frac{2}{a}\{5 \cos[2\pi(x) - 0.47] + 2.5 \cos[2\pi(2x) - 0.17]$$

$$+ 4 \cos[2\pi(3x) - 0.19]\} \qquad (6.42)$$

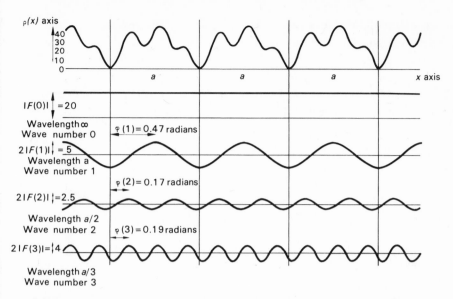

FIGURE 6.5. Analysis of a periodic one-dimensional function of repeat length $a$ (see Figure 6.1). The waves for $h = 1, 2,$ and 3 have been given twice their amplitudes so as to allow for the factor 2 in (6.41).

where, for convenience, $a$ may be taken as unity. The process of X-ray diffraction corresponds to a Fourier analysis, or breakdown of the object $\rho(x)$ into its constituent terms, with the exception that, in recording the *intensities*, the phase information is lost. Fourier synthesis is equivalent to a summation of the terms in order to reconstruct the object, but it cannot be achieved without regaining the phase information.

We will look at the synthesis (6.42) in terms of Figure 6.5. The zeroth-order term $F(0)$ impresses a constant, positive electron density on to the distribution of $\rho(x)$. It is a term of zero phase, and tells us nothing about the structure, only about the contents of the repeat unit $a$. Note that the zeroth-order term does not lie within the multiplying factor $2/a$.

$F(1)$ shows the magnitude of the period $a$ and introduces a little information about the structure. $F(2)$ and $F(3)$ add successively finer detail about the structure, and $\rho(x)$ is the algebraic sum of these waves, as we can see from Figure 6.5. In general, the more orders of diffracted spectra that are included in the sum, the better is the resolution of the image. However, this sum is correct only because the correct phases are used in the equation. The reader may like to show (very easily with tracing paper) that while a change in phase corresponds only to a movement of the wave along the $x$ axis, an entirely different form of $\rho(x)$ may ensue. Try altering the phase of $F(1)$ from 0.47 to 0.17 and then resumming (6.42).

The methods of structure analysis* seek to extract phase information starting from the $|F_o|$ data. A considerable amount of computation is involved in this process. Fortunately for the modern crystallographer, programs for the calculation of structure factors and electron density fields are now standard "equipment" and become relatively easy on high-speed, large-capacity computers. However, it is very important that the student should appreciate the nature of these calculations and the underlying theory of their use in crystallography.

### 6.3.3   Interpretation of Electron Density Distributions

Electron density is concentrated in the vicinity of atoms, rising to peaks at the electron density maxima (atomic "positions") and falling to relatively low values between them. The wavelengths of X-rays used in crystal structure analysis are too long to reveal the intimate electronic structure of atoms themselves, which are seen, therefore, somewhat blurred in the calculated electron density function. Atoms appear as peaks in this function, and the peak position of a given atom is assumed to correspond to its atomic center, within the limit of experimental errors. In general, the more complete and accurate the experimental $|F|$ data, the better will be the atomic resolution and the more precise the final structure model.

### Peak Heights and Weights

To a first approximation, the heights of the peaks in an electron density distribution of a crystal are proportional to the corresponding atomic numbers. The hydrogen atom, at the extreme low end of the atomic numbers, does not appear in electron density representations; its small electron density merges into the background density that arises from errors in both the data and the structure model. However, hydrogen atoms can be detected by a special difference-Fourier technique, as discussed in a later section (page 252).

A better measure of the electron content of a given atom may be obtained from an integrated peak weight, in which the values of $\rho(x, y, z)$ are summed over the volume occupied by the atom. This technique makes some

---

* An apparent paradox, since it refers to Fourier synthesis.

allowance for the variation of individual atomic temperature factors, high values of which tend to decrease peak heights for a given electron content.

## Computation and Display of Electron Density Distributions

Assuming for the moment that phases are available, the electron density function may be calculated over a grid of chosen values of $x$, $y$, and $z$. For this purpose, the unit cell is divided into a selected number of equal divisions, in a manner similar to that employed in the synthesis of the square-wave function (page 204). Intervals corresponding to about 0.3 Å are satisfactory for most electron density maps. The symmetry of $\rho(x, y, z)$ corresponds to the space group of the crystal under investigation. Consequently, a summation over a volume equal to, or just greater than, that of the asymmetric unit is adequate.

In order to facilitate the interpretation of $\rho(x, y, z)$, it is essential to present the distribution of the numerical values in such a way that the geometric relationships between the peaks are easily inspected. This feature is afforded by first calculating the electron density in sections, each corresponding to a constant value of $x$, $y$, or $z$ using (6.40). Each section consists of a field of figures arranged on a grid, closely true to scale for preference, which may be contoured by lines passing through points of equal electron density, interpolating as necessary (Figure 6.6). The grading of the contour intervals is selected to produce a reasonable number of contours around the higher density areas. The contouring should be carried out with care; this exercise leads to precise peak positions and a desirable familiarity with the problem. Sophisticated map-plotting and peak-searching facilities are available, but they should be treated with caution by the beginner.

The contoured sections are finally transferred to a transparent medium, such as thin perspex or clear acetate sheets, which are then stacked at the requisite spatial intervals and viewed over a diffuse light source. Figure 1.2 is a photograph of such a display, extending through 17 sections.

An alternative method of displaying the results of an electron density calculation is by means of a ball-and-stick model. An example of this form of representation is shown in Figure 6.7.

## Projections

The use of two-dimensional studies in crystallography is fairly restrictive, but, nevertheless, worthy of mention because of the relative ease of

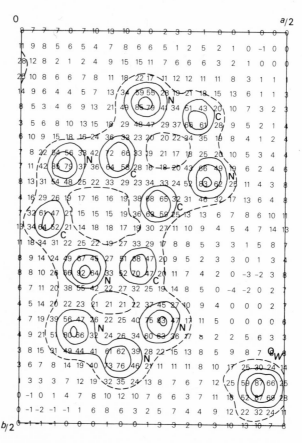

FIGURE 6.6. Two-dimensional electron density projection $\rho(x, y)$ for azidopurine monohydrate, $C_5H_3N_7 \cdot H_2O$ [calculated from the data of Glusker *et al.*, *Acta Crystallographica B* **24**, 359 (1968)]. The isolated peak $(O_w)$ in the lower right-hand region of the map represents the oxygen atom of the water molecule. Hydrogen atom positions are not obtained in a direct electron density synthesis (see page 252). The field figures are $10\rho(x, y)$ in electrons per $Å^2$ contoured at intervals of 20 units.

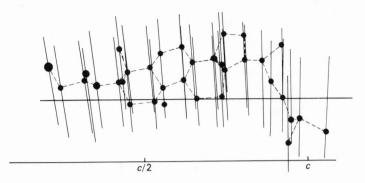

FIGURE 6.7. Three-dimensional model of euphenyl iodoacetate (see Figures 1.2 and 1.3).

calculation and preparation of Fourier maps. For example, the function

$$\rho(x, z) = \frac{2}{A_b} \sum_h \sum_l |F(h0l)| \cos[2\pi(hx + lz) - \phi(h0l)] \qquad (6.43)$$

is calculated with the data from only one level of the reciprocal lattice, the zero level, perpendicular to $b$, and plotted over the area $A_b$ of the $ac$ plane, or the asymmetric portion thereof. The simplification in the calculations is offset, however, by a corresponding complexity in the interpretation of the maps, arising from the superposition of peaks in projection onto the given plane, although this effect is not as severe as in one dimension. Equation (6.43) corresponds to the projection of the electron density along the $b$ axis: it is essential to appreciate the difference between the meaning of $\rho(xz)$ and $\rho(x0z)$, for example; the latter represents the section of the three-dimensional electron density function at $y = 0$. Equations for projections along other principal axes may be written by analogy with (6.43).

Even simple atomic arrangements may appear distorted in projection, with individual molecules overlapping to some degree, but we would not wish to discourage their consideration. We shall restrict their use to examples illustrating various aspects of structure analysis. Practice in the calculation and interpretation of Fourier series is afforded by Problems 6.5 and 6.6 at the end of this chapter (see also Chapter 8).

## 6.4 Methods of Solving the Phase Problem

The set of $|F_o(hkl)|$ data constitute the starting point of all X-ray structure determinations. The approximate number of symmetry-independent reflections measurable may be calculated in the following manner.

### 6.4.1  Number of Reflections in the Data Set

The radius of the limiting sphere (page 134) is 2 RU, and its volume is therefore 33.51 $RU^3$. The number of reciprocal lattice points within the limiting sphere is approximately equal to the number of times the reciprocal unit cell, volume $V^*$, will fit into 33.51; using (2.20), this number is 33.51 $V_c/\lambda^3$. The number of symmetry-independent reflections observable, $N_{max}$, in a given experiment in which $\theta_{max}$ represents the practical upper limit is given by

$$N_{max} = 33.51\, V_c \sin^3\theta_{max}/\lambda^3 Gm \qquad (6.44)$$

where $G$ is the unit-cell translation constant (Table 4.1) for nonzero reflections and $m$ is the number of symmetry-equivalent reflections (the number of general equivalent points in the appropriate Laue group). For zones and rows, $m$ may take different values from that for $hkl$, and a number of systematic absences within the sphere of radius 2 sin $\theta_{max}$ may have to be subtracted.

As an example, consider an orthorhombic crystal of space group $Cmm2$, with unit cell dimensions $a = 9.00$, $b = 10.00$, and $c = 11.00$ Å. For Cu $K\alpha$ radiation ($\bar{\lambda} = 1.542$ Å) and $\theta_{max}$ of 85°, $N_{max}$ is $(33.51 \times 9 \times 10 \times 11 \times \sin^3 85°)/(1.542^3 \times 2 \times 8) = 559$. If Mo $K\alpha$ radiation ($\bar{\lambda} = 0.7107$ Å) had been used instead of Cu $K\alpha$, the number would have been 5710. Such a structure might contain, say, 15 atoms in the asymmetric unit. In the structure analysis, each atom would be determined by three positional parameters $(x_j, y_j, z_j)$ and, say, one isotropic thermal vibration parameter, which, with an overall scale factor, totals 61 variables. Even with Cu $K\alpha$ radiation, there are nine reflections per variable, a situation which, from a mathematical point of view, is considerably overdetermined. This feature is important, since the experimental intensity measurements contain random errors which cannot be eliminated, and the preponderance of data is needed to ensure precision in the structural parameters. We shall consider this situation again in Chapter 7.

### 6.4.2  The Patterson Function

It is interesting to note that although the connection between Fourier theory and X-ray diffraction was recorded first in 1915, it was not until about

1930 that practical use was made of it. Before the advent of computing facilities, the calculation of even a Fourier projection involved considerable time and effort. Add to this the phase problem, which necessitated many such calculations, and it is easy to understand that X-ray analysts were not anxious to become involved with extensive Fourier calculations; many early structure analyses were based on two projections.

In 1934, Patterson reported a new Fourier series which could be calculated directly from the experimental intensity data. However, because phase information is not used in the Patterson series, the result cannot be interpreted as a set of atomic positions, but rather as a collection of interatomic vectors all taken to a common origin. Patterson was led to the formulation of his series from considerations of an earlier theory of Debye on the scattering of X-rays by liquids—a much more difficult problem.

Patterson functions are of considerable importance in X-ray structure analysis, and their application will be considered in some detail. We will study first a one-dimensional function.

## One-Dimensional Patterson Function

The electron density at any fractional coordinate $x$ is $\rho(x)$, and that at the point $(x+u)$ is $\rho(x+u)$. The average product of these two electron densities in a repeat of length $a$, for a given value of $u$, is

$$A(u) = \int_0^1 \rho(x)\rho(x+u)\,dx \qquad (6.45)$$

where the upper limit of integration corresponds to the use of fractional coordinates. Using (6.36) in a form appropriate to a one-dimensional unit repeat, we obtain

$$A(u) = \int_0^1 \frac{1}{a^2} \sum_h \mathbf{F}(h)e^{-i2\pi hx} \sum_{h'} \mathbf{F}(h')e^{-i2\pi h'(x+u)}\,dx \qquad (6.46)$$

The index $h'$ lies within the same range as $h$, but is used to effect distinction between the Fourier series for $\rho(x)$ and $\rho(x+u)$. Separating the parts dependent upon $x$, and remembering that the integral of a sum is the sum of the integrals of the separate terms, we may write

$$A(u) = \frac{1}{a^2} \sum_h \sum_{h'} \mathbf{F}(h)\mathbf{F}(h')e^{-i2\pi h'u} \int_0^1 e^{-i2\pi(h+h')x}\,dx \qquad (6.47)$$

Considering the integral

$$\int_0^1 e^{-i2\pi(h+h')x}\,\mathrm{d}x = \frac{e^{-i2\pi(h+h')x}}{-i2\pi(h+h')}\Bigg]_0^1 \qquad (6.48)$$

$e^{-i2\pi(h+h')}$ is unity, since $h$ and $h'$ are integral, from (4.21), and the integral is, in general, zero. However, for the particular value of $h'$ equal to $-h$, it becomes indeterminate and we must consider making this substitution before integration. Thus,

$$\int_0^1 \mathrm{d}x = 1 \qquad (6.49)$$

Hence, from (6.47), for nonzero value of $A(u)$, where $h' = -h$,

$$A(u) = \frac{1}{a^2}\sum_h\sum_{-h}\mathbf{F}(h)\mathbf{F}(-h)e^{i2\pi hu} \qquad (6.50)$$

Equation (6.50) is not really a double summation, since $h$ and $-h$ cover the same field of the function. Furthermore, $\mathbf{F}(-h)$ is really the conjugate $\mathbf{F}^*(h)$, and using (4.27), we obtain

$$A(u) = \frac{1}{a^2}\sum_h |F(h)|^2 e^{i2\pi hu} \qquad (6.51)$$

where the index $h$ ranges from $-\infty$ to $\infty$. Now using Friedel's law (4.51), we find

$$A(u) = \frac{1}{a^2}\sum_h \left(|F(h)|^2 e^{i2\pi hu} + |F(h)|^2 e^{-i2\pi hu}\right) \qquad (6.52)$$

and from de Moivre's theorem (4.21),

$$A(u) = \frac{2}{a^2}\sum_h |F(h)|^2 \cos 2\pi hu \qquad (6.53)$$

where $h$ now ranges from 0 to $\infty$. The corresponding Patterson function

$P(u)$ is usually defined as

$$P(u) = \frac{2}{a} \sum_h |F(h)|^2 \cos 2\pi h u \qquad (6.54)$$

a trivial difference from the averaging function $A(u)$.

The practical evaluation of $P(u)$ proceeds through (6.54), but its physical interpretation is best considered in terms of (6.45), neglecting the small difference between $P(u)$ and $A(u)$.

Figure 6.8a shows one unit cell of a one-dimensional structure containing two different atoms $A$ and $B$ situated at fractional coordinates $x_A$ and

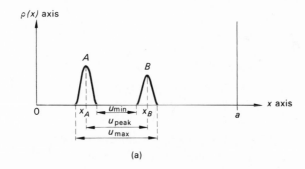

(a)

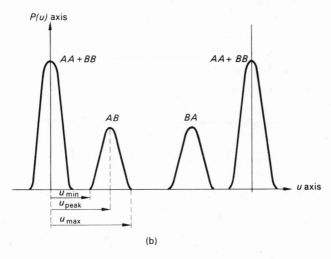

(b)

FIGURE 6.8. Development of a one-dimensional Patterson function for a two-atom structure.

$x_B$, respectively. Equation (6.45) represents the value of the electron density product $\rho(x)\rho(x+u)$, for any constant value of $u$, averaged over the repeat period of the unit cell. The average will be zero if one end of the vector $u$ always lies in zero regions of electron density, small if both ends of the vector encounter low electron densities, large if the electron density products are large, and a *maximum* where $u$ is of such a length that it spans two atomic positions in the unit cell.

For values of $u$ less than $u_{min}$ in Figure 6.8a, no peak will arise from the pair of atoms. As $u$ is increased, however, both ends of the vector will come simultaneously under the electron density peaks, and from (6.45) a finite value of $A(u)$, or $P(u)$, will be obtained. The integration can be simulated by sliding a vector of a given magnitude $u$ along the $x$ axis, evaluating the product $\rho(x)\rho(x+u)$ for all sampling intervals between zero and unit fractional repeat; this process is carried out for all fractional values of $u$ between zero and one. The graph of $P(u)$ as a function of $u$ is similar in appearance to an electron density function, but we must be careful not to interpret it in this way.

As we proceed through the values of $u$, we encounter $u_{peak}$, the interatomic vector $A-B$, which gives rise to the maximum value of $P(u)$, labeled $AB$ in Figure 6.8b. As $u$ increases to $u_{max}$, the electron density product falls to zero and $P(u)$ decreases correspondingly. Since we are concerned with interatomic *vectors*, negative values of $u$ are equally important; $-AB$ is marked off on the negative side of the origin, or at $BA$ within the given unit cell.

If we consider next very small values of $u$, both ends of such vectors will lie inside one and the same electron density peak, and $P(u)$ will be large. In the limit as $u \to 0$, the product involves that of the electron density maximum with itself, which is a local maximum for each atom, and a very large peak at the origin ($u = 0$) is to be expected. Thus the Patterson function is represented as a map of interatomic vectors, including null vectors, all taken to the origin.

The reader should confirm from Figure 6.8, using tracing paper, that the positions of the peaks in Patterson space can be plotted graphically by placing each atom of the structure $\rho(x)$ in turn at the origin of the Patterson map, in parallel orientation, and marking the positions of the other atoms onto the Patterson unit cell. Because of the centrosymmetry of the Patterson function (page 225), it is not strictly necessary to plot vectors lying outside one-half of the unit cell.

## Three-Dimensional Patterson Function

If we replace $\rho(x)$ and $\rho(x+u)$ in (6.45) by the three-dimensional analogs $\rho(x, y, z)$ and $\rho(x+u, y+v, z+w)$ and integrate over a unit fractional volume, we can derive the three-dimensional Patterson function:

$$P(u, v, w) = \frac{2}{V_c} \sum_h \sum_k \sum_l |F(hkl)|^2 \cos 2\pi(hu + kv + lw) \qquad (6.55)$$

where the summations range, in the most general case, over one-half of experimental reciprocal space. This equation should be compared with (6.41): It is a Fourier series with zero phases and $|F|^2$ as coefficients. In practice, it may be handled like the corresponding electron density equation, with $u$, $v$, $w$ replacing $x$, $y$, $z$, but it should be remembered that both functions explore the same field, the unit cell. The roving vector is now specified by three coordinates, $u$, $v$, and $w$, and $P(u, v, w)$ is a maximum where the corresponding vector spans two atoms in the crystal.

## Positions and Weights of Peaks in the Patterson Function

The positions of the peaks in $P(u, v, w)$ may be plotted in three dimensions by placing each atom of the unit cell of a structure in turn at the origin of Patterson space, in parallel orientation, and mapping the positions of all other atoms onto the Patterson unit cell. Examples of this process are illustrated graphically in Figure 6.9; for simplicity the origin peak is not shown in Figure 6.9d. In Figure 6.9a, all atoms and their translation equivalents produce vector peaks lying on the points of a lattice that is identical in shape and size to the crystal lattice. For example, atom 1 at $x, y, z$ and its translation equivalent, 1', at $x, 1+y, z$ give rise to a vector ending at 0, 1, 0 on the Patterson map. Peaks of this nature accumulate at the corners of the Patterson unit cell in exactly the same way as those of the origin peak, $P(0, 0, 0)$. From (6.55), we can derive the height of the origin peak:

$$P(0, 0, 0) = \frac{2}{V_c} \sum_{h=0}^{\infty} \sum_{k, l=-\infty}^{\infty} |F_o(hkl)|^2 \qquad (6.56)$$

In general, (6.56) is equivalent to a superposition at the origin of all $N$ products like $\rho(x_j, y_j, z_j)\rho(x_j, y_j, z_j)$, where $N$ is the number of atoms in the unit cell. Since $\rho(x_j, y_j, z_j)$ is proportional to the atomic number $Z_j$ of the $j$th

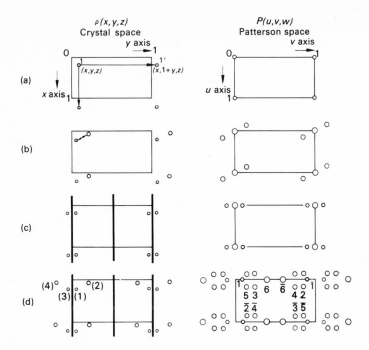

FIGURE 6.9. Effects of symmetry-related and symmetry-independent atoms on the Patterson function. The weights of the peaks are approximately proportional to the diameters of the circles: (a) $P1$ ($N = 1$); (b) $P1$ ($N = 2$)—two atoms per unit cell produce $(2^2 - 2)$ nonorigin peaks; (c) $Pm$ ($N = 2$)—two nonorigin peaks, but with coordinates $\pm\{0, 2y, 0\}$; (d) $Pm$ ($N = 4$)—twelve nonorigin peaks per unit cell; for clarity the origin peak has not been drawn. The *Patterson* space group is $P\bar{1}$ in (a) and (b) and $P2/m$ in (c) and (d). Figure 6.9d is discussed again on pages 226–227.

atom (page 214), we have

$$P(0, 0, 0) \propto \sum_{j=1}^{N} Z_j^2 \tag{6.57}$$

A single vector interaction between two atoms $j$ and $k$ (Figure 6.9b) will have a Patterson peak of height proportional to $Z_j Z_k$. Hence, the height $H(j, k)$ of this peak will be given by

$$H(j, k) \approx P(0, 0, 0) Z_j Z_k \bigg/ \sum_{j=1}^{N} Z_j^2 \tag{6.58}$$

where $P(0, 0, 0)$ is calculated from (6.56). This equation can serve as a useful guide, but overlapping vectors may give rise to misleading indications. The reservations on peak heights already mentioned (page 214) apply also to Patterson peaks.

In a structure with $N$ atoms per unit cell, each atom forms a vector with the remaining $N-1$ atoms. There are, thus, $N(N-1)$ nonorigin peaks. From (6.55), substitution of $-u$, $-v$, $-w$ for $u$, $v$, $w$, respectively, leaves $P(u, v, w)$ unaltered, which is a statement of the centrosymmetry of the Patterson function.

The Patterson unit cell is the same size and shape as the crystal unit cell, but it has to accommodate $N^2$ rather than $N$ "peaks" and is, therefore, correspondingly overcrowded. Thus, peaks in Patterson space tend to overlap when there are many atoms in the unit cell, a feature which introduces difficulties into the process of unraveling the function in terms of the correct distribution of atoms in the crystal.

## Sharpened Patterson Function

In a conceptual point atom, the electrons would be concentrated at a point. The atomic scattering factor curves (Figure 4.6) would be parallel to the abscissa and $f$ would be equal to the atomic number for all values of $(\sin \theta)/\lambda$ and at all temperatures. The electron density for a crystal composed of point atoms would show a much higher degree of resolution than does that for a real crystal. Put another way, the broad peaks representing real atoms (Figure 6.4) would be replaced by peaks of very narrow breadth in the point-atom crystal.

A plot of the mean value of $|F_o|^2$ against $(\sin \theta)/\lambda$ for a typical set of data is shown in Figure 6.10. The radial decrease in $\overline{|F_o|^2}$ can be reduced by modifying $|F_o|^2$ by a function which increases as $(\sin \theta)/\lambda$ increases. The coefficients for a sharpened Patterson synthesis may be calculated by the following equation (see also page 272):

$$|F_{\mathrm{mod}}(hkl)|^2 = \frac{|F_o(hkl)|^2}{\exp[-2B(\sin^2\theta)/\lambda^2]\{\sum_{j=1}^{N} f_j\}^2} \qquad (6.59)$$

$N$ is the number of atoms in the unit cell and $B$ is an overall isotropic temperature factor (pages 153 and 245).

The effect of sharpening on a Patterson synthesis is illustrated in Figure 6.17d, the Harker section $(u, \frac{1}{2}, w)$ for papaverine hydrochloride. It should be compared with Figure 6.17b; the increased resolution is very apparent.

Oversharpening of Patterson coefficients may lead to spurious peaks, particularly where heavy atoms are present, and the technique should not be applied without care. Sometimes the coefficients can be further modified to advantage by multiplication by a function such as $\exp(-m \sin^3 \theta)$, where $m$ is chosen by trial, but might be about 5. This function has the effect of decreasing the magnitude of the $\overline{|F_o|^2}$ curve at the high $\theta$ values. Many other sharpening functions have been proposed, but we shall not dwell on this subject. It is generally useful to calculate both the normal and sharpened Patterson functions for comparison.

### Symmetry of the Patterson Function for a Crystal of Space Group *Pm*

An inspection of Figures 6.9c and 6.9d shows that the peaks on the line $[0, v, 0]$ arise from atom pairs related by the $m$ planes. The vector interactions for case (d) are listed in Table 6.2, and may be easily verified by the reader; the values $z_1 = z_2 = 0.0$ were chosen for convenience only.

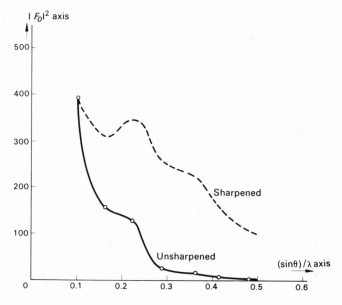

FIGURE 6.10.  Effect of sharpening on the radial decrease of the local average intensity $\overline{|F_o|^2}$.

TABLE 6.2. Vectors Generated by Two Independent Atoms and Their Symmetry Equivalents in Space Group $Pm^a$

| Atom pair | Analytical form of vector | Subtraction of coordinates | | Reduced to one unit cell | | Point in Figure 6.9d |
|---|---|---|---|---|---|---|
| | | $u$ | $v$ | $u$ | $v$ | |
| (1), (3) | $\pm\{0, 2y_1, 0\}$ | 0 | 0.10 | 0 | 0.10 | 1 |
| | | 0 | −0.10 | 0 | 0.90 | $\bar{1}$ |
| (1), (2) | $\pm\{x_1-x_2, y_1-y_2,$ | 0.15 | −0.15 | 0.15 | 0.85 | 2 |
| | $z_1-z_2\}$ | −0.15 | 0.15 | 0.85 | 0.15 | $\bar{2}$ |
| (1), (4) | $\pm\{x_1-x_2, y_1+y_2,$ | 0.15 | 0.25 | 0.15 | 0.25 | 3 |
| | $z_1-z_2\}$ | −0.15 | −0.25 | 0.85 | 0.75 | $\bar{3}$ |
| (2), (3) | $\pm\{x_1-x_2, -y_1-y_2,$ | 0.15 | −0.25 | 0.15 | 0.75 | 4 |
| | $z_1-z_2\}$ | −0.15 | 0.25 | 0.85 | 0.25 | $\bar{4}$ |
| (3), (4) | $\pm\{x_1-x_2, -y_1+y_2,$ | 0.15 | 0.15 | 0.15 | 0.15 | 5 |
| | $z_1-z_2\}$ | −0.15 | −0.15 | 0.85 | 0.85 | $\bar{5}$ |
| (2), (4) | $\pm\{0, 2y_2, 0\}$ | 0 | 0.40 | 0 | 0.40 | 6 |
| | | 0 | −0.40 | 0 | 0.60 | $\bar{6}$ |

$^a$ The coordinates of the four atoms in two sets of general positions are $x, y, z$; $x, \bar{y}, z$ with $x_1 = 0.20$, $y_1 = 0.05$, $x_2 = 0.05$, $y_2 = 0.20$, and $z_1 = z_2 = 0.00$.

The $m$ planes in $Pm$ are carried over into Patterson space, and relate the following pairs of peaks in the vector set:

$$1, \bar{1}; \quad 2, 5; \quad \bar{2}, \bar{5}; \quad 3, 4; \quad \bar{3}, \bar{4}; \quad 6, \bar{6} \qquad (6.60)$$

Furthermore, the introduction of a center of symmetry generates a pattern of $2/m$ symmetry in the Patterson map, which corresponds to the Laue symmetry of all monoclinic crystals. Evidently, the symmetry of the diffraction pattern is impressed onto the Patterson function by the use of $|F|^2$ coefficients in the Patterson–Fourier series. As a consequence, the Patterson synthesis is computed in the primitive space group corresponding to the Laue symmetry of a crystal, and this situation is similar for all space groups.

We can detect the presence of the twofold axis parallel to $b$ in Figure 6.9d through vector peaks such as $5, \bar{2}$ and $3, \bar{4}$. Finally, the symmetry-related pairs of atoms in the crystal, 1,3 and 2,4, give rise to vectors along the line $[0, v, 0]$—the peaks $1, 6, \bar{6}, \bar{1}$ in Patterson space. The presence of a large number of peaks along an axis in a three-dimensional Patterson map may be used as evidence for a mirror plane perpendicular to that axis in the crystal.

This feature is important because an $m$ plane does not give rise to systematic absences in the diffraction pattern (pages 79 and 178). The existence of peaks, arising from symmetry-related atoms, in certain regions of Patterson space was noted first by Harker in 1936. The line $[0, v, 0]$ for $Pm$ is called a Harker line; *planes* containing peaks arising from pairs of symmetry-related atoms are called Harker sections. We shall consider some examples below.

### Vector Interactions in Other Space Groups

We shall consider atoms in general positions in a number of space groups which should be now familiar.

*Space Group P$\bar{1}$.*
General positions:$x, y, z$; $\bar{x}, \bar{y}, \bar{z}$.
Vectors: $\pm\{2x, 2y, 2z\}$.
Harker peaks lie in general positions in Patterson space.

*Space Group P2.*
General positions: $x, y, z$; $\bar{x}, y, \bar{z}$.
Vectors: $\pm\{2x, 0, 2z\}$.
Harker section: $(u, 0, w)$.

It may be noted that for complex structures, not all of the peaks on Harker sections are necessarily true Harker peaks. If in this structure there are two atoms not related by symmetry, which, by chance, have the same or nearly the same $y$ coordinates, the vector between them will produce a peak on the Harker section.

*Space Group P2/m.*
General positions: $x, y, z$; $\bar{x}, \bar{y}, \bar{z}$; $x, \bar{y}, z$; $\bar{x}, y, \bar{z}$.
Vectors:  $\pm\{2x, 0, 2z\}$    double weight    type 1
$\pm\{0, 2y, 0\}$    double weight    type 2
$\pm\{2x, 2y, 2z\}$    single weight    type 3
$\pm\{2x, 2\bar{y}, 2z\}$    single weight    type 4.
Harker section: $(u, 0, w)$.
Harker line: $[0, v, 0]$.

Vector type 1 arises in two ways, once from the pair $x, y, z$; $\bar{x}, y, \bar{z}$ and once from the pair $x, \bar{y}, z$; $\bar{x}, \bar{y}, \bar{z}$. These two interactions give rise to identical vectors, which therefore superimpose in Patterson space and form a double-weight peak. Similar comments apply to type 2, but the centrosymmetrically related atoms give rise to single-weight peaks, types 3 and 4. Figure 6.11 illustrates these vectors, as seen along the $z$ axis. The reader may now

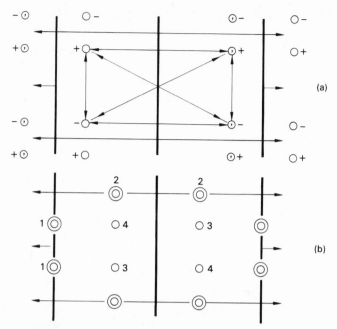

FIGURE 6.11. (a) Vectors between symmetry-related atoms in general equivalent positions in space group $P2/m$. Coordinates like $\bar{x}$ have been treated as $(1-x)$ in drawing the vectors. (b) One unit cell of the Patterson distribution omitting the origin peak. In both diagrams, the twofold axes intersect the $m$ planes in centers of symmetry.

consider how the Patterson function might be used to differentiate among space groups $P2$, $Pm$, and $P2/m$; a clue has already been given on pages 227 and 228. Statistical methods, outlined briefly in Chapter 7, are often employed to verify the results obtained from a study of the vector distribution.

### 6.4.3 Examples of the Use of the Patterson Function in Solving the Phase Problem

In this section, we shall consider how the Patterson function was used in the solution of three quite different structures.

Bisdiphenylmethyldiselenide, $(C_6H_5)_2CHSe_2CH(C_6H_5)_2$

Crystals of this compound form yellow needles, with straight extinction under crossed Polaroids for all directions parallel to the needle axis, and oblique extinction on the section normal to the needle axis. Photographs taken with the crystal oscillating about its needle axis show only a horizontal $m$ line, while zero- and upper-layer Weissenberg photographs show only symmetry 2. The crystals are therefore monoclinic, with $b$ along the needle direction.

*Crystal Data*

System: monoclinic.

Unit-cell dimensions: $a = 18.72, b = 5.773, c = 12.594$ Å, $\beta = 125.47°$.

$V_c$: 1107.1 Å$^3$.

$D_m$: 1.49 g cm$^{-3}$.

$M$: 492.4.

$Z$: 2.02, or 2 to the nearest integer.

Unit-cell contents: 4 Se, 52 C, and 44 H atoms.

Absent spectra:$hkl$: $h + k = 2n$.

Possible space groups: $C2$, $Cm$, $C2/m$.

*Symmetry Analysis.* Where the space group is not determined uniquely by the X-ray diffraction pattern, it may be possible to eliminate certain alternatives at the outset of the structure determination by other means.

Space groups $C2$ and $Cm$ each require four general positions:

$$C2: \quad (0, 0, 0; \ \tfrac{1}{2}, \tfrac{1}{2}, 0) + \{x, y, z; \quad \bar{x}, y, \bar{z}\}$$

$$Cm: \quad (0, 0, 0; \ \tfrac{1}{2}, \tfrac{1}{2}, 0) + \{x, y, z; \quad x, \bar{y}, z\}$$

Since $Z$ is 2, the molecular symmetry is either 2, in $C2$, or $m$, in $Cm$. In both $C2$ and $Cm$, all atoms could satisfy general position requirements, and neither arrangement would be stereochemically unreasonable.

Space group $C2/m$ requires eight general equivalent positions per unit cell. Only special position sets, such as 0, 0, 0 and $\tfrac{1}{2}, \tfrac{1}{2}, 0$ correspond with $Z = 2$. These positions have symmetry $2/m$, but it is not possible to construct the molecule in this symmetry without contradicting known chemical facts. Consequently, we shall regard this space group as highly improbable for the compound under investigation.

*Patterson Studies.* Whatever the answer to the questions remaining from this symmetry analysis, we expect, from chemical knowledge,[*] that the two selenium atoms will be covalently bonded at a distance of about 2.3 Å. This Se—Se interaction will produce a strong peak in the Patterson function at about 2.3 Å from the origin.

* See Bibliography.

The atomic numbers of Se, C, and H are 34, 6, and 1, respectively. Hence, the important vectors in the Patterson function would have single-weight peak heights, from (6.58), as follows:

(a) Se—Se: 1156.
(b) Se—C: 204.
(c) C—C: 36.

Because of the presence of identical vectors arising from the $C$ unit cell, all vectors will be double these values.

Figure 6.12 is the Patterson section $P(u, 0, w)$, calculated with 1053 data of $|F_o(hkl)|^2$, with grid intervals of 50ths along $u$, $v$, and $w$. The origin peak $P(0, 0, 0)$ was scaled to 100 and, from (6.57), $\sum_{j=1}^{N} Z_j^2 = 6540$. Hence, the vector interactions (a), (b), and (c) should have the approximate peak heights of 35, 6, and 1, respectively.

The section is dominated by a large peak of height 39 at a distance of about 2.3 Å from the origin. Making the reasonable assumption that it

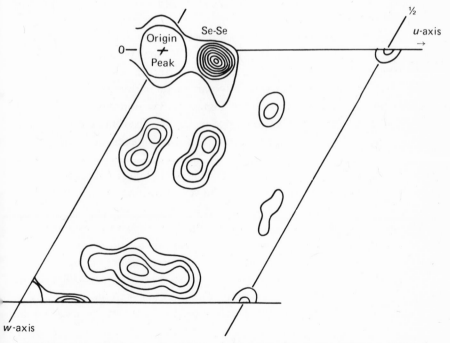

FIGURE 6.12. Patterson section, $P(u, 0, w)$; the origin peak (height = 100) has not been contoured. Contours around the Se—Se peaks are at intervals of 4; elsewhere at intervals of 2.

represents the Se—Se vector, and since there are no significant peaks on the $v$ axis, it follows that the space group cannot be $Cm$, thus leaving $C2$ as the most logical choice.

By measurement on the section, the Patterson coordinates are $u =$ 6.7/50 and $w = 2.2/50$, and from the study of space group $P2$ (pages 85 and 228), it follows that $x_{Se} = 0.067$ and $z_{Se} = 0.022$.

In space group $C2$, the unit cell origin is fixed in the $xz$ plane by the twofold axis. There is no symmetry element that defines the origin in the $y$ direction, which must be fixed by specifying the $y$ coordinate for a selected atom. For convenience, we may set $y_{Se} = 0$, and our analysis so far may be given as the positions

$$Se: \quad 0.067, 0, \quad 0.022$$
$$Se': \quad -0.067, 0, -0.022$$

A space-group ambiguity is not always resolved in this manner. Sometimes it is necessary to proceed further with the structure analysis before confirmation is obtained.

What of the atoms other than selenium? Is it possible to determine the positions of the carbon and hydrogen atoms? We shall find that we can locate the carbon atoms in this structure from the Patterson synthesis. To explain the procedure, we consider first only part of the structure, including one phenyl ring of the asymmetric unit (Figure 6.13a), and neglect all but the C—Se vectors. The vector set generated by the two Se atoms and 14 C atoms in this hypothetical arrangement contains two images of the structure fragment (one per Se atom), which are displaced from each other by the Se—Se vector. The idealized vector set is shown in Figure 6.13b. By shifting one of these images by a *reverse*\* Se—Se vector displacement, it is possible to bring the two images into coincidence. Verify this statement by making a transparent copy of Figure 6.13b and placing its origin over an Se—Se vector position in the same figure, keeping the pairs of $u$ and $w$ axes parallel. Certain peaks overlap, producing a single, displaced image of the structure. Shade the peaks that overlap. This image is displaced with respect to the true space-group origin, which we know to be midway between the two Se atoms. A correctly placed image of the structure can be recovered by inserting the

---

\* If the forward direction of this vector is used, the structure obtained would, in general, be inverted through the origin. This does not happen with the example under study because the molecule possesses twofold symmetry.

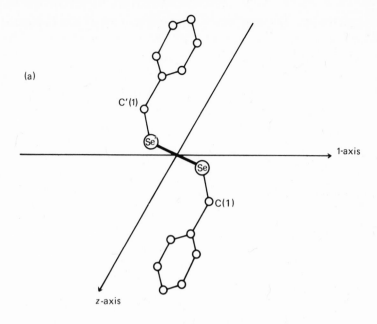

(a)

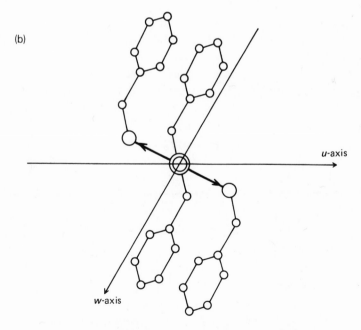

(b)

FIGURE 6.13. (a) Hypothetical structure fragment $C_6H_5CHSe_2CHC_6H_5$; (b) idealized set of Se—Se and Se—C vectors.

true origin position onto the tracing and neglecting any peaks that are not shaded.

The partial vector set was formed from the image of all atoms of the fragment in each Se atom; each image is weighted by $Z_j$, the atomic number of the $j$th atom (carbon, in this example) imaged in Se. The displacement arises because the Patterson synthesis transfers all vectors to a common origin.

*Patterson Superposition.* The technique just described depends upon the recognition of the vector interaction from a given pair of atoms, the two Se atoms in this example. At least a partial unscrambling of the structure images in the Patterson function was effected by correctly displacing two copies of the Patterson map and noting the positions of overlap.

To illustrate the method further and to derive a systematic procedure for its implementation, we return to the Patterson section in Figure 6.12. The two Se atoms have the same $y$ coordinate, which means that the vector shift takes place in this section. Now make two copies on tracing paper of the half unit-cell outline, $x = 0$ to $\frac{1}{2}$ and $z = 0$ to 1, and label them copy 1 and copy 2.

On copy 1 mark in the position $S$ of the point, $-(x_{Se}, z_{Se})$, which is at $-0.067$, $-0.022$, and on copy 2 mark in the position $S'$ of the point $-(x_{Se'}, z_{Se'})$, which is at $0.067, 0.022$. Think of these two unit cells as existing in crystal space, not Patterson space. Place copy 1 over the Patterson $(u, 0, w)$ section, maintaining a parallel orientation, with $S$ over the origin, and trace out the Patterson map (Figure 6.14a). Repeat this procedure with copy 2, placing $S'$ over the Patterson section origin (Figure 6.14b).

Finally, superimpose copy 1 and copy 2. As in the exercise with Figures 6.13a and 6.13b, some peaks overlap and some lie over blank regions in one or the other map. The overlaps correspond to regions of high electron density in the crystal. They are best mapped out by compiling a new diagram which contains the *minimum* value of the vector density between copy 1 and copy 2 for each point, thus eliminating or decreasing in height those regions where one copy has no or only slight overlap. A map prepared in this way is shown in Figure 6.14c.

## Minimum Function

The technique outlined above follows the method of Buerger.* An analytical expression for the minimum function $M_n(x, y, z)$ is given by

* See Bibliography.

(6.61); it may be regarded as an approximation to the electron density $\rho(x, y, z)$.

$$M_n(x, y, z) = \text{Min}[P(u - x_1, v - y_1, w - z_1), P(u - x_2, v - y_2, w - z_2),$$

$$\ldots, P(u - x_n, v - y_n, w - z_n)] \tag{6.61}$$

where $\text{Min}(P_1, P_2, \ldots, P_n)$ is the lowest value at the point $x, y, z$ in the set of superpositions $P_1, P_2, \ldots, P_n$; $n$ corresponds with the number of known or trial atomic positions. The following general comments on the application of the minimum function procedure should be noted:

(a) The $n$ trial atoms should form within themselves a set or sets of points related by the appropriate space-group symmetry.

(b) In a noncentrosymmetric space group, $n$ should be three or more if it is necessary to remove the Patterson center of symmetry.

(c) If the various $n$ trial atoms have different atomic numbers, the corresponding Patterson copies should be weighted accordingly in order to even out the different image strengths.

(d) Incorrectly placed atoms in the trial set tend to confuse the structure image. New atom sites therefore should be added to the model with caution.

Figure 6.15 shows a composite electron density map of the atoms in the asymmetric unit that were revealed by a three-dimensional minimum function $M_2$. This result is quite satisfactory; only C(9), C(10), and C(11) are not yet located. The composite map of the complete structure* and the packing of the molecules in the unit cell are shown in Figure 6.16. In favorable circumstances, the Patterson function can be solved for the majority of the heavier atoms in the crystal structure. The atoms not located by $M_2$ in this example were obtained from an electron density map phased on those atoms that were found, a standard method for attempting to complete a partial structure (see page 248).

## Determination of the Chlorine Atom Positions in Papaverine Hydrochloride

The crystal data for this compound have been given in Chapter 5 (page 184). The calculated origin peak height is approximately 4700, and a single-weight Cl—Cl vector would have a height of about 6% of that of the

---

* H. T. Palmer and R. A. Palmer, *Acta Crystallographica*, **B25**, 1090 (1969). The $|F_o|$ data for this compound may be obtained from one of the authors (R.A.P.).

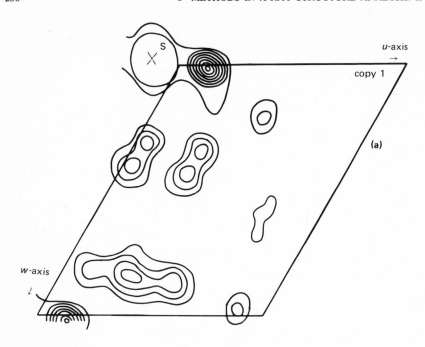

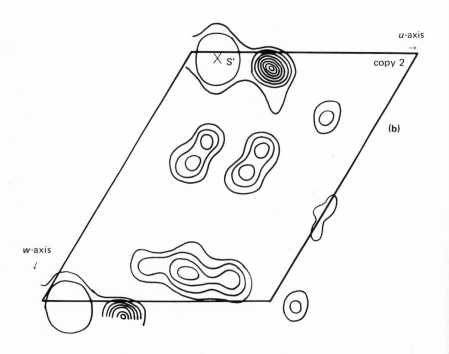

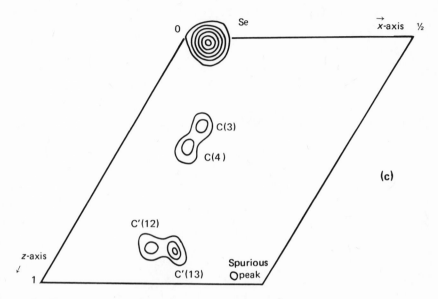

FIGURE 6.14. (a, b) Shifted copies 1 and 2 prepared from the $(u, 0, w)$ section; (c) minimum-function $M_2$ section at $y = 0.0$; C'(12) and C'(13) are symmetry-related to C(12) and C(13) in Figure 6.15.

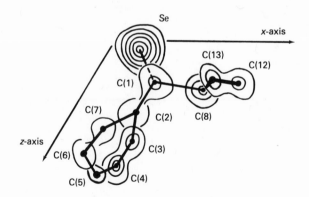

FIGURE 6.15. Composite map of the three-dimensional minimum function $M_2(x, y, z)$.

origin peak. The Cl—Cl vector may not be located as easily as that of Se—Se in the previous example. The general equivalent positions in $P2_1/c$ (see pages 91–92 and problems 2P.7 and 2P.8) give rise to the vectors shown in Table 6.3. The assignment of coordinates to the chlorine atoms follows the recognition of peaks $A$, $B$, and $C$ as Cl—Cl vectors on the Patterson maps (Figures 6.17a–c). Figure 6.17d is the sharpened section, $(u, \frac{1}{2}, w)$. The steps in the solution of the problem are set out in table 6.4.

The results are completely self-consistent, and we may list the Cl coordinates in the unit cell:

$$4\,\text{Cl:} \qquad 0.025, \quad 0.169, \quad 0.038; \qquad 0.025, \quad 0.331, \quad 0.538$$
$$-0.025, -0.169, -0.038; \quad -0.025, -0.331, -0.538$$

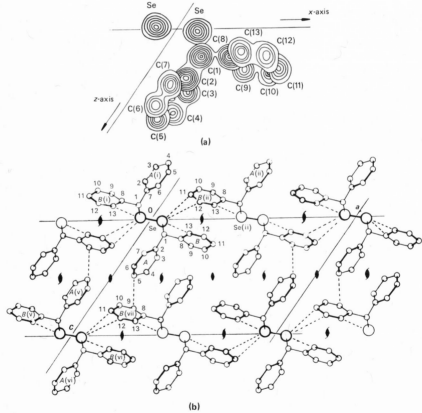

(a)

(b)

FIGURE 6.16. (a) Composite electron density map as seen along $b$; (b) crystal structure as seen along $b$; the dashed lines indicate the closest intermolecular contacts.

TABLE 6.3

| Label | Vector | Peak strength | Harker region |
|-------|--------|---------------|---------------|
| A | $\pm\{0, \frac{1}{2}+2y, \frac{1}{2}\}$ | Double weight | Line: $[0, v, \frac{1}{2}]$ |
| B | $\pm\{2x, \frac{1}{2}, \frac{1}{2}+2z\}$ | Double weight | Section: $(u, \frac{1}{2}, w)$ |
| C | $\pm\{2x, 2y, 2z\}$ | Single weight | General region |
| D | $\pm\{2x, 2\bar{y}, 2z\}$ | Single weight | General region |

TABLE 6.4

| Patterson map | Label | Vector coordinates[a] | Cl coordinates |
|---------------|-------|----------------------|----------------|
| Figure 6.17a, level $v = 8.4/52$ | A | $\frac{1}{2}-2y = 8.4/52$ | $y = 0.169$ |
| Figure 6.17b, level $v = \frac{1}{2}$ | B | $2x = 2.2/44,$ $\frac{1}{2}+2z = 17.3/30$ | $x = 0.025$ $z = 0.038$ |
| Figure 6.17c, level $v = 17.6/52$ | C | $2x = 2.2/44$ $2y = 17.6/52$ $2z = 2.3/30$ | $x = 0.025$ $y = 0.169$ $z = 0.038$ |

[a] The Patterson synthesis was computed with the intervals of subdivision 44, 52, and 30 along $u$, $v$, and $w$, respectively.

For simplicity, peak $A$ was assigned as $-(\frac{1}{2}+2y)$, which is crystallographically the same as $\frac{1}{2}-2y$, in order to obtain $y \leqslant \frac{1}{2}$. For a similar reason, $B$ was retained as $\frac{1}{2}+2z$.

The specification of the peak parameters in this manner is, to some extent, dependent on the observer. A different choice, for example, $\frac{1}{2}+2y$ in $A$, merely results in a set of atomic positions located with respect to one of the other centers of symmetry as origin. In space groups where the origin might be defined with respect to other symmetry elements, similar arbitrary peak specification may be possible.

## Determination of the Mercury Atom Positions in $KHg_2$

This example illustrates the application of the Patterson function to the determination of the coordinates of atoms in special positions of space group *Imma*.

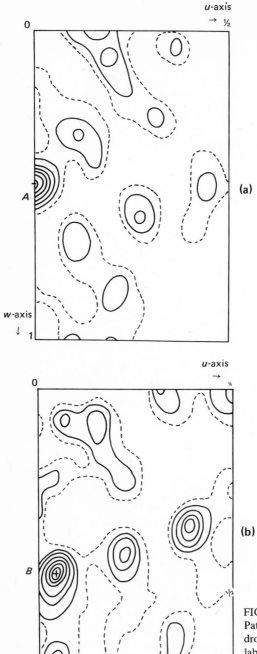

FIGURE   6.17. Three-dimensional Patterson sections for papaverine hydrochloride; the Cl—Cl vectors are labeled $A$, $B$, and $C$: (a) $v = 8.4/52$, (b) $v = \frac{1}{2}$, (c) $v = 17.6/52$, (d) $v = \frac{1}{2}$ (sharpened section).

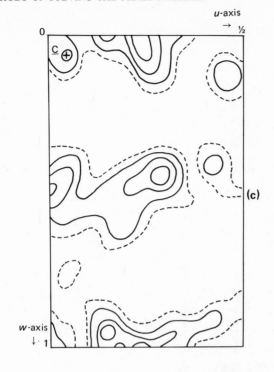

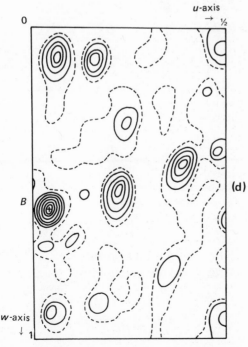

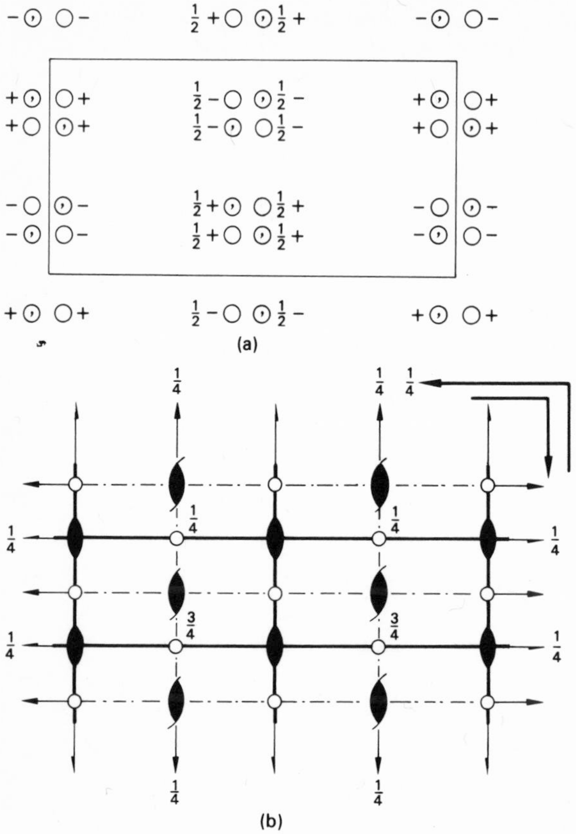

(a)

(b)

FIGURE 6.18. Space group
*Imma*.

*Crystal Data**

System: orthorhombic.

Unit-cell dimensions: $a = 8.10$, $b = 5.16$, $c = 8.77$ Å.

$V_c$: 366.6 Å$^3$.

$D_m$: 7.95 g cm$^{-3}$.

$M$: 440.3.

$Z$: 3.99, or 4 to the nearest integer.

Unit-cell contents: 4 K and 8 Hg atoms.

Absent spectra:   $hkl$: $h + k + l = 2n + 1$

$hk0$: $h = 2n + 1$, $(k = 2n + 1)$.

Possible space groups: *Im2a*, *I2ma*, or *Imma*.

In the absence of further information on the space group, we shall proceed with the analysis in *Imma* (Figures 6.18a and 6.18b). The reader

* E. J. Duwell and N. C. Baenziger, *Acta Crystallographica* **8**, 705 (1955).

TABLE 6.5. Special Positions in *Imma*

| | | | |
|---|---|---|---|
| 4 | (a) | $2/m$ | $0,0,0;\quad \frac{1}{2},0,0;\quad \frac{1}{2},\frac{1}{2},\frac{1}{2};\quad 0,\frac{1}{2},\frac{1}{2}$ |
| 4 | (b) | $2/m$ | $0,\frac{1}{2},0;\quad \frac{1}{2},\frac{1}{2},0;\quad \frac{1}{2},0,\frac{1}{2};\quad 0,0,\frac{1}{2}$ |
| 4 | (c) | $2/m$ | $\frac{1}{4},\frac{1}{4},\frac{1}{4};\quad \frac{1}{4},\frac{3}{4},\frac{1}{4};\quad \frac{3}{4},\frac{3}{4},\frac{3}{4};\quad \frac{3}{4},\frac{1}{4},\frac{3}{4}$ |
| 4 | (d) | $2/m$ | $\frac{1}{4},\frac{3}{4},\frac{1}{4};\quad \frac{1}{4},\frac{1}{4},\frac{3}{4};\quad \frac{3}{4},\frac{1}{4},\frac{1}{4};\quad \frac{3}{4},\frac{3}{4},\frac{1}{4}$ |
| 4 | (e) | $mm2$ | $\frac{1}{4},0,z;\quad \frac{3}{4},0,\bar{z};\quad \frac{3}{4},\frac{1}{2},\frac{1}{2}+z;\quad \frac{1}{4},\frac{1}{2},\frac{1}{2}-z$ |
| 8 | (f) | $2$ | $\pm\{0,y,0;\quad \frac{1}{2},y,0;\quad \frac{1}{2},\frac{1}{2}+y,\frac{1}{2};\quad 0,\frac{1}{2}+y,\frac{1}{2}\}$ |
| 8 | (g) | $2$ | $\pm\{x,\frac{1}{4},\frac{1}{4};\quad x,\frac{3}{4},\frac{1}{4};\quad \frac{1}{2}+x,\frac{3}{4},\frac{1}{4};\quad \frac{1}{2}+x,\frac{1}{4},\frac{3}{4}\}$ |
| 8 | (h) | $m$ | $\pm\{x,0,z;\quad \frac{1}{2}-x,0,z;\quad \frac{1}{2}+x,\frac{1}{2},\frac{1}{2}+z;\quad \bar{x},\frac{1}{2},\frac{1}{2}+z\}$ |
| 8 | (i) | $m$ | $\pm\{\frac{1}{4},y,z;\quad \frac{1}{4},\bar{y},z;\quad \frac{3}{4},\frac{1}{2}+y,\frac{1}{2}+z;\quad \frac{3}{4},\frac{1}{2}-y,\frac{1}{2}+z\}$ |

may like to consider how easily these diagrams may be derived from *Pmma* (origin on $\bar{1}$) $+I$.

*Symmetry and Packing Analyses.* Since $Z$ is 4 and there are 16 general equivalent positions in *Imma*, all atoms must lie in special positions. Table 6.5 lists these positions for this space group, with a center of symmetry (equivalent to $2/m$) as origin.*

This list presents a quite formidable number of alternatives for examination. The eight Hg atoms could lie in (f), (g), (h), or (i). However, further consideration of sets (f), (g), and (i) and sets (c) and (d) shows that they would all involve pairs of Hg atoms being separated by distances less than $b/2$ (2.58 Å). This value is much shorter than known Hg—Hg bond distances in other structures, and we shall reject these sets. (The positions in these sets may be plotted to scale in order to verify the spatial limitations.)

Of the remaining sets, (a) and (b) together would again place neighboring Hg atoms too close to one another. There are three likely models:

Model I: four Hg in (a) + four Hg in (e).
Model II: four Hg in (b) + four Hg in (e).
Model III: eight Hg in (h).

The Patterson function enables us to differentiate among these alternative models.

*Vector Analysis of the Alternative Hg Positions.* Model I would produce, among others, an Hg—Hg vector at $u = \frac{1}{2}$, $w = 0$, from the atoms in set (a). The $b$ axis Patterson projection (Figure 6.19) shows no peak at that position, and we eliminate model I. For a similar reason, with the atoms of

---

* In the original paper, the origin in *Imma* was chosen on a center of symmetry displaced by $\frac{1}{4}, \frac{1}{4}, \frac{1}{4}$ from this origin.

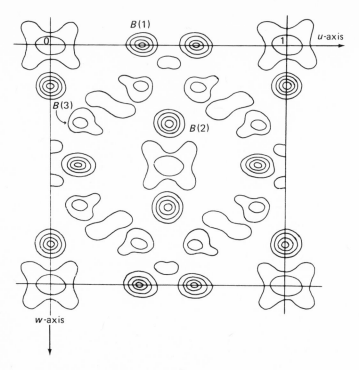

FIGURE 6.19. Patterson projection $P(u, w)$ for $KHg_2$; the origin peak has not been contoured.

set (b), model II is rejected. It is necessary to show next that model III is consistent with the Patterson function. The $a$-axis projection is shown in Figure 6.20.

*Interpretation of $P(u, w)$.* In this projection, no reference is made to the $y$ coordinates, and we look for vectors of the type $\pm\{\frac{1}{2} + 2x, 0\}$ and $\pm\{\frac{1}{2}, 2z\}$, and four vectors related by $2mm$ symmetry $\pm\{2x, 2z\}$ and $\pm\{2\bar{x}, 2z\}$. The double-weight peak labeled $B(1)$ is on the line $w = 0$, and $B(2)$ is on the line $u = 1/2$. Hence, $x_{Hg} = 0.064$ and $z_{Hg} = 0.161$. These values are corroborated by measurements from the single-weight peak $B(3)$.

*Interpretation of $P(v, w)$.* Vectors like $A$ (Figure 6.20) are of the type $\pm\{0, 2z\}$. We deduce $z_{Hg} = 0.161$, in excellent agreement with the value obtained from the $b$-axis projection.

Superposition techniques applied to the $a$-axis projection indicate that the K atoms are in special positions (b), but this result is not supported by the

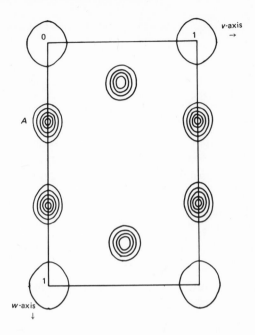

FIGURE 6.20. Patterson projection $P(v, w)$ for $KHg_2$; the origin peak has not been contoured.

$b$-axis projection. Evidently, the Patterson results can give only a partial structure, and supplementary methods are needed to carry the analysis to completion. In summary, we have determined the positions of the mercury atoms to be in set (h),* Table 6.5, with $x = 0.064$, $z = 0.161$.

### 6.4.4  Absolute Scale of $|F_o|$ and Overall Temperature Factor

In any structure analysis it soon becomes necessary to calculate structure factors for a trial structure for comparison with the experimental data To do this successfully we need a temperature factor for the atomic scattering factors and a scale factor for $|F_o|$. They can be derived approximately by Wilson's method.† He showed that for a unit cell containing a

---

* In the work of Duwell and Baenziger (*loc. cit.*), the positions listed are 8 (i), with $x = 0.186$ and $z = 0.089$, each being $-1/4$ *plus* the value given here (see footnote to page 243).

† A. J. C. Wilson, *Nature* **150**, 152 (1942).

random distribution of $N$ identical atoms, the local average value of $|F(hkl)|^2$ is given by

$$\overline{|F(hkl)|^2} \approx \sum_{j=1}^{N} g_j^2(hkl) \tag{6.62}$$

This approximation has been found to hold satisfactorily for a wide range of structures, provided that the values of $|F|^2$ are averaged over small ranges (local) of $(\sin \theta)/\lambda$, so that $f$ is not varying rapidly within any range.

We assume an isotropic temperature factor $B$ for all atoms and a scale factor $K$ for the experimental $|F_o|$ data [see (4.37) and (4.50)]. Hence, from (6.62)

$$K^2 \overline{|F_o(hkl)|^2} = \exp[-2B(\sin^2 \theta_r)/\lambda^2] \sum_{j=1}^{N} f_{j,\,\theta_r}^2 \tag{6.63}$$

where $\theta_r$ is a representative value of each range of $(\sin \theta)/\lambda$. Rearranging, and taking $\log_e$ we obtain,

$$\log_e q_r = 2\log_e K + 2B(\sin^2\theta_r)/\lambda^2 \tag{6.64}$$

where $q_r$ is given by

$$q_r = \left( \sum_{j=1}^{N} f_{j,\theta_r}^2 \right) \bigg/ \overline{|F_o(hkl)|^2}\big|_{\theta_r} \tag{6.65}$$

If $\log_e q_r$ is plotted against $(\sin^2\theta)/\lambda^2$ and the best straight line drawn, the slope is equal to $2B$ and the intercept on the ordinate is equal to $2\log_e K$.

This graph is often called a Wilson plot, and may be carried out in the following way.

(a) Three-dimensional reciprocal space is divided into spherical shells of approximately equal volume, with 80–100 reflections per shell; the shells decrease in thickness with increasing values of $(\sin \theta)/\lambda$.

(b) The average value of $|F_o(hkl)|^2$ is calculated over the available reciprocal space by expanding the data to include the symmetry-related reflections with their correct multiplicities. It is necessary to allocate values to "unobserved" reflections. Wilson has shown that $0.5|F_m|$ for centrosymmetric crystals and $0.67|F_m|$ for noncentrosymmetric crystals are the most probable values; $|F_m|$ is the minimum value of $|F_o|$ in the immediate locality of the shell under consideration. Systematically absent reflections

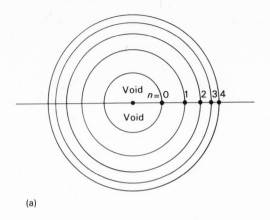

(a)

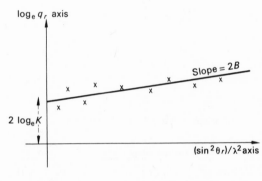

(b)

FIGURE 6.21. Scale and temperature factor: (a) division of reciprocal space into spherical shells, (b) Wilson plot.

and those in a region up to about the second-order reflection on each axis are ignored in taking the averages, as they are atypical of the general distribution of intensities.

(c) The mean value of $(\sin^2\theta)/\lambda^2$ is given, without sensible error, by

$$(\sin^2\theta_r)/\lambda^2 = (\sin^2\theta_n - \sin^2\theta_{n+1})/2\lambda^2 \qquad (6.66)$$

where $n + 1$ is the number of the outer shell defining the $n$th range, starting at $n = 0$, where $\sin\theta = \sin\theta_{\min}$ (Figure 6.21a).

(d) $\sum f_{j,\theta_r}^2$ is calculated for each shell, using tabulated $f$ data.

(e) The Wilson plot is drawn and $B$ and $K$ determined (Figure 6.21b).

### 6.4.5  Heavy-Atom Method and Partial Fourier Synthesis

The heavy-atom method was conceived originally as a method for determining the positions of light atoms in a structure containing a relatively small number of heavier atoms. However, the technique can be applied to most situations where a partial structure analysis has been effected, provided that certain conditions are met.

Imagine a situation where $N_k$ of the $N$ atoms in a unit cell have been located; $N_k$ may be only one atom, if it is a heavy atom. There will be $N_u$ atoms remaining to be located, and we may express the structure factor (4.40) in terms of known and unknown atoms:

$$\mathbf{F}(hkl) = \sum_{j=1}^{N_k} g_j \exp[i2\pi(hx_j + ky_j + lz_j)]$$

$$+ \sum_{u=1}^{N_u} g_u \exp[i2\pi(hx_u + ky_u + lz_u)] \qquad (6.67)$$

or

$$\mathbf{F}(hkl) = \mathbf{F}_c(hkl) + \mathbf{F}_u(hkl) \qquad (6.68)$$

In practice, $|F_o|$ data, appropriately scaled, replace $|F(hkl)|$, and $\mathbf{F}_c(hkl)$ refers to the known ($N_k$) atomic positions. As more of the structure becomes known, the values of $|F_c(hkl)|$ approach $|F_o(hkl)|$ and the phase angle $\phi_c$ approaches the unobservable but required value $\phi(hkl)$. Figure 6.22 illustrates this argument for any given reflection. The values of $\phi_c$ may provide

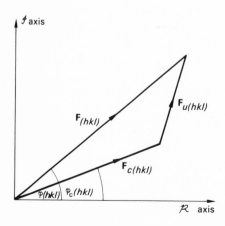

FIGURE 6.22. Partial-structure phasing; $\mathbf{F}(hkl)$ is the true structure factor of modulus $|F_o(hkl)|$ and phase $\phi(hkl)$.

sufficiently reasonable approximations to $\phi(hkl)$ for an electron density map to be calculated with some confidence. The nearer $\mathbf{F}_c$ is to $\mathbf{F}(hkl)$, the better the value of the phase angle, and this is clearly dependent upon the percentage of the scattering power which is known. As a guide to the effective phasing power of a partial structure, the quantity $r$ is calculated:

$$r = \sum_{j=1}^{N_k} Z_j^2 \Big/ \sum_{u=1}^{N_u} Z_u^2 \qquad (6.69)$$

where $Z$ refers to the atomic number of a species. A value of $r$ near unity is considered to provide a useful basis for application of the heavy-atom method. However, values of $r$ quite different from unity have produced successful results, because for a given reflection the important quantity is really $r'$, given by

$$r' = \sum_{j=1}^{N_k} g_j^2 \Big/ \sum_{u=1}^{N_u} g_u^2 \qquad (6.70)$$

If $r$ is large, however, the heavy-atom contributions tend to swamp those from the lighter atoms, which may then not be located very easily from electron density maps. On the other hand, if $r$ is small, the calculated phase will deviate widely from the desired value, and the resulting electron density map may be very difficult to interpret. These extreme situations are found in bisdiphenylmethyldiselenide ($r = 2.4$) and papaverine hydrochloride ($r = 0.28$), based, in each case, on the heavy atoms alone in $N_k$.

The underlying philosophy of the heavy-atom method depends on the acceptance of calculated phases, even if they contain errors, for the computation of the electron density synthesis. Large phase errors give rise to high background features, which mask the image of the correct structure. The calculated phases $\phi_c$ contain errors arising from inadequacies in the model, but the $|F_o|$ data, although subject to experimental errors, hold information on the complete structure. Phase errors may be counteracted to some extent by weighting the Fourier coefficients according to the degree of confidence in a particular phase. For centrosymmetric structures, the weight $w(hkl)$ by which $|F_o(hkl)|$ is multiplied is given by*

$$w(hkl) = \tanh(\chi/2) \qquad (6.71)$$

* M. M. Woolfson, *Acta Crystallographica* **9**, 804 (1956).

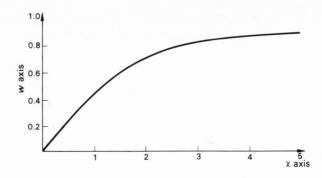

FIGURE 6.23. Weighting factor $w(hkl)$ as a function of $\chi$ in noncentrosymmetric crystals.

where $\chi$ is given by

$$\chi = 2|F_o|\,|F_c|/\sum g_u^2 \qquad (6.72)$$

The subscripts $c$ and $u$ refer, respectively, to the known and unknown parts of the structure. In noncentrosymmetric structures, $w(hkl)$ can be obtained from the graph in Figure 6.23.* Weighting factors should be applied to $|F_o|$ values that have been placed on an absolute, or approximately absolute, scale.

Bearing these points in mind, it follows that the best electron density map one can calculate with phases determined from a partial structure is given by

$$\rho(x, y, z) = \frac{2}{V_c} \sum_{h=0} \sum_{k-\infty}^{\infty} \sum_{l} w(hkl)|F_o(hkl)| \cos[2\pi(hx + ky + lz) - \phi_c(hkl)] \qquad (6.73)$$

where

$$\phi_c(hkl) = \tan^{-1}[B'_c(hkl)/A'_c(hkl)] \qquad (6.74)$$

$A'_c(hkl)$ and $B'_c(hkl)$ are the real and imaginary components, respectively, of the calculated structure factor, which is included in the right-hand side of (6.68).

* G. A. Sim, *Acta Crystallographica* **13**, 511 (1960).

Electron density maps calculated from partial-structure phasing contain features which characterize both the true structure and the partial, or trial, structure. This fact may be illustrated by writing the $x$-, $y$-, and $z$-independent terms from the electron density synthesis (6.37) as $(w|F_o|/|F_c|)|F_c|\exp[i\phi_c(hkl)]$. The terms $|F_c|\exp[i\phi_c(hkl)]$ alone would synthesize the partial structure. This synthesis is modified, in practice, by the term $w|F_o|/|F_c|$; evidently, the electron density synthesis is biased toward the trial structure.

If the model includes atoms in reasonably accurate positions, we can expect two important features in the electron density map: (a) Atoms of the trial structure should appear, possibly in corrected positions, and (b) additional atoms should be revealed by the presence of peaks in stereochemically sensible positions.

If neither of these features is observed in the electron density synthesis, it may be concluded that the trial structure contains very serious errors. Correspondingly, there would be poor agreement in the pattern of relationship between $|F_o|$ and $|F_c|$.

## Pseudosymmetry in Electron Density Maps

The electron density map calculated with phases derived from the heavy-atom positions may not exhibit the true space-group symmetry, but rather that of a space group of higher symmetry. As an example, consider a structure having space group $P2_1$ with one heavy atom per asymmetric unit. The origin is defined with respect to the $x$ and $z$ axes by the $2_1$ axis along $[0, y, 0]$, but the $y$ coordinate of the origin is determined with respect to an arbitrarily assigned $y$ coordinate for *one* of the atoms. Consider the heavy atoms at $x, y, z$ and, symmetry-related, at $\bar{x}, \frac{1}{2}+y, \bar{z}$. This arrangement has the symmetry of $P2_1/m$, with the $m$ planes cutting the $y$ axis at whatever $y$ coordinate is chosen for the heavy atom, say $y_H$, and at $\frac{1}{2}+y_H$. If $y_H = \frac{1}{4}$, a center of symmetry is at the origin, and the calculated phases will be 0 or $\pi$. This situation is illustrated in Figure 6.24, which indicates that an unscrambling of the images must be carried out. If the heavy atom is given any general value for $y_H$, $B'(hkl)$ will not be zero and the phase angles will not be 0 or $\pi$, but the pseudosymmetry will still exist.

## Successive Fourier Refinement

A single application of the method described above does not usually produce a complete set of atomic coordinates. It should lead to the inclusion

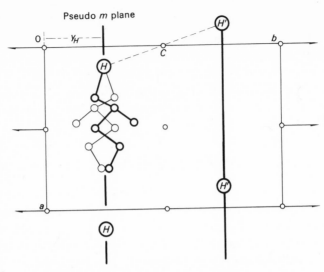

FIGURE 6.24. Introduction of pseudosymmetry into space group $P2_1$ by single heavy-atom phasing. $H$ is the heavy atom and $C$ is a center of symmetry introduced between $H$ and its $P2_1$ equivalent $H'$. The space group (for the heavy atoms alone) thus appears as $P2_1/m$ with mirror (and $\bar{1}$) pseudosymmetry. The electron density map, phased on the $H$ and $H'$ species, will contain contours for two mirror-related images in the asymmetric unit, with a certain degree of confusion between them.

of more atoms into subsequent structure factor calculations and so to a better electron density map, and so on. This iterative process of Fourier refinement should, after several cycles, result in the identification of all nonhydrogen atoms in the structure to within about 0.1 Å of their true positions. Further improvement of the structure would normally be carried out by the method of least squares, which is described in Chapter 7.

### 6.4.6  Difference-Fourier Synthesis

Some errors present in the trial structure may not be revealed by Fourier synthesis. In particular, the following situations are important.

(i) Atoms in completely wrong positions tend to be returned by the Fourier process with the same fractional coordinates, but sometimes with a comparatively low electron density.

(ii) Correctly placed atoms may have been assigned either the wrong atomic number, for example, C for N, or an incorrectly estimated temperature factor.

(iii) Small corrections to the fractional coordinates may be difficult to assess from the Fourier map.

In these circumstances, a difference-Fourier synthesis is valuable. We shall symbolize the Fourier series with $|F_o|$ coefficients as $\rho_o(x, y, z)$ and the corresponding synthesis with $|F_c|$ instead as $\rho_c(x, y, z)$; the difference-Fourier synthesis $\Delta\rho(x, y, z)$ may be obtained in a single-stage calculation from the equation

$$\Delta\rho(x, y, z) = \frac{2}{V_c} \sum_h \sum_k \sum_l (|F_o| - |F_c|) \cos[2\pi(hx + ky + lz) - \phi_c] \quad (6.75)$$

Since the phases are substantially correct at this stage, it is, in effect, a subtraction, point by point, of the "calculated," or trial, Fourier synthesis from that of the "observed," or experimentally based, synthesis. The difference synthesis has the following useful properties.

(a) Incorrectly placed atoms correspond to regions of high electron density in $\rho_c(x, y, z)$ and low density in $\rho_o(x, y, z)$; $\Delta\rho(x, y, z)$ is therefore negative in these regions.

(b) A correctly placed atom with either too small an atomic number or too high a temperature factor shows up as a small positive area in $\Delta\rho$. The converse situations produce negative peaks in $\Delta\rho$.

(c) An atom requiring a small positional correction tends to lie in a negative area at the side of a small positive peak. The correction is applied by moving the atom into the positive area.

(d) Very light atoms, such as hydrogen, may be revealed by a $\Delta\rho$ synthesis when the phases are essentially correct, after least-squares refinement has been carried out. There may be some advantage in using only reflections for which $(\sin\theta)/\lambda$ is less than about 0.35. Hydrogen scatters negligibly above this value.

(e) As one final test of the validity of a refined structure, the $\Delta\rho$ synthesis should be effectively featureless within 2–3 times the standard deviation of the electron density (page 292).

## 6.4.7 Limitations of the Heavy-Atom Method

The Patterson and heavy-atom techniques are effective for structures containing up to about 100 atoms in the asymmetric unit. It is sometimes necessary to introduce heavy atoms artificially into structures. This process may not be desirable because a possible structural interference may arise and there will be a loss in the accuracy of the light-atom positions. An introduction to "direct methods," capable of solving the phase problem for such structures, is given in the next chapter. Very large structures, such as proteins, containing many more than 100 atoms in the asymmetric unit, may be investigated by a variation of the heavy-atom method.

### 6.4.8  Isomorphous Replacement

A common feature of biologically important substances is their high molecular weight. Proteins and enzymes, for example, are polymers built up from various amino acid residues and forming very large assemblies with molecular weights greater than 5000. The study of the conformations of these giant molecules is necessary for the understanding of their biological functions, and the principal method of obtaining structural detail is by X-ray analysis.

Because of their high molecular weight, protein structures do not yield to analysis by the heavy-atom method. The value of $r$, from (6.69), is typically 0.03 for a protein molecule of molecular weight 5000 containing one mercury atom. This value of $r$ is too small to be useful. Another difficulty is that most proteins and enzymes contain neither very heavy atoms nor easily replaceable groups to facilitate the introduction of heavy atoms. In spite of these difficulties, if a heavy-atom derivative of a large molecule can be prepared, it may be possible to induce it to crystallize in a similar size of unit cell and with the same space group as the native compound. Such pairs of compounds are said to be isomorphous.

The structure factor of the heavy-atom derivative may be expressed vectorially as $\mathbf{F}_{PH}$, where

$$\mathbf{F}_{PH} = \mathbf{F}_P + \mathbf{F}_H \tag{6.76}$$

$\mathbf{F}_P$ and $\mathbf{F}_H$ are the structure factors for the parent protein and the heavy atoms alone, respectively, for the same reflection. This relationship is shown in Figure 6.25.

Assuming that the positions of the $N_H$ heavy atoms in the unit cell can be determined, their contribution can be calculated:

$$\mathbf{F}_H \approx \sum_{j=1}^{N_H} Z_j' \exp[-b_j(\sin^2\theta)/\lambda^2]\exp[i2\pi(hx_j + ky_j + lz_j)] \tag{6.77}$$

where $Z_j'$ is an effective atomic number, treated as a variable because only a

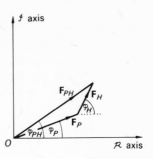

FIGURE 6.25. Graphical interpretation of the isomorphous replacement equation. In practice, the phases $\phi_P$ and $\phi_{PH}$ are unknown initially. $F_H$ may be known with a fair degree of accuracy if the heavy-atom positions in the isomorphous derivative are known. This enables a solution, as illustrated in Figure 6.26, to be obtained.

fraction of the molecules in a protein crystal take up heavy atoms; $b_j$ is a combined atomic scattering factor and temperature factor, both of which are assumed to decrease exponentially with $(\sin \theta)/\lambda$.

To obtain an idea of the effect of a heavy atom on the intensities of X-ray reflections from a protein, we shall carry out a simple calculation for a crystal containing one protein molecule per unit cell in space group $P1$. Assuming that it has a molecular weight of about 13,000, about 1000 nonhydrogen atoms would comprise the molecule; we shall assume that they are all carbon ($Z_c = 6$). Accepting Wilson's approximation (6.62),

$$\overline{|F_P|^2} \approx \sum_{j=1}^{1000} f_C^2 \tag{6.78}$$

At $\sin \theta = 0$, $\overline{|F_P|^2}$ is 36,000. If the derivative contains one mercury atom ($Z_{Hg} = 80$),

$$\overline{|F_{PH}|^2} \approx \sum_{j=1}^{1000} f_C^2 + \sum_{j=1}^{N_H} Z_j'^2 \tag{6.79}$$

which has the value 42,400 at $\sin \theta = 0$. Hence, the maximum change in intensity is about 18%, which is a surprisingly high value.

Experimentally, two sets of data $|F_P(hkl)|$ and $|F_{PH}(hkl)|$ are obtained, and, because of the comparative nature of the phase-determining procedure with isomorphous compounds, they must be placed on the same relative scale, which can be achieved by Wilson's method.

Rewriting (6.76), we have

$$|F_P| \exp(i\phi_P) = |F_{PH}| \exp(i\phi_{PH}) - \mathbf{F}_H \tag{6.80}$$

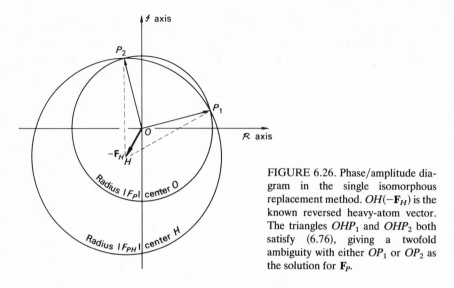

FIGURE 6.26. Phase/amplitude diagram in the single isomorphous replacement method. $OH(-\mathbf{F}_H)$ is the known reversed heavy-atom vector. The triangles $OHP_1$ and $OHP_2$ both satisfy (6.76), giving a twofold ambiguity with either $OP_1$ or $OP_2$ as the solution for $\mathbf{F}_P$.

This equation involves two unknown quantities, $\phi_P$ and $\phi_{PH}$, and cannot yield a unique solution. However, Figure 6.26 shows that only two solutions for $\phi_P$ are real, corresponding to the vectors $OP_1$ and $OP_2$, one of which is the true $\mathbf{F}_P$ vector. Another isomorphous derivative, $PH'$, with a *different set* of heavy-atom positions, will also have two solutions for $\mathbf{F}_P$, say, $OP_3$ and

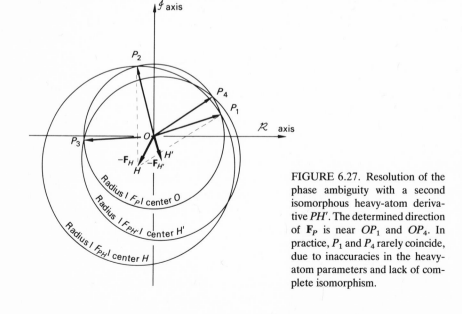

FIGURE 6.27. Resolution of the phase ambiguity with a second isomorphous heavy-atom derivative $PH'$. The determined direction of $\mathbf{F}_P$ is near $OP_1$ and $OP_4$. In practice, $P_1$ and $P_4$ rarely coincide, due to inaccuracies in the heavy-atom parameters and lack of complete isomorphism.

$OP_4$, one of which should agree with either $OP_1$ or $OP_2$ within experimental error, thus resolving the ambiguity (Figure 6.27). With a more extensive series of isomorphous derivatives, it is possible to obtain phases capable of yielding interpretable electron density maps. Many protein structures have been investigated successfully by this technique.

## Centrosymmetric Projections

Proteins always crystallize in noncentrosymmetric space groups because the amino acid residues in the polypeptide have "left-handed" configurations about the $\alpha$-carbon atoms. Amino acid residues with "right-handed" configurations are very rare in nature. Although noncentrosymmetric structures usually present more difficulties than centrosymmetric structures, there is a compensation in the relative ease of determination of the space group; ambiguities such as $P2_1$ and $P2_1/m$ do not exist for the protein crystallographer. Most noncentrosymmetric space groups have at least one centrosymmetric zone. In such a case, (6.76) becomes

$$s_{PH}|F_{PH}| = s_P|F_P| + s_H|F_H| \tag{6.81}$$

where $s$ refers to the sign of the structure factor, and is $\pm 1$ (see page 168).

Unless both $|F_{PH}|$ and $|F_P|$ are very small compared with $|F_H|$, it is unlikely that $s_{PH}$ will differ from $s_P$. Generally $\mathbf{F}_P$ and $\mathbf{F}_{PH}$ are pointing in the same direction. Accepting this statement, we may substitute $s_P$ for $s_{PH}$ in (6.81):

$$s_P(|F_{PH}| - |F_P|) = s_H|F_H| \tag{6.82}$$

or

$$s_P = s_H|F_H|/\Delta F \tag{6.83}$$

where $\Delta F = |F_{PH}| - |F_P|$. Since we are interested only in the signs, (6.83) may be rewritten as

$$s_P = s_H s_\Delta \tag{6.84}$$

where $s_\Delta$ is $+1$ if $|F_{PH}| > |F_P|$ and $-1$ if $|F_{PH}| < |F_P|$. In this way, signs can often be determined for centric reflections in a protein crystal with only a single isomorphous derivative, and we shall illustrate the method by the following example.

TABLE 6.6. $h0l$ Data for Ribonuclease

| | Observed data | | | | Calculated data | | Deduced sign |
|---|---|---|---|---|---|---|---|
| $hkl$ | $|F_P|$ | $|F_{PH}|$ | $|\Delta F|$ | $s_\Delta$ | $|F_H|$ | $s_H$ | $s_P = s_H s_\Delta$ |
| 003 | 437 | 326 | 111 | −1 | 50 | +1 | −1 |
| 006 | 59 | 48 | 11 | −1 | 27 | −1 | +1 |
| 007 | 182 | 109 | 73 | −1 | 90 | −1 | +1 |
| $10\overline{17}$ | 144 | 196 | 52 | +1 | 31 | −1 | −1 |
| $10\overline{13}$ | 146 | 82 | 64 | −1 | 52 | +1 | −1 |
| $10\overline{9}$ | 97 | 165 | 68 | +1 | 55 | −1 | −1 |
| 106 | 183 | 242 | 59 | +1 | 45 | +1 | +1 |
| $30\overline{4}$ | 746 | 861 | 115 | +1 | 72 | +1 | +1 |
| 405 | 103 | 57 | 46 | −1 | 56 | +1 | −1 |

## Sign Determination for Centric Reflections in Protein Structures

We shall consider data for both the enzyme ribonuclease and a heavy-atom derivative prepared by soaking pregrown crystals of the enzyme in $K_2[PtCl_6]$ solution.*

*Crystal Data for Ribonuclease.*

System: monoclinic.

Unit-cell dimensions: $a = 30.31$, $b = 38.26$, $c = 52.91$ Å, $\beta = 105.9°$.

$M$: 13,500 (ribonuclease).

$Z_P$: two molecules of ribonuclease plus an unknown number of water molecules.

$Z_{PH}$: as for $Z_P + N_H [PtCl_6]^{2-}$ groups per unit cell.

Absent spectra: $0k0$: $k = 2n + 1$.

Space group: $P2_1$. The $h0l$ zone is centrosymmetric.

Table 6.6 shows how the signs for some $h0l$ reflections have been determined. Notice that experimental errors in $|F_P|$ and $|F_{PH}|$, together with errors in the calculated $|F_H|$ arising from inaccuracies in the heavy-atom model, are reflected in the inequality of $\Delta F$ and $|F_H|$. The validity of (6.84) is upheld by these data.

## Location of Heavy-Atom Positions in Proteins

In a centrosymmetric zone, it follows from (6.83) that

$$|F_H| = |\Delta F| \tag{6.85}$$

* C. H. Carlisle *et al.*, *Journal of Molecular Biology* **85**, 1 (1974).

where $|\Delta F| = \| F_{PH}| - |F_P \|$. A Patterson function calculated with $|F_H|^2$ as coefficients would give the vector set of the substituted heavy atoms in the protein molecule. Since $|F_H|$ cannot be observed, the next best procedure is to calculate a Patterson map with $(\Delta F)^2$ as coefficients. If the experimental errors in $|F_P|$ and $|F_{PH}|$ are not significant, and not too many sign "crossovers" with $s_P$ and $s_{PH}$ occur, then the $(\Delta F)^2$ Patterson projection would be expected to reveal the heavy-atom vectors (in projection). In the case of general noncentrosymmetric reflections, we note from (6.80) that

$$|F_H|^2 = \| F_{PH}| \exp(i\phi_{PH}) - |F_P| \exp(i\phi_P) \|^2 \tag{6.86}$$

Since $\phi_{PH} \neq \phi_P$, the relation (6.85) does not hold, but instead

$$|F_H| \approx |\Delta F| \cos \delta \tag{6.87}$$

where $\delta$ is an unknown angle ($\phi_H - \phi_P$ in Figure 6.25). Hence

$$|F_H|^2 \approx (\Delta F)^2 \cos^2 \delta \tag{6.88}$$

In practice, the angle $\delta$ is undeterminable, and the best one can do is to calculate a Patterson function with $(\Delta F)^2$ coefficients as for centrosymmetric reflections, but, as an added precaution to ensure that $(\phi_{PH} - \phi_P)$ is small, use only those terms for which both $|F_P|$ and $|F_{PH}|$ are large. Although the noncentrosymmetric $(\Delta F)^2$ is not a true Patterson function, it has been used successfully to determine the heavy-atom distribution in proteins.

The most useful derivatives contain a small number of highly substituted sites. Unlike the structure analysis of smaller molecules, it is not known initially how many heavy-atom sites have been incorporated into the molecule.

As an example, we shall consider the $(\Delta F)^2$ Patterson map for the Pt derivative of ribonuclease (space group $P2_1$). The vectors between symmetry-related atoms occur on the Harker section $(u, \frac{1}{2}, w)$. Eight peaks occur on the Harker section and four at $v = 0$ (Figures 6.28a and 6.28b). This result suggests that there is more than one heavy-atom site per protein molecule. The most obvious choice is two, since four heavy atoms per unit cell would give rise to 12 nonorigin peaks. If the two sites are labeled 1 and 2, their Harker peaks will be of the form $\pm\{2x_1, \frac{1}{2}, 2z_1\}$ and $\pm\{2x_2, \frac{1}{2}, 2z_2\}$.

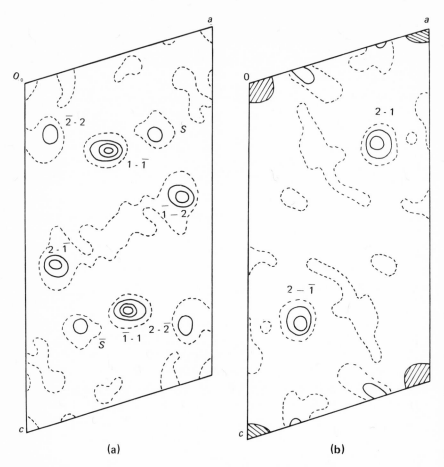

FIGURE 6.28.   $(\Delta F)^2$ Patterson sections for the Pt derivative of ribonuclease: (a) $P(u, \frac{1}{2}, w)$, (b) $P(u, 0, w)$.

Interpretation of the Patterson function is best undertaken in terms of the Harker section, assuming that the peaks represent nonoverlapping vectors and ignoring the possibility that some peaks could be non-Harker peaks. Since the true Harker peaks are of the form $2x, 2z$, values of $x$ and $z$ can be obtained by dividing by 2 the fractional coordinates on the Harker section.

This analysis may be carried out graphically. The peak positions from the Harker section are replotted, on tracing paper, on a unit-cell projection in which the $a$ and $c$ dimensions are each reduced by a factor of $\frac{1}{2}$. This procedure results in one quadrant of Figure 6.29a. The diagram is com-

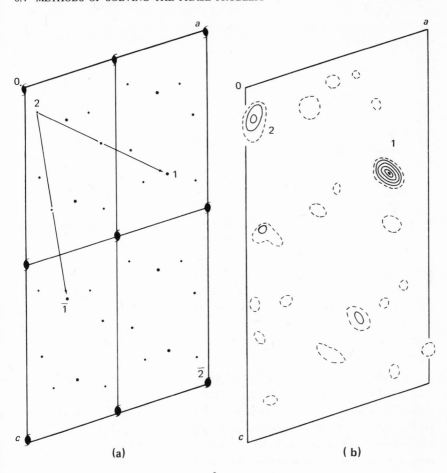

FIGURE 6.29. Interpretation of the $(\Delta F)^2$ Paterson sections: (a) implication diagram, (b) electron density map showing the Pt atom sites.

pleted by operating on the first quadrant with the translation of $a/2$, and then on both quadrants by $c/2$, thus completing an area the size of the true unit-cell projection.

All points marked on this map locate potential $(x, z)$ coordinates for the heavy atoms. In fact, it contains four equivalent solutions with respect to the four unique $2_1$ axes in the unit cell, marked ◗. Cross-vector peaks are found by moving this implication diagram* to other sections of the Patterson function, using pairs of potential sites to generate potential vectors. To see

* See Bibliography.

FIGURE 6.30. Stereoviews of the polypeptide chain in ribonuclease; the main site in the Pt derivative is shown as a simulated octahedrally coordinated group.

how this mechanism operates, place the site marked 2 on the tracing paper over the origin of the section $v = 0$ and note the coincidence of site 1 with the peak 2-1. Similarly, the peak 2-$\bar{1}$ and others on the section $v = \frac{1}{2}$ can be generated from the sites 1, 2, $\bar{1}$, and $\bar{2}$ on the implication diagram. Peaks $S$ and $\bar{S}$ are not explained in this way; they may be assumed to be spurious [remember $(\Delta F)^2$ is not a true representation of $|F_H|^2$].

Figure 6.29b shows a composite electron density map of the Pt atom sites which were prepared by an independent method, and confirm the Patterson analysis. The $y$ coordinates of the two heavy-atom sites are almost equal, which accounts for the presence of the non-Harker peaks $\pm(2\text{-}\bar{1})$ on the Harker section. Figure 6.30 is a stereo pair showing the course of the polypeptide chain in ribonuclease and the position of the main site in the Pt derivative.

## Bibliography

### General Structure Analysis

BUERGER, M. J., *Vector Space*, New York, Wiley.

LIPSON, H. S., *Crystals and X-Rays*, London, Wykeham Publications.

STOUT, G. H., and JENSEN, L. H., *X-Ray Structure Determination—A Practical Guide*, New York, Macmillan.

WOOLFSON, M. M., *An Introduction to X-Ray Crystallography*, Cambridge, University Press.

## Protein Crystallography

EISENBERG, D., *X-Ray Crystallography and Enzyme Structure*, in *The Enzymes*, Vol. I (1970), Academic Press.

PHILLIPS, D. C., *Advances in Structure Research by Diffraction Methods* **2**, 75 (1966).

## Chemical Data

SUTTON, L. E. (Editor), *Tables of Interatomic Distances and Configuration in Molecules and Ions*, The Chemical Society, London (1958; supplement, 1965).

## Problems

**6.1.** Write down the symmetry-equivalent amplitudes of $|F(hkl)|$, $|F(0kl)|$, and $|F(h0l)|$ in (a) the triclinic, (b) the monoclinic, and (c) the orthorhombic crystal systems. Friedel's law may be assumed.

**6.2.** (a) Determine the orientations of the Harker lines and sections in $Pa$, $P2/a$, and $P222_1$.
(b) A monoclinic, noncentrosymmetric crystal shows concentrations of peaks on $(u, 0, w)$ and $[0, v, 0]$. How might this situation arise?

**6.3.** Diphenyl sulfoxide, $(C_6H_5)_2SO$, is monoclinic, with $a = 8.90$, $b = 14.08$, $c = 8.32$ Å, $\beta = 101.12°$, and $Z = 4$. The conditions limiting possible X-ray reflections are as follows.

$$hkl: \quad \text{none}; \qquad h0l: \quad h+l=2n; \qquad 0k0: \quad k=2n$$

(a) What is the space group?
(b) Figures 6P.1a–c are the Patterson sections at $v = \frac{1}{2}$, $v = 0.092$, and $v = 0.408$, respectively, and contain the relevant S—S vector peaks. Determine the fractional coordinates of the four S atoms in the unit cell. Plot these atomic positions as seen along the $b$ axis, with an indication of the heights of the atoms with respect to the plane of the diagram.

**6.4.** Figure 6P.2 shows an idealized vector set for a structure $C_6H_5S$ in space group $P2$ with $Z = 2$, projected down the $b$ axis. Only the S—S and S—C vector interactions are considered.

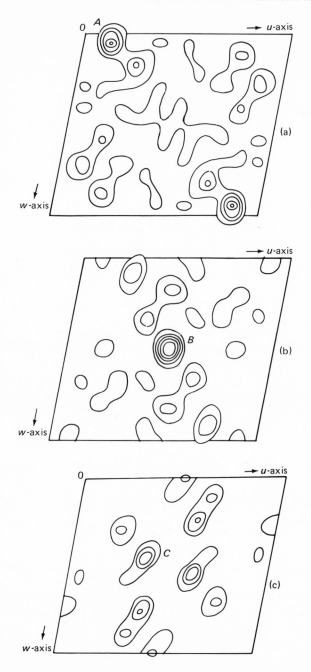

FIGURE 6P.1.  Patterson sections at (a) $v = \frac{1}{2}$, (b) $v = 0.092$, (c) $v = 0.408$.

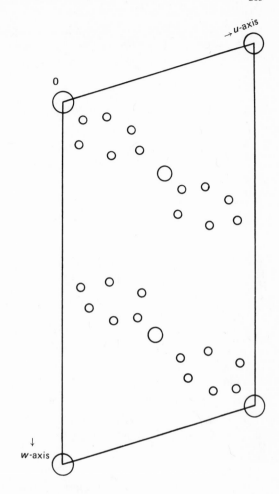

FIGURE 6P.2.

(a) Determine the $x$ and $z$ coordinates for the S atoms, and plot them to the scale of this projection.

(b) Use the Patterson superposition method to locate the carbon atom positions in the same projection.

**6.5.** Hafnium disilicide, $HfSi_2$, is orthorhombic, with $a = 3.677$, $b = 14.55$, $c = 3.649$ Å, and $Z = 4$. The space group is $Cmcm$, and the Hf and Si atoms occupy three sets of special positions of the type

$$\pm\{0, y, \tfrac{1}{4}; \quad \tfrac{1}{2}, \tfrac{1}{2}+y, \tfrac{1}{4}\}$$

The contributions from the Hf atoms dominate the structure factors.

Show that the geometric structure factor $A(0k0)$ is proportional to $\cos 2\pi y_{Hf}$. The $|F(0k0)|$ data are listed below, from which the values of $|F(0k0)|^2$, divided by 10 and rounded to the nearest integer, have been derived.

| $0k0$ | 020 | 040 | 060 | 080 | 010,0 | 012,0 | 014,0 | 016,0 |
|---|---|---|---|---|---|---|---|---|
| $|F(0k0)|$ | 7 | 14 | 18 | 13 | 12 | <1 | 20 | <1 |
| $|F(0k0)|^2$ | 5 | 20 | 32 | 17 | 14 | 0 | 40 | 0 |

(a) Calculate the one-dimensional Patterson function $P(v)$, using the equation

$$P(v) = \sum_k |F(0k0)|^2 \cos 2\pi k v$$

The multiplying factor $2/b$ and the $F(000)$ term have been omitted to simplify the calculation; they can never change the form of the synthesized function, although the neglect of the term involving $F(000)$ gives rise to negative values in the calculated $P(v)$.

One-dimensional summations may be carried out conveniently by means of Fourier summation tables,[*] a technique similar to that first put forward in 1936 by Lipson and Beevers.[†] Each line contains the values of (amplitude) $\cos 2\pi$(index)(interval), evaluated to the nearest integer. The intervals are 60ths of the repeat unit, and the lines run from 0 to 30/60. If there is reflection symmetry at $\frac{1}{4}$ along the repeat (i.e., the index is even only), as well as at $\frac{1}{2}$, then only the values 0–15 need be summed. The columns are added vertically to form the sum over the index. (See Table 6P.1.)

Plot the function, extend it to one repeat unit, interpret the four highest nonorigin peaks, and determine $y_{Hf}$.

(b) Using the value of $y_{Hf}$ and the form of the geometric structure factor $A(0k0)$, determine the signs for the $0k0$ reflections. Hence, compute the electron density:

$$\rho(y) = \sum_k \pm |F(0k0)| \cos 2\pi k y$$

---

[*] A set of tables in the form of a booklet can be obtained from one of the authors (M.F.C.L). They enable summations to be calculated with relative ease for one-dimensional, centrosymmetric structures. The index range is 0–15 and the amplitude range is 0–20. The amplitude range can be extended through multiplication by positive or negative integers.

[†] H. Lipson and C. A. Beevers, *Proceedings of the Physical Society* **48**, 772 (1936).

TABLE 6P.1

| Amplitude | Index | $\frac{0}{60}$ | $\frac{1}{60}$ | $\frac{2}{60}$ | $\frac{3}{60}$ | $\frac{4}{60}$ | $\frac{5}{60}$ | $\frac{6}{60}$ | $\frac{7}{60}$ | $\frac{8}{60}$ | $\frac{9}{60}$ | $\frac{10}{60}$ | $\frac{11}{60}$ | $\frac{12}{60}$ | $\frac{13}{60}$ | $\frac{14}{60}$ | $\frac{15}{60}$ |
|---|---|---|---|---|---|---|---|---|---|---|---|---|---|---|---|---|---|
| 5 | 2 | 5 | 5 | 5 | 4 | 3 | 2 | 2 | 1 | $\bar{1}$ | $\bar{2}$ | $\bar{2}$ | $\bar{3}$ | $\bar{4}$ | $\bar{5}$ | $\bar{5}$ | $\bar{5}$ |
| 20 | 4 | 20 | 18 | 13 | 6 | $\bar{2}$ | $\overline{10}$ | $\overline{16}$ | $\overline{20}$ | $\overline{20}$ | $\overline{16}$ | $\overline{10}$ | $\bar{2}$ | 6 | 13 | 18 | 20 |
| 32 | 6 | 32 | 26 | 10 | $\overline{10}$ | $\overline{26}$ | $\overline{32}$ | $\overline{26}$ | $\overline{10}$ | 10 | 26 | 32 | 26 | 10 | $\overline{10}$ | $\overline{26}$ | $\overline{32}$ |
| 17 | 8 | 17 | 11 | $\bar{2}$ | $\overline{14}$ | $\overline{17}$ | 8 | 5 | 16 | 16 | 5 | 8 | $\overline{17}$ | $\overline{14}$ | $\bar{2}$ | 11 | 17 |
| 14 | 10 | 14 | 7 | $\bar{7}$ | $\overline{14}$ | $\bar{7}$ | 7 | 14 | 7 | $\bar{7}$ | $\overline{14}$ | $\bar{7}$ | 7 | 14 | 7 | $\bar{7}$ | $\overline{14}$ |
| 40 | 14 | 40 | 4 | $\overline{39}$ | $\overline{12}$ | 37 | 20 | $\overline{32}$ | $\overline{27}$ | 27 | 32 | $\overline{20}$ | $\overline{37}$ | 12 | 39 | $\bar{4}$ | $\overline{40}$ |

Add for $\sum\limits_k$

$P(v) \rightarrow 128 \cdots$

TABLE 6P.2

| | | | | | | | | | | | | | | | | | |
|---|---|---|---|---|---|---|---|---|---|---|---|---|---|---|---|---|---|
| 7 | 2 | 7 | 7 | 6 | 6 | 5 | 3 | 2 | 1 | $\bar{1}$ | $\bar{2}$ | $\bar{3}$ | $\bar{5}$ | $\bar{6}$ | $\bar{6}$ | $\bar{7}$ | $\bar{7}$ |
| 14 | 4 | 14 | 13 | 9 | 4 | $\bar{1}$ | $\bar{7}$ | $\overline{11}$ | $\overline{14}$ | $\overline{14}$ | $\overline{11}$ | $\bar{7}$ | $\bar{1}$ | 4 | 9 | 13 | 14 |
| 18 | 6 | 18 | 15 | 6 | $\bar{6}$ | $\overline{15}$ | $\overline{18}$ | $\overline{15}$ | $\bar{6}$ | 6 | 15 | 18 | 15 | 6 | $\bar{6}$ | $\overline{15}$ | $\overline{18}$ |
| 13 | 8 | 13 | 9 | $\bar{1}$ | $\overline{11}$ | $\overline{13}$ | $\bar{6}$ | 4 | 12 | 12 | 4 | $\bar{6}$ | $\overline{13}$ | $\overline{11}$ | $\bar{1}$ | 9 | 13 |
| 12 | 10 | 12 | 6 | $\bar{6}$ | $\overline{12}$ | $\bar{6}$ | 6 | 12 | 6 | $\bar{6}$ | $\overline{12}$ | $\bar{6}$ | 6 | 12 | 6 | $\bar{6}$ | $\overline{12}$ |
| 20 | 14 | 20 | 2 | $\overline{20}$ | $\bar{6}$ | 18 | 10 | $\overline{16}$ | $\overline{13}$ | 13 | 16 | $\overline{10}$ | $\overline{18}$ | 6 | 20 | $\bar{2}$ | $\overline{20}$ |

Again the $2/b$ factor and $F(000)$ have been omitted. Use the summation table given in Table 6P.2; if a negative amplitude is indicated for any reflection, the signs of all the numbers, except the index, in the corresponding line are changed. Plot the function and determine $y_{Hf}$. What can be deduced about the positions of the Si atoms? In the light of your results, study $P(v)$ again.

*6.6. The alums, $MAl(NO_4)_2 \cdot 12H_2O$, where $M = NH_4$, K, Rb, Tl, and $N = S$, Se, are isomorphous. They crystallize in the cubic centrosymmetric space group $Pa3$, with the unit-cell side $a$ in the range 12.2–12.4 Å and $Z = 4$. A symmetry analysis leads to the following atomic positions:

4 $M$: $\quad 0, 0, 0; \quad 0, \frac{1}{2}, \frac{1}{2}; \quad \frac{1}{2}, 0, \frac{1}{2}; \quad \frac{1}{2}, \frac{1}{2}, 0$

4 Al: $\quad \frac{1}{2}, \frac{1}{2}, \frac{1}{2}; \quad \frac{1}{2}, 0, 0; \quad 0, \frac{1}{2}, 0; \quad 0, 0, \frac{1}{2}$

8 $N$: $\quad \pm\{x, x, x; \quad \frac{1}{2}+x, \frac{1}{2}-x, x; \quad \bar{x}, \frac{1}{2}+x, \frac{1}{2}-x; \quad \frac{1}{2}-x, \bar{x}, \frac{1}{2}+x\}$

TABLE  6P.3.  $|F(hhh)|$ for Isomorphous Alums

| hkl | NH$_4^+$ (10 electrons) | K$^+$ (18 electrons) | Rb$^+$ (36 electrons) | Tl$^+$ (80 electrons) |
|-----|-------------------------|----------------------|-----------------------|-----------------------|
| 111 | 86  | 38  | 19  | 113 |
| 222 | 0   | 19  | 79  | 195 |
| 333 | 111 | 125 | 158 | 236 |
| 444 | 25  | 6   | 55  | 125 |
| 555 | 24  | 49  | 64  | 131 |
| 666 | 86  | 86  | 122 | 164 |
| 777 | 53  | 34  | 0   | 18  |
| 888 | 0   | 16  | 22  | 56  |

The $N$ atoms lie on cube diagonals, and $x_N$ may be obtained by a one-dimensional Fourier synthesis along the line [111], using $F(hhh)$ data. Table 6P.3 lists these data for four alums ($N = S$). Tl may be assumed to be sufficiently heavy to make all $F$ values positive in this derivative. The same sites in each crystal are occupied by the replaceable atoms.

(a) Use the isomorphous character to determine the signs of the reflections in Table 6P.3.

(b) Compute $\rho[111]$ for K alum, using the following equation:

$$\rho(D) = \sum_h \pm |F(hhh)| \cos 2\pi hD$$

where $D$ is the sampling interval along [111], again in 60ths. Table 6P.4 is a summation table for the range 0–30/60; the signs of the terms must be adjusted in accordance with your findings in the table of $|F(hhh)|$ data. Plot the function and determine a probable value for $x_S$.

(c) The $hhh$ data for the isomorphous K/Se alum are listed below. Calculate and plot $\rho(D)$ for these data. compare the two electron density plots and comment upon the results. Table 6P.5 is the appropriate summation table.

| hkl | 111 | 222 | 333 | 444 | 555 | 666 | 777 | 888 |
|-----|-----|-----|-----|-----|-----|-----|-----|-----|
| $|F|$ | −48 | −52 | 64 | 0 | 116 | 100 | −16 | 0 |

## TABLE 6P.4

| F | h | | | | | | | | | | | | | | | | |
|---|---|--|--|--|--|--|--|--|--|--|--|--|--|--|--|--|--|
| 38 | 1 | 38 | 37 | 36 | 35 | 33 | 31 | 28 | 25 | 22 | 19 | 15 | 12 | 8 | 4 | 0 | $\bar{4}$ | $\bar{8}$ |
| 19 | 2 | 19 | 17 | 15 | 13 | 9 | 6 | 2 | $\bar{2}$ | $\bar{6}$ | $\bar{9}$ | $\bar{13}$ | $\bar{15}$ | $\bar{17}$ | $\bar{19}$ | $\bar{19}$ | $\bar{19}$ | $\bar{19}$ |
| 125 | 3 | 119 | 101 | 74 | 39 | 0 | $\bar{39}$ | $\bar{74}$ | $\bar{101}$ | $\bar{119}$ | $\bar{125}$ | $\bar{119}$ | $\bar{101}$ | $\bar{74}$ | $\bar{39}$ | 0 | 39 | 74 |
| 49 | 5 | 42 | 24 | 0 | $\bar{24}$ | $\bar{42}$ | $\bar{49}$ | $\bar{42}$ | $\bar{24}$ | 0 | 24 | 42 | 49 | 42 | 24 | 0 | $\bar{24}$ | $\bar{42}$ |
| 86 | 6 | 70 | 27 | $\bar{27}$ | $\bar{70}$ | $\bar{86}$ | $\bar{70}$ | $\bar{27}$ | 27 | 70 | 86 | 70 | 27 | $\bar{27}$ | $\bar{70}$ | $\bar{86}$ | $\bar{70}$ | $\bar{27}$ |
| 34 | 7 | 25 | 4 | $\bar{20}$ | $\bar{33}$ | $\bar{29}$ | $\bar{11}$ | 14 | 31 | 32 | 17 | 7 | 28 | 34 | 23 | 0 | 23 | 34 |
| 16 | 8 | 11 | 2 | $\bar{13}$ | $\bar{16}$ | $\bar{8}$ | 5 | 15 | 15 | 5 | $\bar{8}$ | $\bar{16}$ | $\bar{13}$ | 2 | 11 | 16 | 11 | 2 |

## TABLE 6P.5

| F | h | | | | | | | | | | | | | | | | |
|---|---|--|--|--|--|--|--|--|--|--|--|--|--|--|--|--|--|
| 48 | 1 | $\bar{48}$ | $\bar{46}$ | $\bar{44}$ | $\bar{42}$ | $\bar{39}$ | $\bar{36}$ | $\bar{32}$ | $\bar{28}$ | $\bar{24}$ | $\bar{20}$ | $\bar{15}$ | $\bar{10}$ | $\bar{5}$ | 0 | 5 | 10 |
| 52 | 2 | $\bar{51}$ | 48 | 42 | 35 | 26 | 16 | 5 | $\bar{5}$ | $\bar{16}$ | $\bar{26}$ | $\bar{35}$ | $\bar{42}$ | $\bar{48}$ | 51 | 52 | 51 | 48 |
| 64 | 3 | 61 | 52 | 38 | 20 | 0 | $\bar{20}$ | $\bar{38}$ | $\bar{52}$ | $\bar{61}$ | $\bar{64}$ | $\bar{61}$ | $\bar{52}$ | $\bar{38}$ | 20 | 0 | 20 | 38 |
| 116 | 5 | 101 | 58 | 0 | $\bar{58}$ | $\bar{101}$ | $\bar{116}$ | $\bar{101}$ | $\bar{58}$ | 0 | 58 | 101 | 116 | 101 | 58 | 0 | $\bar{58}$ | $\bar{101}$ |
| 100 | 6 | 81 | 31 | $\bar{31}$ | $\bar{81}$ | $\bar{100}$ | $\bar{81}$ | $\bar{31}$ | 31 | 81 | 100 | 81 | 31 | $\bar{31}$ | $\bar{81}$ | $\bar{100}$ | $\bar{81}$ | $\bar{31}$ |
| 16 | 7 | 12 | 2 | $\bar{9}$ | 16 | 14 | 5 | $\bar{7}$ | $\bar{15}$ | $\bar{15}$ | $\bar{8}$ | 3 | 13 | 16 | 11 | 0 | $\bar{11}$ | $\bar{16}$ |

# Some Further Topics

## 7.1 Introduction

In this chapter we shall consider direct (phase probability) methods of solving the phase problem, and certain other techniques which may be involved in the overall investigation of crystal and molecular structure.

## 7.2 Direct Methods of Phase Determination

Direct methods of solving the phase problem have come to the fore in recent years. They are particularly important for their ability to yield good phase information for structures containing no heavy atoms. Our discussion will be confined to centrosymmetric structures.

One feature common to the structure-determining methods that we have encountered so far is that values for the phases of X-ray reflections are derived initially by structure factor calculations, albeit on only part of the structure. Since the data from which the best phases are ultimately derived are the $|F_o|$ values, we may imagine that the phases are somehow locked within these quantities, even though their actual values are not recorded experimentally. This philosophy led to the search for analytical methods of phase determination, which are independent of trial structures, and initiated the development of direct methods, or phase probability techniques.

### 7.2.1 Normalized Structure Factors

A simplification in direct phase-determining formulae results by replacing $|F(hkl)|$ by the corresponding normalized structure factor $|E(hkl)|$, which is given by the equation*

$$|E_o(hkl)|^2 = \frac{K^2|F_o(hkl)|^2}{\varepsilon \sum\limits_{j=1}^{N} g_j^2} \tag{7.1}$$

The **E** values have properties similar to those of the **F** values derived for a point-atom model (page 225); they are largely compensated for the fall-off of $f$ with $\sin \theta$. High-order reflections with comparatively small $|F|$ values can have quite large $|E|$ values, an important fact in the application of direct methods. We may note in passing that $|E|^2$ values can be used as coefficients in a sharpened Patterson function, and, since $\overline{|E|^2} = 1$[cp. (6.62)], the coefficients ($|E|^2 - 1$) produce a sharpened Patterson function with the origin peak removed. This technique is useful because, in addition to the general sharpening effect, vectors of small magnitude which are swamped by the origin peak may be revealed.

### $\varepsilon$-Factor

Because of the importance of individual reflections in direct phasing methods, care must be taken to obtain the best possible $|E|$ values. The factor $\varepsilon$ in the denominator of (7.1) takes account of the fact that reflections in certain reciprocal lattice zones or rows may have an average intensity which is greater than that for the general reflections. The $\varepsilon$-factor depends upon the crystal class. For example, in space group $P2$, using the methods developed in Chapter 4, we obtain

$$|F(hkl)|^2 = 4\left[\sum_{j=1}^{N/2} g_j \cos 2\pi(hx_j + lz_j) \cos 2\pi k y_j\right]^2$$

$$+ 4\left[\sum_{j=1}^{N/2} g_j \cos 2\pi(hx_j + lz_j) \sin 2\pi k y_j\right]^2 \tag{7.2}$$

The series within the pairs of square brackets will, upon expansion, contain terms such as

$$g_j^2 \cos^2 Y_j \cos^2 \Psi_j + 2g_jg_i \cos Y_j \cos \Psi_j \cos Y_i \cos \Psi_i + g_j^2 \cos^2 Y_j \sin^2 \Psi_j$$

$$+ 2g_jg_i \cos Y_j \sin \Psi_j \cos Y_i \sin \Psi_i \tag{7.3}$$

---

* We shall generally omit the subscript $o$ in the $|E|$ symbol, since such terms almost always refer to observed data.

TABLE 7.1. $\varepsilon$-Factors for Some Crystal Classes

| System | Class | $hkl$ | $0kl$ | $h0l$ | $hk0$ | $h00$ | $0k0$ | $00l$ |
|--------|-------|-------|-------|-------|-------|-------|-------|-------|
| Triclinic | 1 | 1 | 1 | 1 | 1 | 1 | 1 | 1 |
| | $\bar{1}$ | 1 | 1 | 1 | 1 | 1 | 1 | 1 |
| Monoclinic | 2 | 1 | 1 | 1 | 1 | 1 | 2 | 1 |
| | $m$ | 1 | 1 | 2 | 1 | 2 | 1 | 2 |
| | $2/m$ | 1 | 1 | 2 | 1 | 2 | 2 | 2 |
| Orthorhombic | 222 | 1 | 1 | 1 | 1 | 2 | 2 | 2 |
| | $mm2$ | 1 | 2 | 2 | 1 | 2 | 2 | 4 |
| | $mmm$ | 1 | 2 | 2 | 2 | 4 | 4 | 4 |

where $\Upsilon = 2\pi(hx + lz)$, $\Psi = 2\pi ky$, and $j \neq i$. Provided that the $N$ species in the unit cell are similar and randomly distributed, and that $N$ itself is not small, the summation of the cross terms in (7.3) will be negligible, because each is as likely to be negative as positive. Hence,

$$|F(hkl)|^2 = 4 \sum_{j=1}^{N/2} g_j^2 \cos^2 \Upsilon_j \qquad (7.4)$$

For the $0k0$ reflections, $\Upsilon = 0$ and

$$|F(0k0)|^2 = 4 \sum_{j=1}^{N/2} g_j^2 = 2 \sum_{j=1}^{N} g_j^2 \qquad (7.5)$$

which is twice the average for the $hkl$ reflections, given by (6.62). In other words, $\varepsilon = 2$ for the $0k0$ reflections in any crystal of class 2. The values of $\varepsilon$ for some crystal classes are listed in Table 7.1.

## $|E|$ Statistics

The distribution of $|E|$ values holds useful information about the space group of a crystal. Theoretical quantities derived for equal-atom structures in space groups $P1$ and $P\bar{1}$ are listed in Table 7.2, together with the experimental results for two crystals.

Crystal 1 is pyridoxal phosphate oxime dihydrate, $C_8H_{11}N_2O_6P \cdot 2H_2O$, which is triclinic. The values in Table 7.2 favor the centric distribution $C$, and the structure analysis* confirmed the assignment of space

---

* A. N. Barrett and R. A. Palmer, *Acta Crystallographica* **B25**, 688 (1969).

TABLE 7.2.  Some Theoretical and Experimental Values Related to $|E|$ Statistics

| | Theoretical values | | Experimental values | |
|---|---|---|---|---|
| Mean values | $P\bar{1}\,(C)$ | $P1\,(A)$ | Crystal 1 | Crystal 2 |
| $|E|^2$ | 1.00 | 1.00 | 0.99 | 0.98 |
| $|E|$ | 0.80 | 0.89 | 0.85 $A/C$ | 0.84 $A/C$ |
| $||E|^2-1|$ | 0.97 | 0.74 | 0.91 $C$ | 0.82 $A$ |
| Distribution | % | % | % | % |
| $|E|>3.0$ | 0.30 | 0.01 | 0.20 $C$ | 0.05 $A$ |
| $|E|>2.5$ | 1.24 | 0.19 | 0.90 $C$ | 0.98 $C$ |
| $|E|>2.0$ | 4.60 | 1.80 | 2.70 $A/C$ | 2.84 $A/C$ |
| $|E|>1.75$ | 8.00 | 4.71 | 7.14 $C$ | 6.21 $A/C$ |
| $|E|>1.5$ | 13.4 | 10.5 | 12.9 $C$ | 10.5 $A$ |
| $|E|>1.0$ | 32.0 | 36.8 | 33.7 $C$ | 37.1 $A$ |

group $P\bar{1}$. Crystal 2 is a pento-uloside sugar; the results correspond, on the whole, to an acentric distribution $A$, as expected for a crystal of space group $P2_12_12_1$.*

It should be noted that the experimentally derived quantities do not always have a completely one-to-one correspondence with the theoretical values, and care should be exercised in using these statistics to select a space group.

### 7.2.2  Structure Invariants and Origin-Fixing Reflections

The formulae used in direct phasing require, initially, the use of a few reflections with phases known, either uniquely or symbolically. In centrosymmetric crystals, the origin† is taken on one of the eight centers of symmetry in the unit cell, and we speak of the sign $s(hkl)$ of the reflection; $s(hkl)$ is $F(hkl)/|F(hkl)|$ and is either $+1$ or $-1$. We shall show next that, in any primitive, centrosymmetric space group in the triclinic, monoclinic, or orthorhombic systems, arbitrary signs can be allocated to three reflections in order to specify the origin at one of the centers of symmetry. These signs form a basic set to which a "snowballing" technique allows more and more reflections to be included as the analysis proceeds.

From (4.62) it follows that

$$F(hkl)_{0,0,0} = \sum_{j=1}^{N} g_j \cos 2\pi(hx_j + ky_j + lz_j) \qquad (7.6)$$

* H. T. Palmer, Personal communication (1973).
† The origin of the $x$, $y$, $z$ coordinates of the structure.

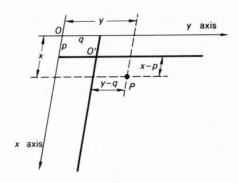

FIGURE 7.1. Transformation of the point $P(x, y)$, with respect to two-dimensional axes, by moving the origin from $O$ to $O'(p, q)$; the transformed coordinates of $P$ are $(x - p, y - q)$.

where $F(hkl)_{0,0,0}$ indicates an origin of coordinates at the point $0, 0, 0$. If this origin is moved to a center of symmetry at $\frac{1}{2}, \frac{1}{2}, 0$, the point that was originally $x_j, y_j, z_j$ becomes $x_j - \frac{1}{2}, y_j - \frac{1}{2}, z_j$ (Figure 7.1), with $p = q = \frac{1}{2}$. The structure factor equation is now

$$F(hkl)_{1/2,1/2,0} = \sum_{j=1}^{N} g_j \cos 2\pi[(hx_j + ky_j + lz_j) - (h + k)/2] \quad (7.7)$$

Expanding the cosine term, and remembering that $\sin[2\pi(h + k)/2]$ is zero, we obtain

$$F(hkl)_{1/2,1/2,0} = (-1)^{h+k} F(hkl)_{0,0,0} \quad (7.8)$$

Equation (7.8) demonstrates that $|F(hkl)|$ is invariant under change of origin, as would be expected, but that a change of sign may occur, depending on the parity of the indices $hkl$. The complete results are listed in Table 7.3.

TABLE 7.3. Effect of a Change of Origin of Coordinates, among Centers of Symmetry, on the Sign of a Structure Factor

| Parity group | 1 | 2 | 3 | 4 | 5 | 6 | 7 | 8 |
|---|---|---|---|---|---|---|---|---|
| Centers of symmetry | $h$ even $k$ even $l$ even | $h$ odd $k$ even $l$ even | $h$ even $k$ odd $l$ even | $h$ even $k$ even $l$ odd | $h$ even $k$ odd $l$ odd | $h$ odd $k$ even $l$ odd | $h$ odd $k$ odd $l$ even | $h$ odd $k$ odd $l$ odd |
| $0, 0, 0$ | + | + | + | + | + | + | + | + |
| $\frac{1}{2}, 0, 0$ | + | − | + | + | + | − | − | − |
| $0, \frac{1}{2}, 0$ | + | + | − | + | − | + | − | − |
| $0, 0, \frac{1}{2}$ | + | + | + | − | − | − | + | − |
| $0, \frac{1}{2}, \frac{1}{2}$ | + | + | − | − | + | − | − | + |
| $\frac{1}{2}, 0, \frac{1}{2}$ | + | − | + | − | − | + | − | + |
| $\frac{1}{2}, \frac{1}{2}, 0$ | + | − | − | + | − | − | + | + |
| $\frac{1}{2}, \frac{1}{2}, \frac{1}{2}$ | + | − | − | − | + | + | + | − |

The use of this table will be illustrated by the following examples. Reflection 312 belongs to parity group 7 (ooe, in short). If $s(312)$ is given a plus sign, the origin could be regarded as being restricted to one from the following list:

$$0, 0, 0; \quad 0, 0, \tfrac{1}{2}; \quad \tfrac{1}{2}, \tfrac{1}{2}, 0; \quad \tfrac{1}{2}, \tfrac{1}{2}, \tfrac{1}{2}$$

Similarly, if $s(322)$, parity group 2 (oee), is also given a plus sign, the possible origins are

$$0, 0, 0; \quad 0, \tfrac{1}{2}, 0; \quad 0, 0, \tfrac{1}{2}; \quad 0, \tfrac{1}{2}, \tfrac{1}{2}$$

Combining these two sign allocations, the common origins are

$$0, 0, 0; \quad 0, 0, \tfrac{1}{2}$$

In order to fix the origin uniquely at, say, 0, 0, 0, we select another reflection with a plus sign with respect to 0, 0, 0 and a minus sign with respect to $0, 0, \tfrac{1}{2}$. Reference to Table 7.3 shows that parity groups 4, 5, 6, and 8 each meet this requirement.

Parity groups 1 and 3 are excluded from the choice as the third origin-specifying reflection. Group 1 is a special case discussed below. Group 3 (eoe) is related to groups 2 and 7 through an addition of indices:

$$312 + 322 \rightarrow 634 \qquad (7.9)$$

or, more generally,

$$\text{ooe} + \text{oee} \rightarrow \text{eoe} \qquad (7.10)$$

since $o + o$ or $e + e \rightarrow e$, and $e + o \rightarrow o$.

Parity groups 2, 3, and 7 are said to be linearly related, and cannot be used together in defining the choice or origin.

Structure factors belonging to parity group 1 do not change sign on change of origin, as is evident from both the development of (7.8) and Table 7.3. Reflections in this group are called structure invariants; their signs depend on the actual structure and cannot be chosen at will.

### 7.2.3  Sign-Determining Formulae

Over the past twenty years, many equations have been proposed which are capable of providing sign information for centrosymmetric crystals. Two of these expressions have proved to be outstandingly useful, and it is to them that we turn our attention.

## Triple-Product Sign Relationship

In 1952, Sayre† derived a general formula for structures containing identical resolved atoms. For centrosymmetric crystals, it may be given in the form

$$s(hkl)s(h'k'l')s(h-h', k-k', l-l') \approx +1 \qquad (7.11)$$

where the sign $\approx$ means "is probably equal to." The vectors associated with these reflections, $d^*(hkl)$, $d^*(h'k'l')$, and $d^*(h-h', k-k', l-l')$ form a closed triangle, or vector triplet, in reciprocal space. In practice, it may be possible to form several such vector triplets for a given $hkl$; Figure 7.2a shows two triplets for the vector 300. If two of the signs in (7.11) are known, the third can be deduced, and we can extend the sign information beyond that given in the starting set.

A physical meaning can be given to equation (7.11) by drawing the traces, in real space, of the three planes that form a vector triplet in reciprocal space (Figures 7.2a and 7.2b). For a centrosymmetric crystal, (6.41) becomes

$$\rho(x, y, z) = \frac{2}{V_c} \sum_h \sum_k \sum_l \pm |F(hkl)| \cos 2\pi(hx + ky + lz) \qquad (7.12)$$

The $|F(hkl)|$ terms in this equation are *positive* if the traces of the corresponding planes pass through the origin, like the full lines in Figure 7.2b and *negative* if they lie midway between these positions, like the dashed lines in Figure 7.2b. The combined contributions from the three planes in question will thus have maxima at the points of their mutual intersections, which are therefore potential atomic sites, and correspond to regions of high electron density.

This argument is particularly strong if the three planes have high $|E|$ values. It may be seen from the diagram that triple intersections occur only at points where either three full lines $(+ + +)$ meet, or two dashed lines and one full line meet (some combination of $+ - -$). This result is in direct agreement with (7.11). It is interesting to note that the structure

† D. Sayre, *Acta Crystallographica* **5**, 60 (1952).

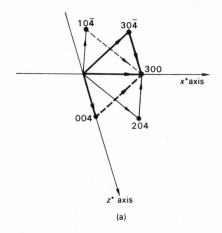

(a)

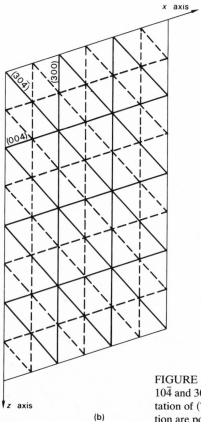

(b)

FIGURE 7.2. (a) Vector triplets 300, 204, $10\overline{4}$ and 300, $30\overline{4}$, 004; (b) physical interpretation of (7.11); the points of triple intersection are possible atomic sites.

of hexamethylbenzene was solved in 1929 by Lonsdale* through drawing the traces of three high-order, high-intensity reflection planes, $7\bar{3}0$, $340$, and $4\bar{7}0$, and placing atoms at their intersections. These planes form a vector triplet, and this structure determination contained, therefore, the first, but apparently inadvertent, use of direct methods.

## $\Sigma_2$ Formula

Hauptman and Karle† have given a more general form of (7.11):

$$s[E(hkl)] \approx s\left[ \sum_{h'k'l'} E(h'k'l')E(h-h', k-k', l-l') \right] \qquad (7.13)$$

the summation being over all vector pairs with known signs which form a triplet with $hkl$. The probability associated with (7.13) is given by

$$P_+(hkl) = \tfrac{1}{2} + \tfrac{1}{2}\tanh[(\sigma_3/\sigma_2^{3/2})\alpha'] \qquad (7.14)$$

where $\alpha'$ is given by

$$\alpha' = |E(hkl)| \sum_{h'k'l'} E(h'k'l')E(h-h', k-k', l-l') \qquad (7.15)$$

and $\sigma_n$ by

$$\sigma_n = \sum_j Z_j^n \qquad (7.16)$$

where $Z_j$ is the atomic number of the $j$th atom.

For a structure containing $N$ identical atoms, $\sigma_3/\sigma_2^{3/2}$ is equal to $N^{-1/2}$. From (7.15), we see that the probability is strongly dependent upon the magnitudes of the $|E|$ values. Furthermore, unless glide-plane or screw-axis symmetry is present [see, for example, (7.18) and (7.19)], or there exists some other means of generating negative signs, (7.13) will produce only positive signs for all $E(hkl)$. Such a situation would correspond to a structure with a very heavy atom at the origin, and would, in general, correspond to an incorrect solution.

* K. Lonsdale, *Proceedings of the Royal Society A* **123**, 494 (1929).

† H. Hauptman and J. Karle, *Solution of the Phase Problem, I. The Centrosymmetric Crystal,* American Crystallographic Association Monograph No. 3 (1953).

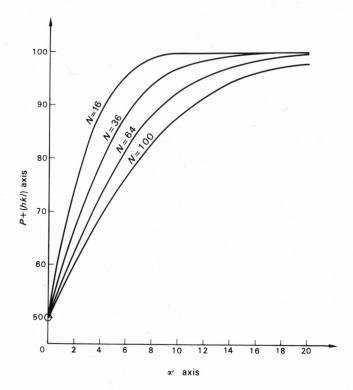

FIGURE 7.3. Percentage probability of a single triple-product
($\Sigma_2$) sign relationship as a function of $\alpha'$ for different numbers $N$ of
atoms in a unit cell.

If the combination of signs under the summation in (7.15) produces a
large and negative value for $\alpha'$, the corresponding value of $P_+(hkl)$ may tend
to zero. This result indicates that $s(hkl)$ is negative, with a probability that
tends to unity.

Probability curves for different numbers $N$ of atoms in the unit cell as a
function of $\alpha'$ are shown in Figure 7.3. Since the most reliable signs from
(7.13) are, from (7.15), associated with large $|E|$ values, we can now add to
the origin-specifying criteria (page 274) the requirements of both large $|E|$
values and a large number of $\Sigma_2$ interactions for each reflection in the
starting set. In this way, strong and reliable sign propagation is encouraged.

To illustrate the operation of the $\Sigma_2$ relationship, we shall consider the
two vector triplets in Figure 7.2. The sign to be determined is $s(300)$, the
others are assumed to be known. The data are tabulated as follows.

| $hkl$ | $|E(hkl)|$ | | | | | | |
|-------|-----------|---|---|---|---|---|---|
| 300 | 2.40 | | | | | | |
| $h'k'l'$ | $E(h'k'l')$ | $h-h', k-k', l-l'$ | $E(h-h', k-k', l-l')$ | $\alpha'$ | $s(hkl)$ | $P_+(hkl)$, % |
| $10\bar{4}$ | +2.03 | 204 | −2.22 | | −1 | |
| 004 | −1.95 | $30\bar{4}$ | +1.81 | 19.3 | −1 | 0.4 |

Assuming that $N$ is 64, the indication given is that $s(300)$ is negative with a probability of 99.6%.

## 7.2.4 Amplitude Symmetry and Phase Symmetry

In space group $P\bar{1}$, the only symmetry-related structure factors are $F(hkl)$ and $F(\bar{h}\bar{k}\bar{l})$. According to Friedel's law (4.51), the intensities and, hence, the amplitudes* of these structure factors are equal, and, in centrosymmetric space groups, $s(hkl) = s(\bar{h}\bar{k}\bar{l})$. Thus, the amplitude symmetry and the phase symmetry follow the same law, but this will not necessarily be true in other space groups.

From the geometric structure factor for space group $P2_1/c$, (4.109), and (4.110),

$$|F(hkl)| = |F(\bar{h}\bar{k}\bar{l})| = |F(h\bar{k}l)| = |F(\bar{h}k\bar{l})| \qquad (7.17)$$

while for the signs there are two possibilities:

$$k+l = 2n: \qquad s(hkl) = s(\bar{h}\bar{k}\bar{l}) = s(h\bar{k}l) = s(\bar{h}k\bar{l}) \qquad (7.18)$$

$$k+l = 2n+1: \qquad s(hkl) = s(\bar{h}\bar{k}\bar{l}) = -s(h\bar{k}l) = -s(\bar{h}k\bar{l}) \qquad (7.19)$$

These relationships provide enhanced opportunities for $\Sigma_2$ relationships to be developed, and in this way space-group symmetry can improve the chances of successful phase determination. The amplitude symmetry and phase symmetry for all space groups are contained in the *International Tables for X-Ray Crystallography*, Vol. I.†

## 7.2.5 $\Sigma_2$ Listing

Because of both the increased probability in relationships developed for reflections with high $|E|$ values and the existence of many vector triplets

---

* **E** and **F** follow the same symmetry relationships in all space groups.
† N. F. M. Henry and K. Lonsdale (Editors), *International Tables for X-Ray Crystallography*, Vol. I, Birmingham, Kynoch Press.

in a complete set of data, the initial application of direct methods is limited to reflections with large $|E|$ values, say, greater than 1.5. As a more stringent condition, however, it must be remembered that an electron density map needs about eight symmetry-independent reflections per atom in the asymmetric unit in order to provide reasonable resolution of the electron density image.

A $\Sigma_2$ listing is prepared by considering each value of $|E(hkl)|$ greater than the preset limit, in order of decreasing magnitude, as a basic $hkl$ vector, and searching the data for all interactions with $h'k'l'$ and $h-h', k-k', l-l'$. Some reflections will enter into many such interactions, while others will produce only a small number.

## 7.2.6  Symbolic-Addition Procedure

Karle and Karle* have described a technique for the systematic application of the $\Sigma_2$ formula for building up a self-consistent sign set. The various steps involved are outlined below, using results obtained with pyridoxal phosphate oxime dihydrate.†

Crystal Data

Formula: $C_8H_{11}N_2O_6P \cdot 2H_2O$.
System: triclinic.
Unit-cell dimensions: $a = 10.94$, $b = 8.06$, $c = 9.44$ Å, $\alpha = 57.18$, $\beta = 107.68$, $\gamma = 116.53°$.
$V_c$: 627 Å$^3$.
$D_m$: 1.57 g cm$^{-3}$.
$M$: 261.
$Z$: 2.01, or 2 to the nearest integer.
Absent spectra: none.
Possible space groups: $P1$ or $P\bar{1}$. $P\bar{1}$ was chosen on the basis of intensity statistics (Table 7.2).
All atoms are in general positions.

Sign Determination

(a) A total of 163 reflections for which $|E| \geqslant 1.5$ were arranged in descending order of magnitude and a $\Sigma_2$ listing was obtained using a computer program.

* J. Karle and I. L. Karle, *Acta Crystallographica* **21**, 849 (1966).
† Barrett and Palmer, *op. cit.* (see footnote, page 273).

TABLE 7.4. Starting Set for the Symbolic-
Addition Procedure

| hkl | $|E|$ | Sign |
|-----|-------|------|
| $9\bar{1}\bar{4}$ | 2.97 | $+$ |
| $8\bar{1}\bar{5}$ | 3.00 | $+$ } $^a$ |
| $\bar{1}40$ | 2.38 | $+$ |
| 020 | 4.50 | $A$ |
| 253 | 2.24 | $B$ |
| 822 | 2.71 | $C$ } $^b$ |
| 303 | 2.69 | $D$ |
| 023 | 2.28 | $E$ |

$^a$ Origin-fixing reflections.
$^b$ Letter symbols, each representing $+1$ or $-1$.

(b) From a study of the $\Sigma_2$ listing, three reflections were allocated $+$ signs (Table 7.4); they are the origin-fixing reflections, selected according to the procedures already discussed.

(c) Equation (7.13) was used by searching, initially, between members of the origin-fixing set and other reflections. To maintain a high probability, only the highest $|E|$ values were used. For example, $9\bar{5}\bar{5}$ ($|E| = 2.31$) is generated by the combination of $8\bar{1}\bar{5}$ and $\bar{1}40$:

$$s(9\bar{5}\bar{5}) \approx s(8\bar{1}\bar{5})s(\bar{1}40) = (+1)(+1) = +1 \qquad (7.20)$$

From (7.14), $\alpha'$ is 16.5, and Figure 7.3 tells us ($N = 38$, excluding hydrogen) that the probability of this indication is about 99.7%. The new sign was accepted and used to generate more signs. This process was continued until no new signs could be developed with high probability.

(d) At this stage, it is usually found that the number of signs developed with confidence is small. This situation was found with pyridoxal phosphate oxime dihydrate, and the $\Sigma_2$ formula was then applied to reflections with symbolic signs. In this technique, a reflection was selected, again by virtue of its high $|E|$ value and long $\Sigma_2$ listing, and allocated a letter symbol (Table 7.4). As a rule, five or less symbolic phases are sufficient, and there are no necessary restrictions on the parities of these reflections. However, it is desirable that there are no redundancies in the complete starting set, that is, no three reflections in the set should themselves be related by a triple product relationship.

As a symbol became involved in a sign of a reflection, it was written into the $\Sigma_2$ listing. As an example, Table 7.5 shows a $\Sigma_2$ entry for $9\bar{8}\bar{6}$. Reading

TABLE 7.5. $\Sigma_2$ listing for the Reflection $98\bar{6}$ of Pyridoxal Oxime Phosphate with Appropriate Phase Symbols Added[a]

| $h'k'l'$ | $s(h'k'l')$ | $\dfrac{|E_2|}{|E(h'k'l')|}$ | $h-h', k-k', l-l'$ | | | $s(h-h', k-k', l-l')$ | $\dfrac{|E_3|}{E(h-h', k-k', l-l')}$ | $|E_1||E_2||E_3|$ | $s(98\bar{6})$ | $P_+(98\bar{6})$, % |
|---|---|---|---|---|---|---|---|---|---|---|
| $1\,\bar{5}\,0$ | $BD$ | 2.16 | 8 | $\bar{3}$ | $\bar{6}$ | $A$ | 1.63 | 6.40 | $ABD$ | 90 |
| $10,\bar{2}\,\bar{2}$ | $AB$ | 2.04 | 1 | 6 | 4 | $D$ | 1.88 | 6.97 | $ABD$ | 91 |
| $10,\bar{7}\,\bar{1}$ | $D$ | 1.87 | 1 | 1 | 5 | $AB$ | 1.63 | 5.54 | $ABD$ | 87 |
| $4\,\bar{8}\,\bar{3}$ | $D$ | 1.83 | 5 | 0 | $\bar{3}$ | $ECD$ | 1.58 | 5.25 | $EC$ | 85 |
| $3\,\bar{9}\,\bar{4}$ | | 1.76 | 6 | 1 | $\bar{2}$ | | 1.58 | 5.03 | | |
| $3\,\bar{5}\,\bar{6}$ | | 1.70 | 6 | $\bar{3}$ | 0 | | 1.51 | 4.66 | | |
| $6\,\bar{7}\,\bar{2}$ | | 1.68 | 3 | $\bar{1}$ | $\bar{4}$ | | 1.63 | 4.98 | | |
| $10,\bar{4}\,\bar{2}$ | $B$ | 1.62 | 1 | 4 | 4 | $AD$ | 1.67 | 4.93 | $ABD$ | 84 |
| $0\,\bar{2}\,0$ | $-A$ | 4.50 | 9 | $\bar{6}$ | $\bar{6}$ | | 1.73 | 14.08 | | |
| $0\,\bar{8}\,0$ | $+$ | 2.48 | 9 | 0 | $\bar{6}$ | | 1.85 | 8.30 | | |

[a] $|E(98\bar{6})| = |E_1| = 1.89$.

across the table, sign combinations are seen to be generated by multiplying $s(h'k'l')$ by $s(h-h', k-k', l-l')$, which are then written as $s(9\bar{8}\bar{6})$ in the penultimate column. Recurring combinations, such as $ABD$, gave rise to consistent indications. If the probability that $s(9\bar{8}\bar{6}) = s(ABD)$ is sufficiently large, this sign value is entered for $s(9\bar{8}\bar{6})$ wherever these indices occur. In the final column of the table, the probability of each sign indication is listed. Although they are small individually, the combined probability that $s(9\bar{8}\bar{6})$ was $ABD$ is 100% [see (7.14)].

(e) When this process, too, had been exhausted with all letter signs, the results were examined for agreement among sign relationships. For example, in Table 7.5 there is a weak (85%) indication that $ABD = EC$. The most significant relationships found overall were

$$AC = E \qquad (7.21)$$

$$C = EB \qquad (7.22)$$

$$B = ED \qquad (7.23)$$

$$AD = E \qquad (7.24)$$

$$AB = CD \qquad (7.25)$$

Multiplying (7.21) by (7.22), and (7.21) by (7.24), and remembering that products such as $A^2$ equal $+1$, reduces this list to

$$A = B \qquad (7.26)$$

$$C = D \qquad (7.27)$$

$$E = AC \qquad (7.28)$$

The five symbols were reduced, effectively, to two, $A$ and $C$. The sign determination was rewritten in terms of signs and the symbols $A$ and $C$; reflections with either uncertain or undetermined signs were rejected from the first electron density calculation.

### 7.2.7 Calculation of $E$ Maps

The result of the above analysis meant that four possible sign sets could be generated by the substitutions $A = \pm 1$, $C = \pm 1$. The set with $A = C = +1$ was rejected because this phase assignment implies a very heavy atom at the origin of the unit cell. The three other sign combinations were used to calculate $E$ maps. These maps are obtained by Fourier calculations, using (7.12), but with $|E|$ replacing $|F|$ as the coefficients. The "sharp" nature of $E$ [see (7.1)] is advantageous when using a limited number of data to resolve atomic peaks in the electron density map.

The sign combination for pyridoxal phosphate oxime dihydrate that led to an interpretable $E$ map was $A = C = -1$. The atomic coordinates from this map (Figure 7.4a) were used in a successful refinement of the structure, and Figure 7.4b shows the conformation of this molecule.

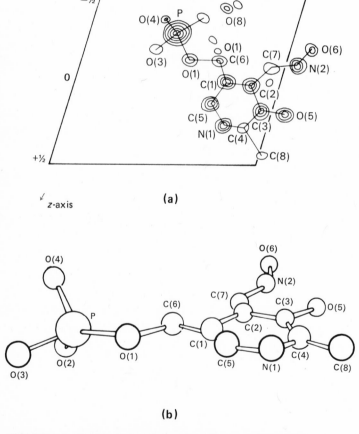

FIGURE 7.4. Pyridoxal phosphate oxime dihydrate: (a) composite three-dimensional $E$ map on the $ac$ plane; (b) molecular conformation viewed at about 10° to the plane of the six-membered ring.

Direct methods have been extended to noncentrosymmetric crystals, and computer methods have been developed to cope with the enormity of three-dimensional calculations involved. Their discussion, however, lies outside the scope of this book.*

## 7.3  Least-Squares Refinement

In Chapter 1, we used the equation of a line in two-dimensional space:

$$Y = mX + b \qquad (7.29)$$

If we have two error-free pairs of values for $X$ and $Y$ for measurements which are related by this equation, we can obtain a unique answer for the constants $m$ and $b$. Sometimes, as in the Wilson plot (page 247), we have several pairs of values which contain random errors, and we need to obtain those values of $m$ and $b$ that best fit the set of observations.

In practical problems, we have often a situation in which the errors in, say, the $X$ values are negligible compared with those in $Y$. Let the best estimates of $m$ and $b$ under these conditions be $m_o$ and $b_o$. Then, the error of fit in the $i$th observation is

$$e_i = m_o X_i + b_o - Y_i \qquad (7.30)$$

The principle of least squares states that the best-fit parameters are those that minimize the sum of the squares of the errors. Thus,

$$\sum_i e_i^2 = \sum_i (m_o X_i + b_o - Y_i)^2 \qquad (i = 1, 2, \ldots, N) \qquad (7.31)$$

has to be minimized over the number $N$ of observations. This condition corresponds to differentiating partially with respect to $m_o$ and $b_o$, in turn, and equating the derivatives to zero. Hence,

$$m_o \sum_i X_i^2 + b_o \sum_i X_i = \sum_i X_i Y_i \qquad (7.32)$$

$$m_o \sum_i X_i + b_o N = \sum_i Y_i \qquad (7.33)$$

* J. Karle, *Advances in Chemical Physics* **16**, 131 (1969); M. M. Woolfson, *Reports on Progress in Physics* **34**, 369 (1971).

which constitute a pair of simultaneous equations, easily solved for $m_o$ and $b_o$.

In a crystal structure analysis, we are always manipulating more observations than there are unknown quantities; the system is said to be overdetermined. We shall consider in outline two important applications of the method of least squares.

### 7.3.1  Unit-Cell Dimensions

In Chapter 3, we considered methods for obtaining unit-cell dimensions with moderate accuracy. Generally, we need to enhance the precision of these measurements, which may be achieved by a least-squares analysis. Consider, for example, a monoclinic crystal for which the $\theta$ values of a number of reflections, preferably high-order, of known indices, have been measured to the nearest $0.01°$. In the monoclinic system, $\sin \theta$ is given, through Table 2.4 and (3.32), by

$$4 \sin^2\theta = h^2 a^{*2} + k^2 b^{*2} + l^2 c^{*2} + 2hla^*c^* \cos \beta^* \qquad (7.34)$$

In order to obtain the best values of $a^*$, $b^*$, $c^*$, and $\cos \beta^*$, we write, following (7.31),

$$\sum_i (h_i^2 a^{*2} + k_i^2 b^{*2} + l_i^2 c^{*2} + 2h_i l_i a^*c^* \cos \beta^* - 4 \sin^2\theta_i)^2 \qquad (7.35)$$

and then minimize this expression, with respect to $a^*$, $b^*$, $c^*$, and $\cos \beta^*$, over the number of observations. The procedure is a little more involved numerically; we obtain four simultaneous equations to be solved for the four variables, but the principles are the same as those involved with the straight line, (7.29)–(7.33).

### 7.3.2  Atomic Parameters

Correct trial structures are usually refined by least-squares methods. In essence, this process involves adjusting the positional and temperature parameters of the atoms in the unit cell so as to obtain the best agreement between the experimental $|F_o|$ values and the derived $|F_c|$ quantities. The difference between $|F_o|$ and $|F_c|$ is the function in the numerator of the

$R$-factor, (5.11). We minimize $R'$, given by

$$R' = \sum_{hkl} w\left(|F_o| - \frac{1}{K}|F_c|\right)^2 \tag{7.36}$$

where $w$ is a weight, which varies inversely with the error in $|F_o|$, $K$ is the scale factor, and the summation extends over the symmetry-independent reflections.

In a crystal with $N$ atoms in the asymmetric unit, in order to refine the $x$, $y$, $z$, and $B$ parameters for each atom, and an overall scale factor, we must adjust simultaneously $4N + 1$ variables.[*] The method is complicated in this application by the fact that $|F_c|$ is not a simple function of the variables. Fortunately, the problem is highly overdetermined. There could be about $50N$ reflections, and a modern high-speed computer can handle the necessary calculation. The overdetermined nature of the structure analysis is important, since it enables the analysis to be carried out with adequate precision.

After the least-squares refinement, a difference-Fourier synthesis (page 252) should be calculated. It should show a zero field within experimental error, as already discussed. This check is important, for it should be borne in mind that while the least-squares procedure will produce the best fit to the model which it is given, a difference-Fourier map may reveal some feature previously unsuspected, such as a molecule of water of crystallization.

## 7.4 Molecular Geometry

When the formal structure analysis is complete, we need to express our results in terms of molecular geometry and packing. This part of the analysis includes the determination of bond lengths, bond angles, and intermolecular contact distances, with measures of their precision.

Consider three atoms with fractional coordinates $x_1, y_1, z_1, x_2, y_2, z_2$, and $x_3, y_3, z_3$ in a unit cell of sides $a$, $b$, and $c$ (Figure 7.5). The vector from the origin O to atom 1 is given by

$$\mathbf{O1} = x_1\mathbf{a} + y_1\mathbf{b} + z_1\mathbf{c} \tag{7.37}$$

---

[*] Thermal motion is frequently anisotropic, and the atomic scattering factor is corrected by an expression of the form $e^{-T}$, where $T = 2\pi^2\lambda^{-2}(h^2a^{*2}U_{11} + k^2b^{*2}U_{22} + l^2c^{*2}U_{33} + 2klb^*c^*U_{23} + 2lhc^*a^*U_{31} + 2hka^*b^*U_{12})$. In this case, the six anisotropic temperature factors $U$ are refined, and the number of variables is increased to $9N + 1$.

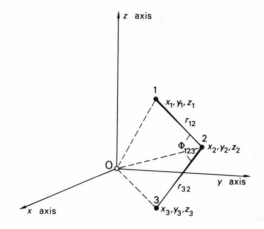

FIGURE 7.5. Geometry of the calculation of
interatomic distances and angles.

The vector $\mathbf{r}_{12}$ is given by

$$\mathbf{r}_{12} = \mathbf{O2} - \mathbf{O1} \qquad (7.38)$$

or, using (7.37),

$$\mathbf{r}_{12} = (x_2 - x_1)\mathbf{a} + (y_2 - y_1)\mathbf{b} + (z_2 - z_1)\mathbf{c} \qquad (7.39)$$

Forming the dot product of each side with itself, remembering that

$$\mathbf{p} \cdot \mathbf{q} = pq \cos \widehat{pq} \qquad (7.40)$$

we obtain

$$r_{12}^2 = (x_2 - x_1)^2 a^2 + (y_2 - y_1)^2 b^2 + (z_2 - z_1)^2 c^2$$
$$+ 2(y_2 - y_1)(z_2 - z_1)bc \cos \alpha + 2(z_2 - z_1)(x_2 - x_1)ca \cos \beta$$
$$+ 2(x_2 - x_1)(y_2 - y_1)ab \cos \gamma \qquad (7.41)$$

This equation may be simplified for crystal systems other than triclinic. Thus,
if the atoms are referred to a tetragonal unit cell,

$$r_{12}^2 = [(x_2 - x_1)^2 + (y_2 - y_1)^2]a^2 + (z_2 - z_1)^2 c^2 \qquad (7.42)$$

In a similar manner, we can evaluate $r_{32}$ (Figure 7.5). Using (7.40), for the tetragonal system,

$$\cos \Phi_{123} = \frac{\{[(x_2-x_1)(x_2-x_3)+(y_2-y_1)(y_2-y_3)]a^2+(z_2-z_1)(z_2-z_3)c^2\}}{(r_{12}r_{32})}$$

(7.43)

where $r_{12}$ and $r_{32}$ are evaluated following (7.41). Similar equations enable any distance or angle to be calculated, in any crystal system, in terms of the atomic coordinates and the unit-cell dimensions.

## 7.5 Accuracy

Closely related to the calculations of bond lengths and angles is the expression of the probable accuracy in these quantities. We state, without proof here, that the least-squares refinement procedures establish values for an estimated standard deviation (esd) in each of the variables used in these calculations. Thus, a fractional coordinate of 0.3712 might have a standard deviation of 0.0003, for instance, written as 0.3712(3).

We need to know further how errors are propagated in a quantity which is a function of several variables, each of which contains some uncertainty arising from random errors. The answer is provided by the statistical principle of superposition of errors.

Let $q$ be a function of several variables $p_i$ $(i = 1, 2, 3, \ldots, N)$, with known standard deviations $\sigma(p_i)$. Then the esd in $q$ is given by

$$\sigma^2(q) = \sum_{i=1}^{N} \left[ \frac{\partial q}{\partial p_i} \sigma(p_i) \right]^2$$

(7.44)

A simple example may be given, through (7.39) and (7.44), for a bond between two atoms lying along the $c$ edge of a tetragonal unit cell. Let $c$ be 10.06(1) Å, $z_1$ be 0.3712(3), and $z_2$ be 0.5418(2). From (7.39),

$$r_{12} = (z_2 - z_1)c = (0.5418 - 0.3712)10.06 = 1.716 \text{ Å}$$

(7.45)

and from (7.44),

$$\sigma^2(r_{12}) = (0.5418 - 0.3712)^2 (0.01)^2$$
$$+ (10.06)^2 (0.0002)^2 + (10.06)^2 (0.0003)^2$$

(7.46)

Thus, $\sigma(r_{12})$ is 0.004 Å and we write $r_{12} = 1.716(4)$ Å. Similar calculations may be used for all distance and angle calculations in all crystal systems, but the general equations are quite involved numerically and are best handled by computer methods.

## 7.6  Correctness of a Structure Analysis

At this stage we may summarize four criteria of correctness of a good structure analysis. If we can satisfy these conditions in one and the same structure model, we shall have a high degree of confidence in it.

(a) There should be good agreement between $|F_o|$ and $|F_c|$, expressed through the $R$-factor (page 193). Ultimately, $R$ depends upon the quality of the experimental data. At best, it will probably be 1–2% greater than the average standard deviation in $|F_o|$.

(b) The electron density map should show neither unaccountable positive nor negative density regions, other than Fourier series termination errors (Figure 6.3).

(c) The difference-Fourier map should be relatively flat. This map eliminates series termination errors as they are present in both $\rho_o$ and $\rho_c$. Random errors produce small fluctuations on a difference map, but they should lie within 2.5–3 times the standard deviation of the electron density $\sigma(\rho_o)$:

$$\sigma(\rho_o) = \frac{1}{V_c} \left[ \sum_{hkl} (\Delta F)^2 \right]^{1/2} \qquad (7.47)$$

where $\Delta F = |F_o| - |F_c|$ and the sum extends over all symmetry-independent reflections.

(d) The molecular geometry should be chemically sensible, within the limits of current structural knowledge.* Abnormal bond lengths and angles may be correct, but they must be supported by strong evidence of their validity in order to gain acceptance. Normally, a deviation of less than three times the appropriate standard deviation is not considered to be statistically significant.

---

* See Bibliography.

## 7.7 Anomalous Scattering

Friedel's law (page 156) is not an exact relationship, and tends to become less so as the atomic numbers of the constituent atoms in a crystal increase. The law breaks down severely if X-rays are used that have a wavelength just less than that of an absorption edge (page 344) of an atom in the crystal. However, this criterion is not essential for anomalous scattering to be used in two important aspects of crystal structure analysis, namely, the determination of absolute stereochemical configurations and the phasing of reflections.

Anomalous scattering introduces a phase change into the given atomic scattering factor, which becomes complex:

$$\mathbf{f} = \mathbf{f}_o + \Delta\mathbf{f}' + i\,\Delta\mathbf{f}'' \tag{7.48}$$

$\Delta\mathbf{f}'$ is a real correction, usually negative, and $\Delta\mathbf{f}''$ is an imaginary component which is rotated anticlockwise through $90°$ in the complex plane with respect to $\mathbf{f}_o$ and $\Delta\mathbf{f}'$.

A possible situation is illustrated in Figure 7.6. In Figure 7.6a, atom $A$ is assumed to be scattering in accordance with Friedel's law, and it is clear that $|F(\mathbf{h})| = |F(\bar{\mathbf{h}})|$, where $\mathbf{h}$ stands for $hkl$. In Figure 7.6b, atom $A$ is represented as an anomalous scatterer, with its three components, according to (7.48). In this situation, $|F(\mathbf{h})| \neq |F(\bar{\mathbf{h}})|$, and intensity measurements of Friedel pairs of reflections produce different values.

A given set of atomic coordinates can be used effectively to calculate $|F_c|$ for a number of Friedel pairs, because the coordinates of the enantiomorph may be obtained, for example, by inversion through the origin. One of the enantiomorphs will show the better fit, and thus the correct absolute configuration can be deduced. Some typical results are listed in Table 7.6, from which it may be deduced that the structure giving $|F_c|_{\bar{x},\bar{y},\bar{z}}$ corresponds to the absolute configuration. An equivalent procedure would be to measure the values of both $|F(hkl)|$ and $|F(\bar{h}\bar{k}\bar{l})|$ and compare them with $|F_c|_{x,y,z}$. A detailed statistical analysis of this method has been given by Hamilton.[*]

The technique can be used only with crystals which are noncentrosymmetric, but this limitation is not important because molecules which crystallize with a single enantiomorph cannot do so in a space group containing any form of inversion symmetry.

[*] W. C. Hamilton, *Acta Crystallographica* **18**, 502 (1965).

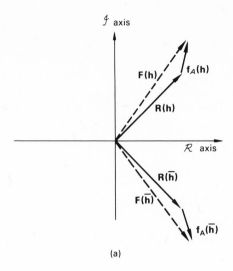

(a)

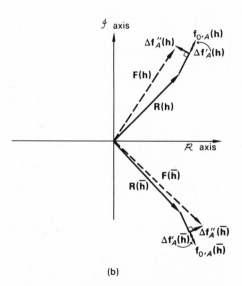

(b)

FIGURE 7.6. Anomalous scattering of atom $A$ with respect to the rest of the structure $R$; (a) normal case—$|F(\mathbf{h})| = |F(\bar{\mathbf{h}})|$; (b) anomalous case—$|F(\mathbf{h})| \neq |F(\bar{\mathbf{h}})|$.

TABLE 7.6.  Example of Some Friedel Pairs and the
Corresponding $|F|$ Values

| hkl | $|F_o|$ | $|F_c|_{x,y,z}$ | $|F_c|_{\bar{x},\bar{y},\bar{z}}$ |
|-----|------|------|------|
| 121 | 17.0 | 19.1 | 18.3 |
| 122 | 21.2 | 22.9 | 21.9 |
| 123 | 41.4 | 44.4 | 42.8 |
| 341 | 36.7 | 38.7 | 35.5 |
| 342 | 7.8  | 9.5  | 8.2  |
| 413 | 14.2 | 15.3 | 13.5 |

Anomalous scattering can be used in phasing reflections. We shall not describe this method in detail, but merely note that the anomalous scatterer can be considered to act like a second isomorphous derivative (page 254). Consequently, it is of considerable importance in protein crystallography.

## 7.8  Limitations of X-Ray Structure Analysis

There are certain things which X-ray analysis cannot do well, and it is prudent to consider the more important of them.

Liquids and gases lack three-dimensional order, and cannot be used in diffraction experiments in the same way as are crystals. Certain information about the radial distribution of electron density can be obtained, but it lacks the distinctive detail of crystal analysis.

It is not easy to locate light atoms in the presence of heavy atoms. Difference-Fourier maps alleviate the situation to some extent, but the atomic positions are not precise. Least-squares refinement of light-atom parameters is not always successful, because the contributions to the structure factor from the light atoms are relatively small.

Hydrogen atoms are particularly difficult to locate with precision because of their small scattering power and the fact that the center of the hydrogen atom does not, in general, coincide with the maximum of its electron density. Terminal hydrogen atoms have a more aspherical electron density distribution than do hydrogen-bonded hydrogen atoms, and their bond distances, from X-ray studies, often appear short when compared with spectroscopic or neutron diffraction values. For similar reasons, refinement of hydrogen atom parameters in a structure analysis may be imprecise, and the standard deviations in their coordinate values may be as much as ten times greater than those for a carbon atom in the same structure.

In general, bond lengths determined by X-ray methods represent distances between the centers of gravity of the electron clouds, which may not be the same as the internuclear separations. Internuclear distances can be found from neutron diffraction data, because neutrons are scattered by the atomic nuclei. If, for a given crystal, the synthesized neutron scattering density is subtracted from that of the X-ray scattering density, a much truer picture of the electron density can be obtained—but that is another story.

## Bibliography

### Direct Methods

STOUT, G. H., and JENSEN, L. H., *X-Ray Structure Determination—A Practical Guide*, New York, Macmillan.

WOOLFSON, M. M., *An Introduction to X-Ray Crystallography*, Cambridge, University Press.

WOOLFSON, M. M., *Direct Methods in Crystallography*, New York, Oxford University Press.

### Refinement and Molecular Geometry

STOUT, G. H., and JENSEN, L. H., *X-Ray Structure Determination—A Practical Guide*, New York, Macmillan.

WOOLFSON, M. M., *An Introduction to X-Ray Crystallography*, Cambridge, University Press.

### Chemical Data

SUTTON, L. E. (Editor), *Tables of Interatomic Distances and Configurations in Molecules and Ions*, London, The Chemical Society (1958; supplement, 1965).

## Problems

**7.1.** Choose three of the following reflections to fix the origin in space group $P\bar{1}$. Give reasons for your choice.

| $hkl$ | $|E|$ | $hkl$ | $|E|$ |
|-------|-------|-------|-------|
| 705 | 2.2 | $6\bar{1}\bar{7}$ | 3.2 |
| $42\bar{6}$ | 2.7 | 203 | 2.3 |
| $4\bar{3}2$ | 1.1 | $8\bar{1}\bar{4}$ | 2.1 |

Are there any triplets which meet the vector requirements of the $\Sigma_2$ formula?

**7.2.** The geometric structure factor formulae for space group $P2_1$ are

$$A = 2 \cos 2\pi(hx + lz + k/4) \cos 2\pi(ky - k/4)$$

$$B = 2 \cos 2\pi(hx + lz + k/4) \sin 2\pi(ky - k/4)$$

Deduce the amplitude symmetry and the phase symmetry for this space group according to the two conditions $k = 2n$ and $k = 2n + 1$.

**7.3.** In space group $P2_1/c$, two starting sets of reflections for the application of the $\Sigma_2$ formula are proposed:

|     | Origin-fixing | Symbols |
|-----|---------------|---------|
| (a) | $041, 117, \bar{1}23$ | $242, \bar{1}62$ |
| (b) | $223, 012, 13\bar{7}$ | $111, 162$ |

Which starting set would be chosen in practice? Give reasons. What modification would have to be made to the starting set if the space group is $C2/c$?

**7.4.** The following values of $\log_e[\sum_j f_j^2(hkl)/|F_o(hkl)|^2]$ and $\overline{(\sin^2\theta)/\lambda^2}$ were obtained from a set of three-dimensional data for a monoclinic crystal. Use the method of least squares to obtain values for the scale $K$ (of $|F_o|$) and temperature factor $B$ by Wilson's method.

| $\log_e[\sum_j f_j^2(hkl)/|F_o(hkl)|^2]$ | $\overline{(\sin^2\theta)/\lambda^2}$ |
|---|---|
| 4.0 | 0.10 |
| 5.6 | 0.20 |
| 6.5 | 0.30 |
| 7.9 | 0.40 |
| 9.4 | 0.50 |

What is the value of the root mean square atomic displacement corresponding to the derived value of $B$?

**7.5.** An orthorhombic crystal contains four molecules of a chloro compound in a unit cell of dimensions $a = 7.210(4)$, $b = 10.43(1)$, $c =$

15.22(2) Å. The coordinates of the Cl atoms are

$$\tfrac{1}{4}, y, z; \quad \tfrac{3}{4}, \bar{y}, z; \quad \tfrac{1}{4}, \tfrac{1}{2}+y, \tfrac{1}{2}+z; \quad \tfrac{3}{4}, \tfrac{1}{2}-y, \tfrac{1}{2}+z$$

with $y = 0.140(2)$ and $z = 0.000(2)$. Calculate the shortest $Cl \cdots Cl$ contact distance and its estimated standard deviation.

**7.6.**  A crystal contains five atoms per unit cell. Four of them contribute together $100e^{i\phi}$ to $F(010)$. The fifth atom has fractional coordinates 0.00, 0.10, 0.00, and its atomic scattering factor components $f_o$, $\Delta f'$, and $\Delta f''$ are 52.2, $-2.7$, and 8.0, respectively. If $\phi = 60°$, determine, graphically or otherwise, $|F(010)|$, $|F(0\bar{1}0)|$, $\phi(010)$, and $\phi(0\bar{1}0)$.

# 8

# Examples of Crystal Structure Analysis

## 8.1 Introduction

In the final chapter of this book, we wish to draw together the material already presented by means of two actual examples of crystal structure determinations. It will be desirable for the reader to refer back to the earlier chapters for the descriptions of the techniques used, since we shall present here mainly the results obtained at each stage of the work. The first example shows the use of the heavy-atom method, and the second illustrates an application of direct methods in structure analysis.

## 8.2 Crystal Structure of 2-Bromobenzo[b]indeno[3,2-e]pyran (BBIP)*

BBIP is an organic compound which is prepared by heating a solution in ethanol of equimolar amounts of 3-bromo-6-hydroxybenzaldehyde (I) and 2-oxoindane (II) under reflux in the presence of piperidine acetate. The two molecules condense with the elimination of two molecules of water. Upon recrystallization of the product from toluene, it has a m.p. of 176.5–177.0°C. Its molecular formula is $C_{16}H_9BrO$, and its classical structural formula is shown by III.

* M. F. C. Ladd and D. C. Povey, *Journal of Crystal and Molecular Structure* **2**, 243 (1972).

### 8.2.1  Preliminary Physical and X-Ray Measurements

The compound was recrystallized slowly as red, acicular (needle-shaped) crystals, with the forms (subsequently named) {100}, {001}, and {011} predominant (Figure 8.1). The red color is characteristic of the chromophoric nature of a conjugated double-bond system.

The density of the crystals was measured by suspending them in aqueous sodium bromide solution in a stoppered measuring cylinder in a thermostat bath at 20°C. Water or concentrated sodium bromide solution, as necessary, was added to the suspension until the crystals neither settled to the bottom of the cylinder nor floated to the surface of the solution. Then, the crystals and solution were of the same density, and the density of the solution was measured with a pyknometer.

Under a polarizing microscope, the crystals showed straight extinction on (100) and (001), and oblique extinction (about 3° to a crystal edge) on a

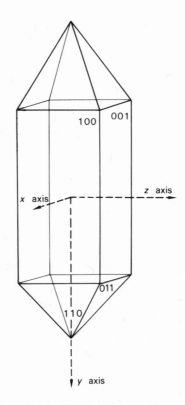

FIGURE 8.1.  Crystal habit of BBIP with the crystallographic axes drawn in.

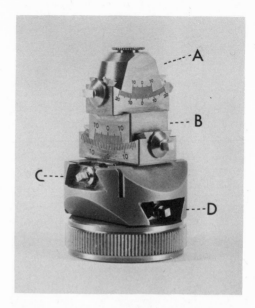

FIGURE 8.2. Standard goniometer head (by courtesy of Stoe et Cie); $A$ and $B$ are two arcs for angular adjustments; $C$ and $D$ are two sledges for horizontal adjustments.

section cut normal to the needle axis ($b$). These observations suggested that the crystals were probably monoclinic.

The crystals chosen for X-ray studies had the approximate dimensions 0.2, 0.4, and 0.3 mm parallel to $a$, $b$, and $c$, respectively. A crystal was mounted on the end of an annealed quartz fiber with "UHU" adhesive thinned with amyl acetate, and the fiber was attached to an X-ray goniometer head (arcs) (Figure 8.2) with dental wax. The arcs were affixed to a single-crystal oscillation camera, and the crystal was set with the needle axis accurately parallel to the axis of oscillation, first by eye and then by X-ray methods. Copper $K\alpha$ radiation ($\bar{\lambda} = 1.5418$ Å) was used throughout the work.

A symmetric oscillation photograph taken about the $b$ axis is shown in Figure 8.3. The horizontal mirror symmetry line indicates that the Laue group of the crystal has an $m$ plane normal to the needle axis. Further X-ray photographs, for example, Figure 8.4, showed that the only axial symmetry was 2 parallel to $b$, thus confirming the monoclinic system for BBIP.

Weissenberg photographs are shown in Figures 8.5–8.7. The indexing of the reflections can be understood with reference to Figures 8.8 and 8.9. There are no systematic absences for the $hkl$ reflections, so that the unit cell is primitive ($P$), but systematic absences do arise for $h0l$ with $l$ odd and for $0k0$ with $k$ odd. These observations confirm the monoclinic symmetry, and the systematic absences lead unambiguously to space group $P2_1/c$.

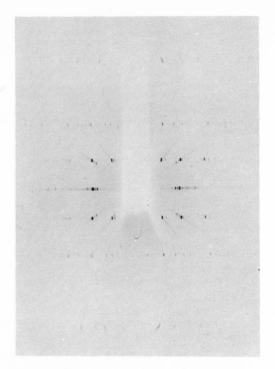

FIGURE 8.3. Symmetric oscillation photograph taken with the X-ray beam normal to $b$; the "shadow" arises from the beam stop and holder.

Measurements on the X-ray photographs gave the approximate unit cell dimensions $a = 7.51$, $b = 5.96$, $c = 26.2$ Å, and $\beta = 92.5°$.

The Bragg angles $\theta$ of 20 high-order reflections of known indices, distributed evenly in reciprocal space, were measured to the nearest $0.01°$ on a four-circle diffractometer. From these data, the unit-cell dimensions were calculated accurately by the method of least squares (page 288). The complete crystal data are listed in Table 8.1. The calculated density $D_c$ is in good agreement with the measured value $D_m$, which indicates a high degree of self-consistency in the parameters involved.

### 8.2.2 Intensity Measurement and Correction

The intensities of about 1700 symmetry-independent reflections with* $(\sin \theta)/\lambda \leq 0.56$ Å$^{-1}$ were measured, and corrections were applied to

---

* Large thermal vibrations of the atoms in the structure led to this rather low practicable upper limit for $(\sin \theta)/\lambda$.

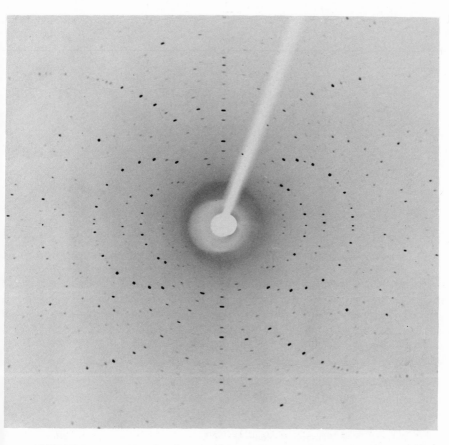

FIGURE  8.4.  Laue photograph taken with the X-ray beam along the twofold ($b$) axis.

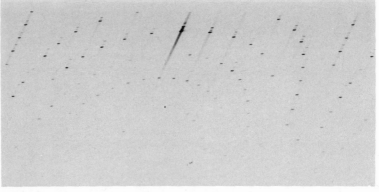

FIGURE 8.5. Weissenberg photograph of the $h0l$ layer. The more intense reflections show spots arising from both Cu $K\alpha$ ($\lambda = 1.542$ Å) and Cu $K\beta$ ($\lambda = 1.392$ Å) radiations. In some areas, spots from W $L\alpha$ radiation ($\lambda = 1.48$ Å) arise due to sputtering of the copper target in the X-ray tube with tungsten from the filament (see Appendix A.4).

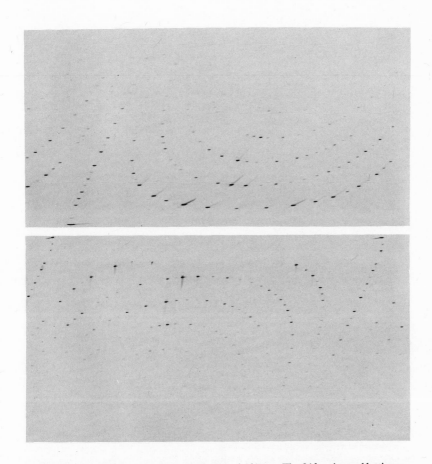

FIGURE 8.6. Weissenberg photograph of the $h1l$ layer. The $01l$ reciprocal lattice row shows evidence of slight mis-setting of the crystal.

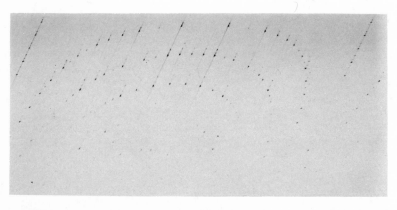

FIGURE 8.7. Weissenberg photograph of the $0kl$ layer; Cu $K\beta$ spots can be seen for the more intense reflections. The $00l$ $(z^*)$ reciprocal lattice row is common to this photograph and that of the $h0l$ layer.

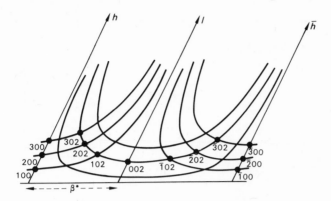

FIGURE 8.8. Sample of indexed reflections on an $h0l$ Weissenberg photograph diagram.

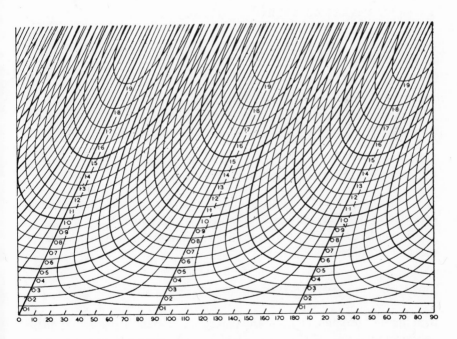

FIGURE 8.9. Weissenberg chart: camera diameter 57.3 mm, 2° rotation per mm travel.
(Reproduced with the permission of the Institute of Physics and the Physical Society, London.)

TABLE 8.1.  Crystal Data for BBIP at 20°C$^a$

| Molecular formula | $C_{16}H_9BrO$ |
|---|---|
| Molecular weight | 297.16 |
| Space group | $P2_1/c$ |
| $a$, Å | 7.508(4) |
| $b$, Å | 5.959(5) |
| $c$, Å | 26.172(6) |
| $\beta$, deg | 92.55(2) |
| Unit cell volume, Å$^3$ | 1169(2) |
| $D_m$, g cm$^{-3}$ | −1.68(1) |
| $D_c$, g cm$^{-3}$ | 1.688(3) |
| $Z$ | 4 |
| $F(000)$ | 592 |

$^a$ The numbers in parentheses are estimated standard deviations, to be applied to the least significant figure.

take account of polarization and Lorentz effects, but not absorption (pages 343, 351). Approximate scale $(K)$ and isotropic temperature $(B)$ factors were obtained by Wilson's method (page 245). The straight line was fitted by a least-squares method, and has the equation

$$\log_e\left\{\sum_j f_j^2/\overline{|F_o|^2}\right\} = -1.759 + 3.480\,\overline{\sin^2\theta} \qquad (8.1)$$

From the slope $(2B/\lambda^2)$ and intercept $(2\log_e K)$, $B = 4.1$ Å$^2$ and $K = 0.13$. The graphical Wilson plot is shown in Figure 8.10 and the scaled $|F(h0l)|$ data in Table 8.2. We are now ready to proceed with the structure analysis.

### 8.2.3  Structure Analysis in the (010) Projection

This projection of the unit cell has the largest area and thus would be expected to show an appreciable resolution of the molecule.

From the data for $P2_1/c$ (page 91), we can associate the coordinates of the general equivalent positions,

$$\pm\{x, y, z;\quad x, \tfrac{1}{2}-y, \tfrac{1}{2}+z\}$$

with the four bromine atoms in the unit cell. In the (010) projection, these coordinates give rise to two repeats within the length $c$, so we shall consider this projection in terms of plane group $p2$ with unit cell edges $a' = a$ and

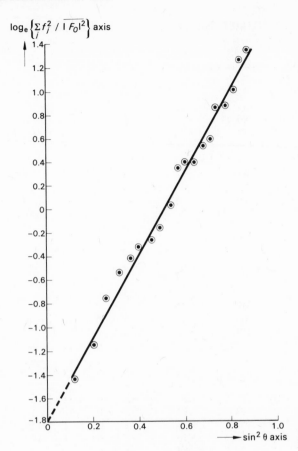

FIGURE 8.10. Wilson plot.

$c' = c/2$. The Patterson projection $P(u, w)$ would be expected to show two Br—Br vectors at $\pm(2x, 2z)$, and it is sufficient to calculate $P(u, w)$ over the range $u = 0$ to $a/2$, $w = 0$ to $c/2$, or one-half of the *projected* unit cell. The projection is shown in Figure 8.11.

The two rows of peaks indicate that, in projection, the molecules lie closely parallel to the $z$ axis; this conclusion is supported by the large magnitude of $|F(200)|$ (Table 8.2). This value may be compared with $F(000)$ in Table 8.1, and, more important, with $\sum_j f_{j,\theta_{200}}$, which is 474. The peak arising from the Br—Br vector is marked $A$, and by direct measurement we obtain the fractional coordinates $x = 0.247$, $z = 0.031$ for the Br atom in the asymmetric unit.

TABLE 8.2. $|F_o(h0l)|$ Data for BBIP on an Approximately Absolute Scale[a]

| h | k | l | $|F_o|$ | $F_{Br}$ | h | k | l | $|F_o|$ | $F_{Br}$ |
|---|---|---|---|---|---|---|---|---|---|
| 0 | 0 | 4 | 34 | 121 | −5 | 0 | 20 | 12 | 19 |
| 0 | 0 | 6 | 57 | 101 | −5 | 0 | 22 | 19 | 13 |
| 0 | 0 | 8 | 53 | 78 | −5 | 0 | 24 | 13 | 8* |
| 0 | 0 | 10 | 80 | 55 | −4 | 0 | 2 | 61 | 66 |
| 0 | 0 | 12 | 52 | 34 | −4 | 0 | 4 | 63 | 57 |
| 0 | 0 | 16 | 6 | 2* | −4 | 0 | 6 | 53 | 46 |
| 0 | 0 | 20 | 19 | −14 | −4 | 0 | 8 | 45 | 33 |
| 0 | 0 | 22 | 68 | −18 | −4 | 0 | 10 | 36 | 20 |
| 0 | 0 | 24 | 7 | −19 | −4 | 0 | 12 | 28 | 8* |
| 1 | 0 | 0 | 25 | 8* | −4 | 0 | 16 | 5 | −8* |
| 2 | 0 | 0 | 303 | −114 | −4 | 0 | 18 | 10 | −14 |
| 3 | 0 | 0 | 4 | −17 | −4 | 0 | 20 | 42 | −17 |
| 4 | 0 | 0 | 61 | 71 | −4 | 0 | 22 | 7 | −18 |
| 5 | 0 | 0 | 15 | 17 | −4 | 0 | 24 | 22 | −17 |
| 6 | 0 | 0 | 17 | −38 | −3 | 0 | 2 | 48 | −34 |
| 7 | 0 | 0 | 18 | −12 | −3 | 0 | 4 | 33 | −49 |
| 8 | 0 | 0 | 11 | 17 | −3 | 0 | 6 | 52 | −59 |
| −8 | 0 | 2 | 10 | 15 | −3 | 0 | 8 | 94 | −65 |
| −8 | 0 | 4 | 6 | 12 | −3 | 0 | 10 | 59 | −66 |
| −8 | 0 | 6 | 10 | 8* | −3 | 0 | 12 | 47 | −63 |
| −8 | 0 | 10 | 6 | 1* | −3 | 0 | 14 | 60 | −57 |
| −7 | 0 | 4 | 20 | −21 | −3 | 0 | 16 | 43 | −49 |
| −7 | 0 | 6 | 15 | −23 | −3 | 0 | 18 | 45 | −40 |
| −7 | 0 | 8 | 12 | −24 | −3 | 0 | 20 | 37 | −31 |
| −7 | 0 | 10 | 20 | −24 | −3 | 0 | 22 | 37 | −22 |
| −7 | 0 | 12 | 8 | −22 | −3 | 0 | 24 | 18 | −15 |
| −7 | 0 | 14 | 15 | −19 | −2 | 0 | 2 | 129 | −108 |
| −7 | 0 | 16 | 26 | −16 | −2 | 0 | 4 | 90 | −95 |
| −6 | 0 | 2 | 9 | −34 | −2 | 0 | 6 | 54 | −78 |
| −6 | 0 | 4 | 13 | −29 | −2 | 0 | 8 | 50 | −59 |
| −6 | 0 | 6 | 30 | −22 | −2 | 0 | 10 | 6 | −39 |
| −6 | 0 | 8 | 23 | −15 | −2 | 0 | 12 | 58 | −22 |
| −6 | 0 | 10 | 16 | −7* | −2 | 0 | 14 | 8 | −6 |
| −6 | 0 | 12 | 10 | −1* | −2 | 0 | 18 | 15 | 13 |
| −6 | 0 | 14 | 11 | 4* | −2 | 0 | 20 | 23 | 18 |
| −6 | 0 | 18 | 7 | 11 | −2 | 0 | 22 | 37 | 20 |
| −6 | 0 | 20 | 34 | 12 | −2 | 0 | 24 | 30 | 21 |
| −5 | 0 | 2 | 16 | 27 | −1 | 0 | 2 | 53 | 32 |
| −5 | 0 | 4 | 36 | 35 | −1 | 0 | 4 | 51 | 53 |
| −5 | 0 | 6 | 18 | 40 | −1 | 0 | 6 | 87 | 68 |
| −5 | 0 | 8 | 43 | 43 | −1 | 0 | 8 | 86 | 77 |
| −5 | 0 | 10 | 50 | 43 | −1 | 0 | 10 | 79 | 80 |
| −5 | 0 | 12 | 18 | 41 | −1 | 0 | 12 | 57 | 77 |
| −5 | 0 | 14 | 40 | 37 | −1 | 0 | 14 | 54 | 70 |
| −5 | 0 | 16 | 34 | 31 | −1 | 0 | 16 | 76 | 61 |
| −5 | 0 | 18 | 19 | 25 | −1 | 0 | 18 | 71 | 51 |

TABLE 8.2—cont.

| h | k | l | $|F_o|$ | $F_{Br}$ | h | k | l | $|F_o|$ | $F_{Br}$ |
|---|---|---|---|---|---|---|---|---|---|
| −1 | 0 | 20 | 63 | 40 | 3 | 0 | 18 | 30 | 39 |
| −1 | 0 | 22 | 37 | 30 | 3 | 0 | 20 | 49 | 33 |
| −1 | 0 | 24 | 17 | 21 | 3 | 0 | 22 | 6 | 26 |
| 1 | 0 | 2 | 25 | −16 | 3 | 0 | 24 | 21 | 19 |
| 1 | 0 | 4 | 62 | −38 | 4 | 0 | 2 | 42 | 72 |
| 1 | 0 | 6 | 57 | −55 | 4 | 0 | 4 | 50 | 69 |
| 1 | 0 | 8 | 53 | −66 | 4 | 0 | 6 | 22 | 62 |
| 1 | 0 | 10 | 76 | −71 | 4 | 0 | 8 | 46 | 53 |
| 1 | 0 | 12 | 28 | −71 | 4 | 0 | 10 | 42 | 42 |
| 1 | 0 | 14 | 67 | −67 | 4 | 0 | 12 | 44 | 31 |
| 1 | 0 | 16 | 89 | −59 | 4 | 0 | 14 | 3 | 20 |
| 1 | 0 | 18 | 58 | −51 | 4 | 0 | 16 | 13 | 11 |
| 1 | 0 | 20 | 67 | −41 | 4 | 0 | 18 | 4 | 3* |
| 1 | 0 | 22 | 21 | −31 | 4 | 0 | 22 | 13 | −6* |
| 1 | 0 | 24 | 23 | −22 | 4 | 0 | 24 | 10 | −8* |
| 2 | 0 | 2 | 51 | −112 | 5 | 0 | 2 | 11 | 7* |
| 2 | 0 | 4 | 60 | −104 | 5 | 0 | 4 | 6 | −3* |
| 2 | 0 | 6 | 76 | −90 | 5 | 0 | 6 | 11 | −12 |
| 2 | 0 | 8 | 60 | −73 | 5 | 0 | 8 | 17 | −20 |
| 2 | 0 | 10 | 80 | −55 | 5 | 0 | 12 | 29 | −28 |
| 2 | 0 | 12 | 41 | −38 | 5 | 0 | 14 | 26 | −28 |
| 2 | 0 | 14 | 6 | −22 | 5 | 0 | 16 | 13 | −27 |
| 2 | 0 | 16 | 19 | −9* | 5 | 0 | 18 | 18 | −25 |
| 2 | 0 | 20 | 9 | 8* | 5 | 0 | 20 | 21 | −21 |
| 2 | 0 | 22 | 47 | 12 | 5 | 0 | 22 | 6 | −17 |
| 2 | 0 | 24 | 16 | 14 | 6 | 0 | 2 | 49 | −39 |
| 3 | 0 | 2 | 3 | 0* | 6 | 0 | 4 | 9 | −39 |
| 3 | 0 | 4 | 30 | 16 | 6 | 0 | 6 | 14 | −36 |
| 3 | 0 | 6 | 20 | 30 | 6 | 0 | 8 | 28 | −32 |
| 3 | 0 | 8 | 35 | 41 | 6 | 0 | 10 | 17 | −26 |
| 3 | 0 | 10 | 30 | 47 | 6 | 0 | 12 | 23 | −20 |
| 3 | 0 | 12 | 29 | 50 | 6 | 0 | 14 | 9 | −14 |
| 3 | 0 | 14 | 63 | 49 | 6 | 0 | 18 | 3 | −4* |
| 3 | 0 | 16 | 50 | 45 | | | | | |

[a] $F_{Br}$ is the calculated contribution of the bromine atoms to $F(h0l)$, assuming a $B$ value of 4.1 Å$^2$. Values marked with asterisks have small amplitudes and their signs may not be correct for the corresponding $F(h0l)$ reflections.

The contribution of the bromine atoms $F_{Br}$ to the $h0l$ reflections is shown in Table 8.2. Values marked with an asterisk have small amplitudes, and their signs may not be correct for the corresponding $F(h0l)$ reflections. The value of the $R$-factor at this stage was 0.35, and there is good agreement in the pattern of $|F(h0l)|$ and $F_{Br}$ in Table 8.2.

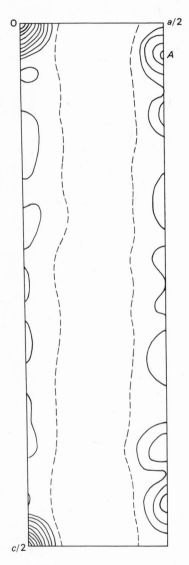

FIGURE 8.11. Asymmetric unit of
$P(u, w)$.

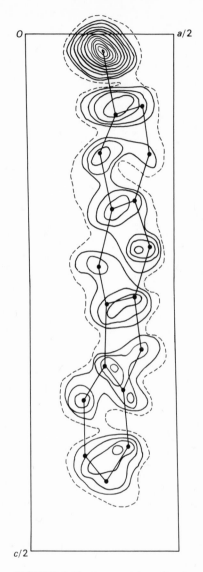

FIGURE 8.12. Asymmetric unit of
$\rho(x, z)$ phased on the bromine atoms;
the probable atomic positions are
marked in.

An electron density map, in this projection, was calculated using the signs given by $F_{Br}$ with the experimental values of $|F(h0l)|$, provided that $F_{Br}$ was not asterisked. Figure 8.12 shows the electron density map with the molecule, fitted with the aid of a model, marked in. The resolution is not yet good, but we can see that we are working along the right lines. Evidently, the molecule is inclined to the plane of this projection, and there will be a limit to the possible improvement of the resolution attainable in projection. Consequently, we shall proceed directly with three-dimensional studies.

### 8.2.4 Three-Dimensional Structure Determination

In order to obtain better resolution in the electron density map and to gain information on all spatial coordinates, we proceeded to a three-dimensional study by calculating a general Patterson map $P(u, v, w)$ section by section normal to the $b$ axis.

The coordinates of the general positions show that the Br—Br vectors in the asymmetric unit will be found at $2x, 2y, 2z$ (single-weight peak, $B$), 0, $\frac{1}{2} - 2y, \frac{1}{2}$ (double-weight peak, $C$), and $2\bar{x}, \frac{1}{2}, \frac{1}{2} + 2z$ (double-weight peak, $D$).

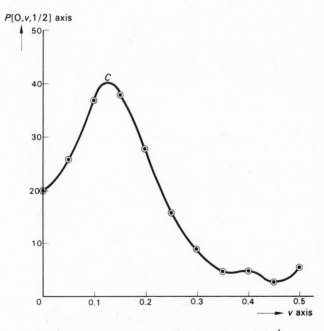

FIGURE 8.13. Patterson function along the line $[0, v, \frac{1}{2}]$.

Consequently, we must study the Patterson map carefully, particularly the line $[0, v, \frac{1}{2}]$ and the section $(u, \frac{1}{2}, w)$. Figures 8.13 and 8.14 show these two regions of Patterson space, and Figure 8.15 illustrates the general section containing the single-weight peak $B$. From the peaks $B$, $C$, and $D$, the coordinates for the bromine atom in the asymmetric unit were found to be

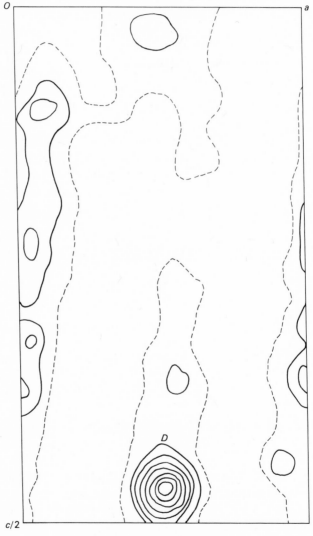

FIGURE 8.14.  Patterson section $(u, \frac{1}{2}, w)$.

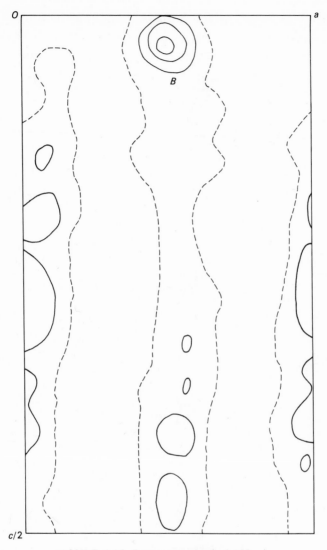

FIGURE 8.15. Patterson section $(u, 0.375, w)$.

0.243, 0.188, 0.030. Repeating the phasing procedure, but now for $hkl$ reflections, and calculating a three-dimensional electron density map produced a good resolution of the complete structure, with the exception of the hydrogen atoms. Figure 8.16 illustrates a composite electron density map, which consists of superposed sections calculated at intervals along $b$.

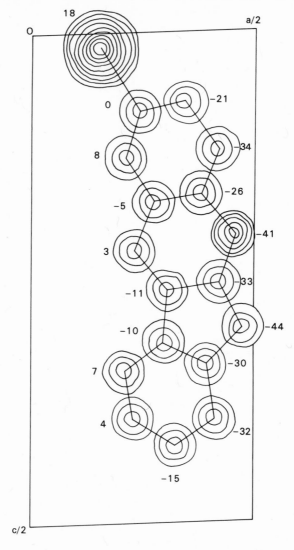

FIGURE 8.16. Composite three-dimensional electron density map with the molecule (excluding H atoms) marked in. The contour of zero electron density is not shown, and the numbers represent $100y$ for each atom.

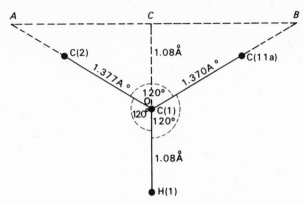

FIGURE 8.17. Calculation of the coordinates of H(1); the configuration of atoms is planar.

The scattering of X-rays by hydrogen is small in magnitude, and these atoms are not resolved by the direct summation of the electron density. Since the molecule is planar, within experimental error, and the rings are aromatic in character, we can calculate the most probable positions for the hydrogen atoms from the geometry of the structure, instead of computing a difference-Fourier synthesis. The principle of a simple calculation is shown by Figure 8.17.

Consider the hydrogen atom H(1) attached to C(1), and assume a C—H bond length* of 1.08 Å, with $sp^2$ geometry at C(1). This atom is transferred, temporarily, to the origin, and the transformed fractional coordinates converted to absolute values in Å. Next, the atoms are, for convenience, referred to orthogonal axes (Figure 8.18). Triangle $OAB$ is isosceles, with $OC = C(1)—H(1)$, and the coordinates of point $C$ are the averages of those at $A$ and $B$. The coordinates of H(1) are now obtained by inverting those of $C$ through $O$ and then transforming back to fractional values in the monoclinic unit cell. In this structure, H(1) has the coordinates 0.809, 0.761, 0.375, which values the reader may like to confirm.

Finally, we arrive at the complete structure of BBIP, shown in Figure 8.19 with a convenient atom numbering scheme.

### 8.2.5 Refinement

During the refinement of the structure, the hydrogen atoms were included, at their calculated positions, in the evaluation of the structure

---

* Idealized value (see page 295), which corresponds to the internuclear distance. The usual experimental X-ray value is closer to 1 Å.

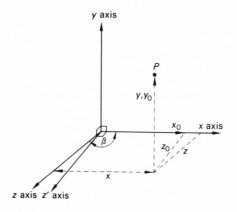

FIGURE 8.18. Orthogonal coordinates in the monoclinic system. The point $P$ has coordinates $X, Y, Z$ (Å) on the monoclinic and $X_0, Y_0, Z_0$ on orthogonal axes, where $X_0 = X + Z \cos \beta$, $Y_0 = Y$, and $Z_0 = Z \sin \beta$.

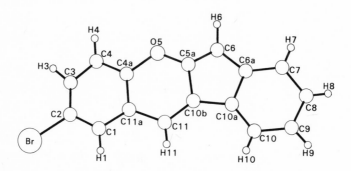

FIGURE 8.19.  Structural formula for BBIP.

factors $F_c$, but no attempt was made to refine the parameters of the hydrogen atoms because the main interest in the problem lay in determining the molecular conformation. The final adjustments of the structural parameters of the Br, O, and C atoms ($x$, $y$, $z$, and anisotropic temperature factors) and the scale factor were carried out by the method of least squares (page 288). The refinement converged with an $R$-factor of 0.070, and a final difference-Fourier synthesis showed no fluctuations in density greater than about 2.5 times $\sigma(\rho_o)$. The analysis was considered to be satisfactory, and the refinement was terminated at this stage.

## 8.2.6 Molecular Geometry

It remained to determine the bond lengths, bond angles, and other features of the geometry of the molecule and its relationship with other molecules in the unit cell.

From the coordinates of the atomic positions (Table 8.3) and using equations such as (7.41) and (7.43), bond lengths and angles were calculated. They are shown on the drawings of the molecule in Figures 8.20 and 8.21. Figure 8.22 illustrates the packing of the molecules in the unit cell, as seen along the $b$ axis; the average intermolecular contact distance is 3.7 Å,

TABLE 8.3. Fractional Atomic Coordinates in BBIP, with esd's in Parentheses[a]

| Atom | $x$ | $y$ | $z$ |
|------|------|------|------|
| C(1) | 0.7820(16) | 0.5813(22) | 0.3789(4) |
| C(2) | 0.7310(16) | 0.5049(24) | 0.4252(4) |
| C(3) | 0.6524(16) | 0.2925(25) | 0.4297(5) |
| C(4) | 0.6214(16) | 0.1587(23) | 0.3871(5) |
| C(4a) | 0.6794(16) | 0.2381(22) | 0.3406(4) |
| O(5) | 0.6520(11) | 0.0949(14) | 0.2995(3) |
| C(5a) | 0.6973(14) | 0.1671(21) | 0.2526(4) |
| C(6) | 0.6714(14) | 0.0603(19) | 0.2077(5) |
| C(6a) | 0.7384(15) | 0.2010(19) | 0.1678(4) |
| C(7) | 0.7401(17) | 0.1770(24) | 0.1150(4) |
| C(8) | 0.8078(18) | 0.3474(24) | 0.0858(4) |
| C(9) | 0.8766(17) | 0.5426(24) | 0.1079(5) |
| C(10) | 0.8731(16) | 0.5732(21) | 0.1605(4) |
| C(10a) | 0.8035(16) | 0.4046(20) | 0.1908(4) |
| C(10b) | 0.7767(14) | 0.3924(21) | 0.2454(5) |
| C(11) | 0.8064(15) | 0.5266(21) | 0.2850(4) |
| C(11a) | 0.7593(14) | 0.4525(20) | 0.3359(5) |
| Br | 0.7602(2) | 0.6848(3) | 0.4848(0) |
| H(1) | 0.809 | 0.761 | 0.375 |
| H(3) | 0.622 | 0.211 | 0.460 |
| H(4) | 0.565 | 0.001 | 0.389 |
| H(6) | 0.630 | 0.879 | 0.208 |
| H(7) | 0.674 | 0.056 | 0.097 |
| H(8) | 0.804 | 0.333 | 0.043 |
| H(9) | 0.886 | 0.639 | 0.076 |
| H(10) | 0.809 | 0.747 | 0.375 |
| H(11) | 0.870 | 0.689 | 0.285 |

[a] There are no esd's for the hydrogen atom coordinates because these parameters were not included in the least-squares refinement.

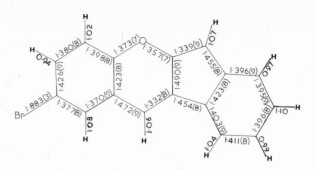

FIGURE 8.20. Bond lengths in BBIP, with their estimated standard deviations in parentheses.

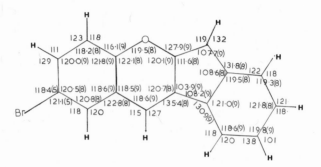

FIGURE 8.21. Bond angles in BBIP, with their estimated standard deviations in parentheses.

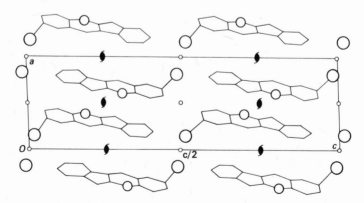

FIGURE 8.22. Molecular packing in the unit cell of BBIP: ◯, bromine; ○, oxygen.

which is typical of the van der Waals forces which link molecules in this type of compound.

## 8.3 Crystal Structure of Potassium 2-Hydroxy-3,4-dioxocyclobut-1-ene-1-olate Monohydrate (KHSQ)*

1,2-Dihydroxy-3,4-dioxocyclobut-1-ene (IV) may be prepared by acid-catalyzed hydrolysis of 1,2-diethoxy-3,3,4,4-tetrafluorocyclobut-1-ene (V). On recrystallization from water, it has a melting point of 293°C, at which temperature it decomposes.

It has been called by the trivial name, squaric acid; the hydrogen atoms in the hydroxyl groups are acidic, and can be replaced by a metal. The potassium hydrogen salt monohydrate (VI), which is the subject of this example, can be obtained by mixing hot, concentrated, equimolar, aqueous solutions of potassium hydroxide and squaric acid and then cooling the reaction mixture.

### 8.3.1 Preliminary X-Ray and Physical Measurements

The compound was recrystallized from water as colorless, prismatic crystals with the forms {001}, {110}, and {100} predominant (Figure 8.23). Under a polarizing microscope, straight extinction was observed on {001} and {100}, and an extinction angle of about 2° was obtained on a section cut normal to $b$. These results suggest strongly that the crystals belong to the monoclinic system.

The crystal specimen chosen for X-ray work had the dimensions 0.5, 0.5, and 0.3 mm parallel to $a$, $b$, and $c$, respectively. The details of the preliminary measurements are similar to those described for the previous example, and we list the crystal data immediately (Table 8.4). Copper $K\alpha$ radiation ($\bar{\lambda} = 1.5418$ Å) was used throughout this work.

* R. J. Bull et al., Crystal Structure Communications 2, 625 (1973).

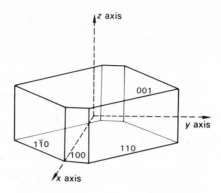

FIGURE 8.23. Crystal habit of KHSQ
with the crystallographic axes drawn in.

TABLE 8.4.  Crystal Data for KHSQ at 20°C

| | |
|---|---|
| Molecular formula | $C_4HO_4^-, K^+, H_2O$ |
| Molecular weight | 170.17 |
| Space group | $P2_1/c$ |
| $a$, Å | 8.641(1) |
| $b$, Å | 10.909(1) |
| $c$, Å | 6.563(2) |
| $\beta$, deg | 99.81(1) |
| Unit-cell volume, Å$^3$ | 609.6(2) |
| $D_m$, g cm$^{-3}$ | 1.839(7) |
| $D_c$, g cm$^{-3}$ | 1.854(1) |
| $Z$ | 4 |
| $F(000)$ | 296 |

## 8.3.2  Intensity Measurement and Correction

About 900 symmetry-independent intensities with $(\sin \theta)/\lambda \leqslant$ 0.57 Å$^{-1}$ were measured. Corrections were applied for polarization and Lorentz effects, but not for absorption. Scale $(K)$ and isotropic temperature $(B)$ factors were deduced by Wilson's method (page 245) and the $|F_o|$ data were converted to $|E|$ values* through (7.1).

In using direct methods, a structure analysis often begins with those reflections for which $|E|$ is greater than some value $E_{min}$, which, typically, is chosen to be about 1.5. For a centrosymmetric crystal, these reflections usually number about 12–15% of the total data. There were 142 of such

---

* Although KHSQ provides a good example of a structure analysis by direct methods, it could have been solved by the heavy-atom technique.

TABLE 8.5. $|E|$ Statistics in KHSQ

|  | Acentric | Centric | This structure |
|---|---|---|---|
| $\overline{\|E\|^2}$ | 1.00 | 1.00 | 1.03 |
| $\overline{\|E\|}$ | 0.89 | 0.80 | 0.82 |
| $\overline{\|E\|^2 - 1}$ | 0.74 | 0.97 | 0.96 |
| % $\geqslant 1.5$ | 10.5 | 13.4 | 14.6 |
| % $\geqslant 1.75$ | 4.7 | 8.0 | 8.4 |
| % $\geqslant 2.0$ | 1.8 | 4.6 | 4.9 |
| % $\geqslant 2.5$ | 0.2 | 1.2 | 1.1 |

reflections for KHSQ, representing 15.8% of the experimental reflection data, and their statistics are shown in Table 8.5. The agreement with the theoretical values for a centric distribution of $|E|$ values is very close.

### 8.3.3 $\Sigma_2$ Listing

The next stage in a direct attack of a structure analysis is the preparation of a $\Sigma_2$ listing (page 281). Symmetry-related reflections now become very important in generating triplet relationships: 300 and $30\bar{4}$ can lead to both 004 and $60\bar{4}$, the latter by replacing $30\bar{4}$ by $\bar{3}04$, taking note of the phase symmetry. We recall the relevant phase symmetry for space group $P2_1/c$ (see page 281), which may be summarized as follows:

$$s(hkl) = s(\bar{h}\bar{k}\bar{l}) \tag{8.2}$$
$$s(hkl) = s(h\bar{k}l)(-1)^{k+l} \tag{8.3}$$

A portion of the $\Sigma_2$ listing is shown in Table 8.6. The numbers in parentheses under each **h** are the total numbers of $\Sigma_2$ triplets for each of the reflections* **h**, and **k**, and **h** − **k** represent vectors forming a triplet with **h**.

### 8.3.4 Specifying the Origin

Following the procedure described on pages 276–280 and using the reflections in Table 8.6, three reflections were chosen and allocated positive signs, in order to fix the origin at 0, 0, 0. The symmetry relationships in the space group of this compound allowed, in all, 12 signs in the origin set (Table 8.7). The reader should check the signs, starting from the first one in each group of four, with equations (8.2) and (8.3).

---

* We use now the convenient notation **h** for the reflection $hkl$, **k** for $h'k'l'$, and **h** − **k** for $h - h'$, $k - k'$, $l - l'$.

TABLE 8.6.  Part of the $\Sigma_2$ Listing for KHSQ

| k | $\lvert E(k) \rvert$ | h−k | $\lvert E(h-k) \rvert$ | h | $\lvert E(h) \rvert$ | $\lvert E(h) \rvert \lvert E(k) \rvert \lvert E(h-k) \rvert$ |
|---|---|---|---|---|---|---|
| 53$\bar{1}$ | 2.6 | 010,4 | 2.8 | 573 | 2.6 | 18.9 |
| (37) |  | 041 | 2.2 | 57$\bar{2}$ | 3.3 | 17.2 |
|  |  | 041 | 2.0 | 570 | 2.7 | 14.0 |
|  |  | 114 | 2.3 | 62$\bar{5}$ | 1.7 | 10.2 |
|  |  | 032 | 1.7 | 56$\bar{3}$ | 2.0 | 8.8 |
| 11$\bar{4}$ | 2.3 | 57$\bar{2}$ | 3.3 | 482 | 1.9 | 14.4 |
| (45) |  | 66$\bar{4}$ | 1.8 | 570 | 2.7 | 11.2 |
|  |  | 68$\bar{1}$ | 1.5 | 573 | 2.6 | 9.0 |
|  |  | 56$\bar{3}$ | 2.0 | 451 | 1.5 | 6.9 |
|  |  | 454 | 1.6 | 540 | 1.5 | 5.5 |
| 032 | 1.7 | 53$\bar{1}$ | 2.6 | 56$\bar{3}$ | 2.0 | 8.8 |
| (54) |  | 57$\bar{2}$ | 3.3 | 540 | 1.5 | 8.4 |
|  |  | 482 | 1.9 | 454 | 1.6 | 5.2 |
|  |  | 451 | 1.5 | 48$\bar{1}$ | 2.0 | 5.1 |
| 11$\bar{2}$ | 2.5 | 57$\bar{2}$ | 3.3 | 66$\bar{4}$ | 1.8 | 14.9 |
| (39) |  | 482 | 1.9 | 570 | 2.7 | 12.8 |
|  |  | 11$\bar{4}$ | 2.3 | 002 | 1.9 | 10.2 |
|  |  | 571 | 1.7 | 68$\bar{1}$ | 1.5 | 6.4 |
| 010,4 | 2.8 | 33$\bar{2}$ | 2.2 | 372 | 1.9 | 11.7 |
| (35) |  | 62$\bar{5}$ | 1.7 | 68$\bar{1}$ | 1.5 | 7.1 |
| 33$\bar{2}$ | 2.2 | 33$\bar{2}$ | 2.2 | 66$\bar{4}$ | 1.8 | 8.7 |
| (46) |  | 11$\bar{4}$ | 2.3 | 242 | 1.7 | 8.6 |
|  |  | 31$\bar{3}$ | 1.8 | 041 | 2.0 | 7.9 |
|  |  | 625 | 1.7 | 313 | 1.8 | 6.7 |
| 002 | 1.9 | 041 | 2.0 | 041 | 2.0 | 7.4 |
| (25) |  | 11$\bar{4}$ | 2.3 | 11$\bar{6}$ | 1.5 | 6.6 |
|  |  | 68$\bar{1}$ | 1.5 | 681 | 1.6 | 4.6 |

### 8.3.5   Sign Determination

The $\Sigma_2$ listing is examined with a view to generating new signs, using (7.13), which may be given in relation to Table 8.6 by

$$s[E(h)] \approx s\left[ \sum_k E(k)E(h-k) \right] \qquad (8.4)$$

where the sum is taken over the several k triplets involved with h. The probability of (8.4) is given by (7.14). If only a single $\Sigma_2$ interaction is considered, (8.4) becomes

$$s[E(h)] \approx s[E(k)]s[E(h-k)] \qquad (8.5)$$

TABLE 8.7.  Origin-Fixing Reflections
and Their Symmetry Equivalents

| $hkl$ | Sign | $|E|$ | No. of $\Sigma_2$ triplets |
|---|---|---|---|
| $53\bar{1}$ | + | 2.6 | 37 |
| $\bar{5}31$ | + | | |
| $5\bar{3}\bar{1}$ | + | | |
| $\bar{5}31$ | + | | |
| | | | |
| $11\bar{4}$ | + | 2.3 | 45 |
| $\bar{1}\bar{1}4$ | + | | |
| $1\bar{1}\bar{4}$ | − | | |
| $\bar{1}14$ | − | | |
| | | | |
| $032$ | + | 1.7 | 54 |
| $0\bar{3}2$ | + | | |
| $0\bar{3}2$ | − | | |
| $03\bar{2}$ | − | | |

TABLE 8.8.  Sign Determination.
Starting from the "Origin Set"

| $\mathbf{k}$ | $\mathbf{h} - \mathbf{k}$ | $\mathbf{h}$ | Indication for $s\,(\mathbf{h})$ |
|---|---|---|---|
| $53\bar{1}+$ | $1\bar{1}\bar{4}-$ | $62\bar{5}$ | − |
| $53\bar{1}+$ | $0\bar{3}\bar{2}-$ | $56\bar{3}$ | − |
| $56\bar{3}-$ | $\bar{1}\bar{1}4+$ | $451$ | − |
| $56\bar{3}-$ | $032+$ | $59\bar{1}$ | − |
| $451-$ | $0\bar{3}\bar{2}-$ | $48\bar{1}$ | + |

[which is equivalent to (7.11)], with a corresponding change in the equation for $P_+(\mathbf{h})$, (7.14).

In our discussion, we shall assume that the values of $P_+(\mathbf{h})$ are sufficiently high for the sign to be accepted as correct; very small or zero values of $P_+(\mathbf{h})$ indicate strongly a negative sign for $\mathbf{h}$. Some examples of the application of (8.5) are given in Table 8.8.*

Use of Sign Symbols

The above process of sign determination was carried out as far as possible throughout the entire $\Sigma_2$ listing, which, although it contained 1276

* It does not matter which reflections in a triplet are labeled $\mathbf{h}$, $\mathbf{k}$.

TABLE 8.9.  Symbolic Signs

| **h** | Sign | $|E|$ | No. of $\Sigma_2$ relationships |
|-------|------|-------|-------------------------------|
| $11\bar{2}$ | $A$ | 2.5 | 39 |
| $\bar{1}\bar{1}2$ | $A$ | | |
| $1\bar{1}\bar{2}$ | $-A$ | | |
| $\bar{1}12$ | $-A$ | | |
| $010,4$ | $B$ | 2.8 | 35 |
| $0\bar{1}0,\bar{4}$ | $B$ | | |
| $0\bar{1}0,4$ | $B$ | | |
| $010,\bar{4}$ | $B$ | | |
| $33\bar{2}$ | $C$ | 2.2 | 46 |
| $\bar{3}\bar{3}2$ | $C$ | | |
| $3\bar{3}\bar{2}$ | $-C$ | | |
| $\bar{3}32$ | $-C$ | | |

triple products, was exhausted after only 24 signs had been found. To enable further progress to be made, three reflections were assigned the symbols $A$, $B$, and $C$, where each symbol represented either a plus or a minus sign. Twelve symbolic signs (Table 8.9) were added to the set, and the sign determination was continued, now in terms of both signs and symbols. It may be noted that although the symbols are given to reflections with large $|E|$ values and large numbers of $\Sigma_2$ interactions, there are not, necessarily, any restrictions on either parity groups or the use of structure invariants.

Some examples of this stage of the process are given in Table 8.10. The values of **h** and **k** are taken from either Tables 8.7 and 8.9, which constitute the "starting set," or as determined through (8.4). The reader is invited to follow through the stages in Table 8.10, working out the correct symmetry-equivalent signs from (8.2) and (8.3) as necessary.

From Table 8.10, we see that six more reflections have been allocated signs, and another 17 are determined in terms of $A$, $B$, and $C$. Multiple indications can now be seen. For example, there are two indications that $s(573) = B$, two indications that $s(570) = -$, and two indications that $s(540) = A$. Three indications for 041 suggest that both $s(041) = -$ and $A = -$.

Continuing in this manner, it was found possible to allot signs and symbols to all 142 $|E|$ values greater than 1.5. The symbols $A$, $B$, and $C$

TABLE 8.10. Further Sign Determinations[a]

| k | s(k) | h−k | s(h−k) | h | Sign indication, s(h) |
|------|------|------|--------|------|------------------------|
| 010,4 | $B$ | $62\bar{5}$ | − | $68\bar{1}$ | $s(68\bar{1}) = B$ |
| 010,4 | $B$ | $33\bar{2}$ | $C$ | 372 | $s(372) = -BC$ |
| $11\bar{2}$ | $A$ | $68\bar{1}$ | $B$ | 571 | $s(571) = AB$ |
| $53\bar{1}$ | + | 010,4 | $B$ | 573 | $s(573) = B$ |
| $11\bar{4}$ | + | $68\bar{1}$ | $B$ | 573 | $s(573) = B$ |
| $33\bar{2}$ | $C$ | $33\bar{2}$ | $C$ | $66\bar{4}$ | $s(66\bar{4}) = CC = +$ |
| $11\bar{4}$ | + | $66\bar{4}$ | + | 570 | $s(570) = -$ |
| $11\bar{2}$ | $A$ | 570 | − | 482 | $s(482) = A$ |
| $11\bar{2}$ | $A$ | $66\bar{4}$ | + | $57\bar{2}$ | $s(57\bar{2}) = -A$ |
| $11\bar{4}$ | + | $57\bar{2}$ | $-A$ | 482 | $s(482) = A$ |
| 032 | + | $57\bar{2}$ | $-A$ | 540 | $s(540) = A$ |
| 032 | + | 482 | $A$ | 454 | $s(454) = -A$ |
| $11\bar{4}$ | + | 454 | $-A$ | 540 | $s(540) = A$ |
| $33\bar{2}$ | $C$ | $11\bar{4}$ | + | 242 | $s(242) = -C$ |
| $11\bar{2}$ | $A$ | $11\bar{4}$ | + | 002 | $s(002) = A$ |
| $11\bar{2}$ | $A$ | 482 | $A$ | 570 | $s(570) = -AA = -$ |
| 570 | − | $53\bar{1}$ | + | 041 | $s(041) = -$ |
| $62\bar{5}$ | − | $33\bar{2}$ | $C$ | $31\bar{3}$ | $s(31\bar{3}) = C$ |
| $31\bar{3}$ | $C$ | $33\bar{2}$ | $C$ | 041 | $s(041) = -CC = -$ |
| $53\bar{1}$ | + | 041 | − | $57\bar{2}$ | $s(57\bar{2}) = +$ |
| 002 | $A$ | 041 | − | 041 | $s(041) = A$ |
| 002 | $A$ | $68\bar{1}$ | $B$ | 681 | $s(681) = AB$ |
| 002 | $A$ | $11\bar{4}$ | + | $11\bar{6}$ | $s(11\bar{6}) = A$ |

[a] Symmetry relations should be employed as necessary.

were involved in 65, 72, and 55 relationships, respectively. Consistent indications, such as those mentioned above for $s(041)$, led finally to the sign relationships $A = AC = B = -$, from which it follows that $C = +$.

### 8.3.6 The E Map

The signs of the 142 $|E|$ values used in this procedure were obtained with a high probability, and, assuming their correctness, an electron density map was computed using the signed $|E|$ values as coefficients. The sections of this map $\rho(x, y, z)$ at $z = 0.15$, 0.20, 0.25, and 0.30 are shown in Figures 8.24–8.27. They reveal the $K^+$ ion and the $C_4O_4^-$ ring system clearly; the oxygen atom of the water molecule was not indicated convincingly at this stage of the analysis. A tilt of the plane of the molecule with respect to (001) can be inferred from Figures 8.25–8.27. Some spurious peaks, marked $S$,

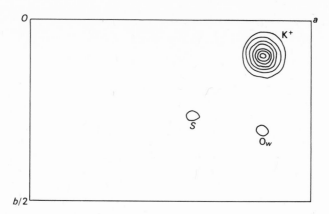

FIGURE 8.24.  $E$ map at $z = 0.15$.

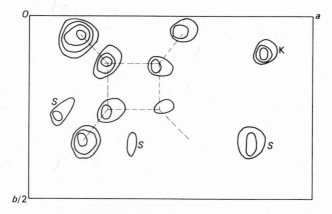

FIGURE 8.25.  $E$ map at $z = 0.20$.

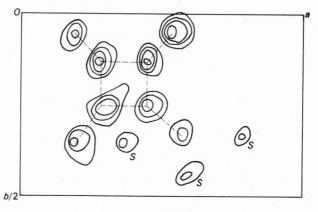

FIGURE 8.26.  $E$ map at $z = 0.25$.

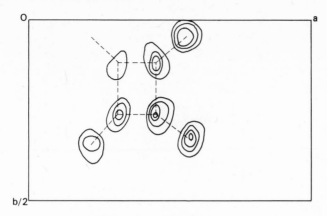

FIGURE 8.27. $E$ map at $z = 0.30$.

may be seen. This is a common feature in $E$ maps. We must remember that a limited data set (142 out of 900) is being used, and that the $|E|$ values are sharpened coefficients corresponding to an approximate point-atom model. The data set is therefore terminated while the coefficients for the Fourier series are relatively large, a procedure which can lead to spurious maxima (page 205). However, such peaks are often of smaller weight than those that correspond to atomic positions.

### 8.3.7 Completion and Refinement of the Structure

Sometimes, all atomic positions are not contained among the peaks in an $E$ map. Those peaks that do correspond to atomic positions may be used to form a trial structure for calculation of structure factors and an $|F_o|$ electron density map (page 250). A certain amount of subjective judgement may be required to decide upon the best peaks for the trial structure at such a stage.

This situation was obtained for KHSQ, although it was not difficult to pick out a good trial structure. The coordinates were obtained for all nonhydrogen atoms except the oxygen atom of the water molecule. The $R$-factor for this trial structure was 0.30, and the composite three-dimensional electron density map obtained is shown in Figure 8.28, which now reveals $O_w$. It may be noted in passing that the small peak labeled $O_w$ in Figure 8.24 corresponds to the position of this atom, but this fact could not be determined at that stage of the analysis.

Further refinement was carried out by the method of least squares, and an $R$-factor of 0.078 was obtained. Figure 8.29 shows a composite three-

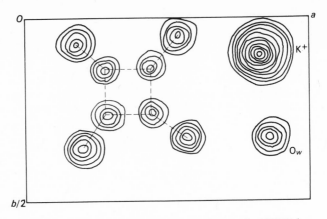

FIGURE 8.28. Composite electron density map for KHSQ (excluding H atoms); the atomic coordinates are listed in Table 8.11.

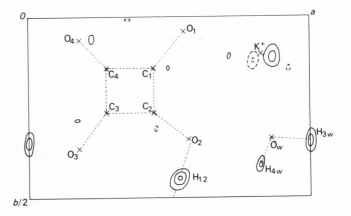

FIGURE 8.29. Composite difference electron density map for KHSQ: Positive contours are solid lines and negative contours are broken lines. Bonds in the squarate ring and those involving hydrogen atoms are shown in dotted lines. Some spurious, small peaks (unlabeled) are shown by this synthesis.

dimensional difference-Fourier map (page 252) of KHSQ. Peaks numerically greater than 0.5 electron per $\text{Å}^3$, representing about 2.5 times the standard deviation in the electron density, are significant, and have been contoured. Some of these peaks indicate areas of small disagreement between the true structure and the model. Three positive peaks, however, are in positions expected for hydrogen atoms. Inclusion of these atoms in the structure factor calculations in the final cycles of least-squares refinement

TABLE 8.11. Fractional Atomic Coordinates for KHSQ

|          | $x$        | $y$        | $z$         |
|----------|------------|------------|-------------|
| $K^+$    | 0.8249(2)  | 0.1040(2)  | 0.1295(3)   |
| C(1)     | 0.4353(9)  | 0.1295(7)  | 0.2572(12)  |
| C(2)     | 0.4495(9)  | 0.2597(7)  | 0.2714(12)  |
| C(3)     | 0.2795(9)  | 0.2714(8)  | 0.2462(11)  |
| C(4)     | 0.2659(9)  | 0.1345(7)  | 0.2305(12)  |
| O(1)     | 0.5399(6)  | 0.0450(5)  | 0.2649(10)  |
| O(2)     | 0.5649(6)  | 0.3346(5)  | 0.2920(10)  |
| O(3)     | 0.1874(7)  | 0.3582(6)  | 0.2386(10)  |
| O(4)     | 0.1578(6)  | 0.0605(5)  | 0.2022(10)  |
| $O_w$    | 0.8789(7)  | 0.3429(6)  | 0.0424(10)  |
| H(12)    | 0.522      | 0.413      | 0.246       |
| H(3$w$)  | 1.000      | 0.346      | 0.075       |
| H(4$w$)  | 0.826      | 0.400      | 0.100       |

had little effect on the $R$-factor, bringing it to its final value of 0.077. The fractional atomic coordinates for the atoms in the asymmetric unit are listed in Table 8.11.

Interatomic distances and angles are shown in Figure 8.30, and a molecular packing diagram, as seen along $c$, is given in Figure 8.31. From

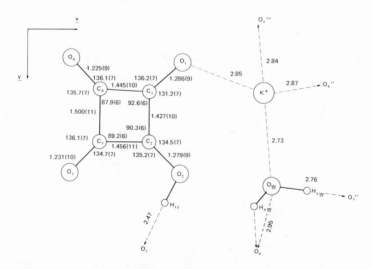

FIGURE 8.30. Bond lengths and bond angles in the crystal-chemical unit of KHSQ; the O—H · · · O distances refer to the overall O · · · O separations. Primes on atom symbols indicate neighboring crystal-chemical units; this diagram should be studied in conjunction with Figure 8.31.

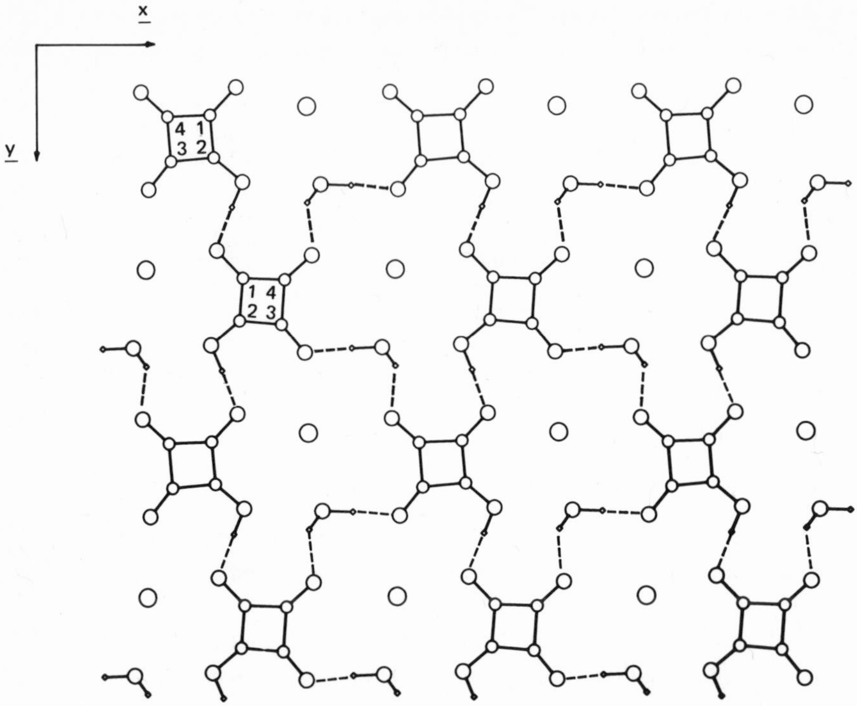

FIGURE 8.31. Molecular packing diagram of one layer as seen along $c$. The circles in order of decreasing size represent K, O, C, and H. The hydrogen-bond network is shown by dashed lines.

the analysis, we find that intermolecular hydrogen bonds exist between O(2) and O(1)′ [2.47(1) Å], between O(3)″ and $O_w$ [2.76(1) Å], and between O(4)′ and $O_w$[2.95(1) Å]; they are responsible for much of the cohesion between molecules in the solid state.* We cannot say at this stage how the negative charge on the HSQ⁻ ion is distributed; a more detailed investigation would be needed in order to answer this question.

## 8.4   Concluding Remarks

No description of crystal structure analysis can be as complete or satisfying as a practical involvement in the subject. In teaching crystallography, extended projects in the analysis of simple structures are becoming an

---

* Single and double primes indicate different neighboring molecules.

important part of the work. The two examples given in this chapter could be followed through in detail with samples of premeasured intensity data* and appropriate computing facilities.

## Bibliography

### Published Structure Analyses

*Acta Crystallographica* (the early issues contain most detail for the beginner).

*Journal of Crystal and Molecular Structure.*

*Zeitschrift für Kristallographie.*

### General Structural Data

WYCKOFF, R. W. G., *Crystal Structures*, Vols. I–VI, New York, Wiley.

KENNARD, O., *et al.* (Editors), *Molecular Structures and Dimensions*, Vols. 1-8, A1, Utrecht, Oosthoek.

## Problems

**8.1.** The unit cell of euphenyl iodoacetate, $C_{32}H_{53}IO_2$, has the dimensions $a = 7.26$, $b = 11.55$, $c = 19.22$ Å, and $\beta = 94.07°$. The space group is $P2_1$ and $Z = 2$. Figure 8P.1 is the Patterson section $(u, \frac{1}{2}, w)$.

(a) Determine the $x$ and $z$ coordinates for the iodine atoms in the unit cell.

(b) Atomic scattering factor data for iodine are tabulated below; temperature factor corrections may be ignored.

| $(\sin \theta)/\lambda$ | 0.00 | 0.05 | 0.10 | 0.15 | 0.20 | 0.25 | 0.30 | 0.35 | 0.40 |
|---|---|---|---|---|---|---|---|---|---|
| $f_I$ | 53.0 | 51.7 | 48.6 | 45.0 | 41.6 | 38.7 | 36.1 | 33.7 | 31.5 |

Determine probable signs for the reflections 001 ($|F_o| = 40$), 0014 ($|F_o| = 37$), 106 ($|F_o| = 33$), and 300 ($|F_o| = 35$). Comment upon the likelihood of the correctness of the signs which you have determined.

---

* Available from one of the authors (M.F.C.L.).

FIGURE 8P.1. Sharpened section, $P(u, \frac{1}{2}, w)$, for euphenyl iodoacetate.

**8.2.** The following two-dimensional $|E|$ values were determined for the [100] zone of a crystal of space group $P2_1/a$. Prepare a $\Sigma_2$ listing, assign an origin, and determine signs for as many reflections as possible, and give reasons for each step which you carry out. In projection, two reflections for which the indices are not both even may be used to specify the origin. It may be assumed that there should be nearly equal numbers of plus and minus signs in the correct data set.

| hkl | $|E|$ | hkl | $|E|$ |
|------|-----|-------|-----|
| 0018 | 2.4 | 0310 | 1.9 |
| 011 | 0.6 | 0312 | 0.1 |
| 021 | 0.1 | 059 | 1.9 |
| 024 | 2.8 | 081 | 2.2 |
| 026 | 0.3 | 0817 | 1.8 |
| 035 | 1.8 | 011,7 | 1.3 |
| 038 | 2.1 | 011,9 | 2.2 |

# Appendix

## A.1 Stereoviews and Crystal Models

### A.1.1 Stereoviews

The representation of crystal and molecular structures by stereoscopic pairs of drawings has become commonplace in recent years. Indeed, some very sophisticated computer programs have been written which draw stereoviews from crystallographic data. Two diagrams of a given object are necessary, and they must correspond to the views seen by the eyes in normal vision. Correct viewing requires that each eye sees only the appropriate drawing, and there are several ways in which it can be accomplished.

   1. A stereoviewer can be purchased for a modest sum from most shops that retail optical instruments or drawing materials. Stereoscopic pairs of drawings may then be viewed directly.

   2. The unaided eyes can be trained to defocus, so that each eye sees only the appropriate diagram. The eyes must be relaxed, and look straight ahead. This process may be aided by placing a white card edgeways between the drawings so as to act as an optical barrier. When viewed correctly, a third (stereoscopic) image is seen in the center of the given two views.

   3. An inexpensive stereoviewer can be constructed with comparative ease. A pair of planoconvex or biconvex lenses each of focal length about 10 cm and diameter 2–3 cm are mounted in a framework of opaque material so that the centers of the lenses are about 60–65 mm apart. The frame must be so shaped that the lenses can be held close to the eyes. Two pieces of cardboard shaped as shown in Figure A.1 and glued together with the lenses in position represents the simplest construction. This basic stereoviewer can be refined in various ways.

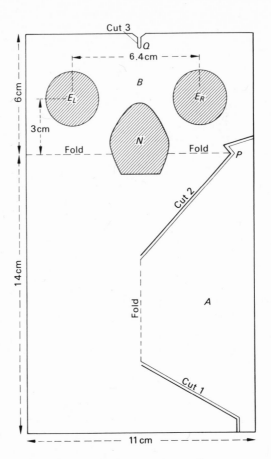

FIGURE A.1. Simple stereoviewer. Cut out two pieces of card as shown and discard the shaded portions. Make cuts along the double lines. Glue the two cards together with the lenses $E_L$ and $E_R$ in position, fold the portions $A$ and $B$ backward, and fix $P$ into the cut at $Q$. View from the side marked $B$. (A similar stereoviewer is marketed by the Taylor–Merchant Corporation, New York.)

## A.1.2 Model of a Tetragonal Crystal

The crystal model illustrated in Figure 1.30 can be constructed easily. This particular model has been chosen because it exhibits a $\bar{4}$ axis, which is one of the more difficult symmetry elements to appreciate from plane drawings.

A good quality paper or thin card should be used for the model. The card should be marked out in accordance with Figure A.2 and then cut out

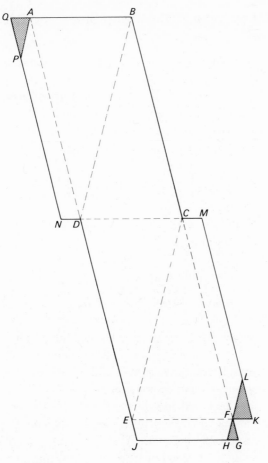

FIGURE A.2. Construction of a tetragonal crystal of point group $\bar{4}2m$:

$NQ = AD = BD = BC = DE = CE = CF = KM$
    $= 10$ cm;

$AB = CD = EF = GJ = 5$ cm;

$AP = PQ = FL = KL = 2$ cm;

$AQ = DN = CM = FK = FG = FH = EJ = 1$ cm.

along the solid lines, discarding the shaded portions. Folds are made in the same sense along all dotted lines, the flaps *ADNP* and *CFLM* are glued internally, and the flap *EFHJ* is glued externally. The resultant model belongs to crystal class $\bar{4}2m$.

## A.2   Crystallographic Point-Group Study and Recognition Scheme

The first step in this scheme is a search for the center of symmetry and mirror plane; they are probably the easiest to recognize. If a model with a center of symmetry is placed on a flat surface, it will have a similar face uppermost and parallel to the supporting surface. For the $m$ plane, a search is made for the left-hand/right-hand relationship in the crystal.

The point groups may be classified into four sections:

(I)   No $m$ and no $\bar{1}$:
1, 2, 222, 3, 32, 4, $\bar{4}$, 422, 6, 622, 23, 432

(II)   $m$ present but no $\bar{1}$:
$m$, $mm2$, $3m$, $4mm$, $\bar{4}2m$, $\bar{6}$, $6mm$, $\bar{6}m2$, $\bar{4}3m$

(III)   $\bar{1}$ present but no $m$:
$\bar{1}$, $\bar{3}$

(IV)   $m$ and $\bar{1}$ both present:

$2/m$, $mmm$, $\bar{3}m$, $4/m$, $\frac{4}{m}mm$, $6/m$, $\frac{6}{m}mm$, $m3$, $m3m$

The further systematic identification is illustrated by means of the block diagram in Figure A.3. Here $R$ refers to the maximum degree of *rotational* symmetry in a crystal, or crystal model, and $N$ is the number of such axes. Questions are given in ovals, point groups in squares, and error paths in diamonds. It may be noted that in sections I, II, and IV, the first three questions (with a small difference in II) are similar. The cubic point groups evolve from question 2 in I, II, and IV.

Readers familiar with computer programming may liken Figure A.3 to a flow diagram. Indeed, this scheme is ideally suited to a computer-aided self-study enhancement of a lecture course on crystal symmetry, and some success with the method has been obtained.*

## A.3   Schoenflies' Symmetry Notation

Theoretical chemists and spectroscopists use the Schoenflies notation for describing point-group symmetry, which is a little unfortunate, because although the crystallographic (Hermann–Mauguin) and Schoenflies notations are adequate for point groups, only the Hermann–Mauguin system is satisfactory for space groups.

* M. F. C. Ladd, *International Journal of Mathematical Education in Science and Technology*, Vol. 7, pp. 395–400 (1976).

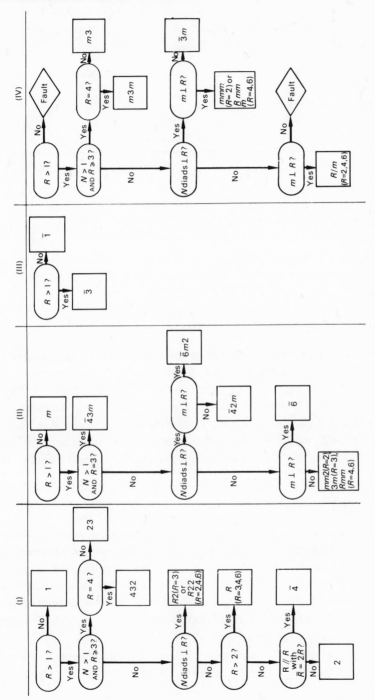

FIGURE A.3. Flow diagram for point-group recognition.

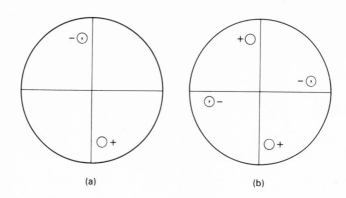

FIGURE A.4. Stereograms of point groups: (a) $S_2$, (b) $S_4$.

The Schoenflies notation uses the rotation axis and mirror plane symmetry elements with which we are now familiar, but introduces the alternating axis of symmetry in place of the inversion axis.

### A.3.1 Alternating Axis of Symmetry

A crystal is said to have an alternating axis of symmetry $S_n$ of degree $n$, if it can be brought into apparent self-coincidence by the combined operation of rotation through $(360/n)$ degrees and reflection across a plane normal to the axis. It must be stressed that this plane is *not* necessarily a mirror plane.* Operations $S_n$ are nonperformable. Figure A.4 shows stereograms of $S_2$ and $S_4$; we recognize them as $\bar{1}$ and $\bar{4}$, respectively. The reader should consider which point groups are obtained if, additionally, the plane of the diagram *is* a mirror plane.

### A.3.2 Notation

Rotation axes are symbolized by $C_n$, where $n$ takes the meaning of $R$ in the Hermann–Mauguin system. Mirror planes are indicated by subscripts $v$, $d$, and $h$; $v$ and $d$ refer to mirror planes containing the principal axis, and $h$ indicates a mirror plane normal to that axis. The symbol $D_n$ is introduced for point groups in which there are $n$ twofold axes in a plane normal to the principal axis of degree $n$. The cubic point groups are represented through

---

* The usual Schoenflies symbol for $\bar{6}$ is $C_{3h}(3/m)$. The reason that $3/m$ is not used in the Hermann–Mauguin system is that point groups containing the element $\bar{6}$ describe crystals that belong to the hexagonal system rather than to the trigonal system; $\bar{6}$ cannot operate on a rhombohedral lattice.

TABLE A.1. Schoenflies and Hermann–Mauguin Point-Group Symbols

| Schoenflies | Hermann–Mauguin[a] | Schoenflies | Hermann–Mauguin[a] |
|---|---|---|---|
| $C_1$ | 1 | $D_4$ | 422 |
| $C_2$ | 2 | $D_6$ | 622 |
| $C_3$ | 3 | $D_{2h}$ | $mmm$ |
| $C_4$ | 4 | $D_{3h}$ | $\bar{6}m2$ |
| $C_6$ | 6 | | |
| $C_i, S_2$ | $\bar{1}$ | $D_{4h}$ | $\dfrac{4}{m}\,mm$ |
| $C_s, S_1$ | $m\ (\bar{2})$ | | |
| $S_6$ | $\bar{3}$ | | |
| $S_4$ | $\bar{4}$ | $D_{6h}$ | $\dfrac{6}{m}\,mm$ |
| $C_{3h}$ | $\bar{6}$ | | |
| $C_{2h}$ | $2/m$ | $D_{2d}$ | $\bar{4}2m$ |
| $C_{4h}$ | $4/m$ | $D_{3d}$ | $\bar{3}m$ |
| $C_{6h}$ | $6/m$ | $T$ | 23 |
| $C_{2v}$ | $mm2$ | $T_h$ | $m3$ |
| $C_{3v}$ | $3m$ | $O$ | 432 |
| $C_{4v}$ | $4mm$ | $T_d$ | $\bar{4}3m$ |
| $C_{6v}$ | $6mm$ | $O_h$ | $m3m$ |
| $D_2$ | 222 | $C_{\infty v}$ | $\infty$ |
| $D_3$ | 32 | $D_{\infty h}$ | $\infty/m\ (\overline{\infty})$ |

[a] $2/m$ is an acceptable way of writing $\frac{2}{m}$, but $4/mmm$ is not as satisfactory as $\frac{4}{m}\,mm$.

the special symbols $T$ and $O$. Table A.1 compares the Schoenflies and Hermann–Mauguin symmetry notations.

## A.4 Generation and Properties of X-Rays

### A.4.1 X-Rays and White Radiation

X-rays are electromagnetic radiations of short wavelength, and are produced by the sudden deceleration of rapidly moving electrons at a target material. If an electron falls through a potential difference of $V$ volts, it acquires an energy of $eV$ electron-volts. If this energy were converted entirely into a quantum $h\nu$ of X-rays, the wavelength $\lambda$ would be given by

$$\lambda = hc/eV \tag{A.1}$$

where $h$ is Planck's constant, $c$ is the velocity of light, and $e$ is the charge on the electron. Substitution of numerical values in (A.1) leads to the equation

$$\lambda = 12.4/V \qquad (A.2)$$

where $V$ is measured in kilovolts (kV).

Generally, an electron does not lose all its energy in this way. It enters into multiple collisions with the atoms of the target material, increasing their vibrations and so generating heat in the target. Thus, (A.2) gives the minimum value of wavelength for a given accelerating voltage. Longer wavelengths are more probable, but very long wavelengths have a small probability and the upper limit is indeterminate. Figure A.5 is a schematic diagram of an X-ray tube, and Figure A.6 shows typical intensity vs. wavelength curves for X-rays. Because of the continuous nature of the spectrum from an X-ray tube, it is often referred to as "white" radiation. The generation of X-rays is a very uneconomical process. Most of the incident electron energy appears as heat in the target, which must be thoroughly water-cooled; about 0.1% of the energy is usefully converted for crystallographic purposes.

## A.4.2  Characteristic X-Rays

If the accelerating voltage applied to an X-ray tube is sufficiently large, the impinging electrons excite inner electrons in the target atoms, which may

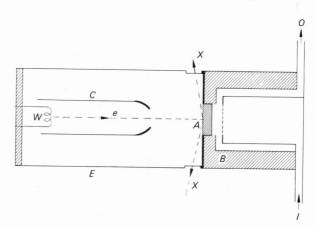

FIGURE A.5. Schematic diagram of an X-ray tube: $W$, heated tungsten filament; $E$, evacuated glass envelope; $C$, accelerating cathode; $e$, electron beam; $A$, target anode; $X$, X-rays (about 6° angle to target surface); $B$, anode supporting block of material of high thermal conductivity; $I$, cooling water in; and $O$, cooling water out.

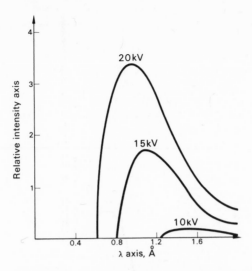

FIGURE A.6. Variation of X-ray intensity with
wavelength $\lambda$.

be expelled from the atoms. Then, other electrons, from higher energy
levels, fall back to the inner levels and their transition is accompanied by the
emission of X-rays. In this case, the X-rays have a wavelength dependent
upon the energies of the two levels involved. If this energy difference is $\Delta E$,
we may write

$$\lambda = hc/\Delta E \qquad (A.3)$$

This wavelength is characteristic of the target material. The white radiation
distribution now has sharp lines of very high intensity superimposed on it
(Figure A.7). In the case of a copper target, very commonly used in X-ray
crystallography, the characteristic spectrum consists of $K\alpha$ ($\lambda = 1.542$ Å)
and $K\beta$ ($\lambda = 1.392$ Å); $K\alpha$ and $K\beta$ are always produced together.

### A.4.3 Absorption of X-Rays

All materials absorb X-rays according to an exponential law:

$$I = I_0 \exp(-\mu t) \qquad (A.4)$$

where $I$ and $I_0$ are, respectively, the transmitted and incident intensities, $\mu$ is
the linear absorption coefficient, and $t$ is the path length through the

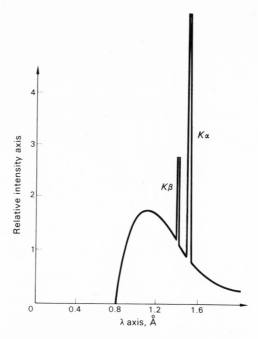

FIGURE A.7. Characteristic $K$ spectrum superposed on the "white" radiation continuum.

material. The absorption of X-rays increases with increase in the atomic
number of the elements in the material.

The variation of $\mu$ with $\lambda$ is represented by the curve of Figure A.8; $\mu$
decreases approximately as $\lambda^3$. At a value which is specific to a given atom in
the material, the absorption rises sharply. This wavelength corresponds to a
resonance level in the atom: A process similar to that involved in the
production of the characteristic X-rays occurs, with the exciting species
being the incident X-rays themselves. The particular wavelength is called
the absorption edge; for metallic nickel it is 1.487 Å.

### A.4.4 Filtered Radiation

If we superimpose Figures A.7 and A.8, we see that the absorption edge
of nickel lies between the $K\alpha$ and $K\beta$ characteristic lines of copper (Figure
A.9). Thus, the effect of passing X-rays from a copper target through a thin
(0.018 mm) nickel foil is that the $K\beta$ radiation is selectively almost completely absorbed. The intensities of both $K\alpha$ and the white radiation are also
reduced, but the overall effect is a spectrum in which the most intense part is

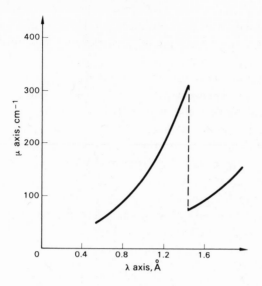

FIGURE A.8.  Variation of $\mu$(Ni) with wavelength
$\lambda$ of X-radiation.

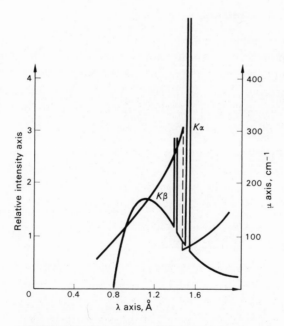

FIGURE A.9.  Superposition of Figures A.7 and A.8 to
show diagrammatically the production of "filtered"
radiation.

the $K\alpha$ line; we speak of filtered radiation, to indicate the production of effectively monochromatic radiation by this process. The copper $K\alpha$ line ($\bar{\lambda} = 1.542$ Å) actually consists of a doublet, $\alpha_1$ ($\lambda = 1.5405$ Å) and $\alpha_2$ ($\lambda = 1.5443$ Å); the doublet is resolved on photographs at high $\theta$ values, but we shall not be concerned here with that feature. The value of 1.542 Å is a weighted mean $(2\alpha_1 + \alpha_2)/3$, the weights being derived from the relative intensities ($2:1$) of the $\alpha_1$ and $\alpha_2$ lines.

The absorption effect is important also in considering the radiation to be used for different materials. We have mentioned that Cu $K\alpha$ is very commonly used, but it would be unsatisfactory for materials containing a high percentage of iron (absorption edge 1.742 Å) since radiation of this wavelength is highly absorbed by iron atoms and re-emitted as characteristic Fe $K$ spectrum. In this case, Mo $K\alpha$ ($\lambda = 0.7107$ Å) is a satisfactory alternative.

## A.5   Crystal Perfection and Intensity Measurement

### A.5.1   Crystal Perfection

In the development of the Bragg equation (3.16), we assumed geometric perfection of the crystal, with all unit cells in the crystal stacked side by side in a completely regular manner. Few, if any, crystals exhibit this high degree of perfection. Figure A.10 shows a family of planes, all in exactly the same orientation with respect to the X-ray beam, at the correct angle for a Bragg reflection. It is clear that the first reflected ray $BC$ is in the correct

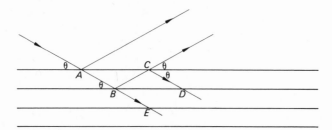

FIGURE A.10. Primary extinction: The phase changes on reflection at $B$ and $C$ are each $\pi/2$, so that between the directions $BE$ and $CD$ there is a total phase difference of $\pi$. Hence, some attenuation of the intensity occurs for the beam incident upon planes deeper in the crystal.

FIGURE A.11. "Mosaic" character in a crystal; the angular misalignment between blocks may vary from 2' to about 30' of arc.

position for a second reflection $CD$, and so on. Since there is a phase change of $\pi/2$ on reflection,* the doubly reflected ray has $\pi$ phase difference with respect to the incident ray $(BE)$. In general, rays reflected $n$ and $n-2$ times differ in phase by $\pi$, and the net result is a reduction in the intensity of the X-ray beam passing through the crystal. This effect is termed primary extinction, and is a feature of geometric perfection of a crystal. In the *ideally perfect* crystal, $I \propto |F|$.

Most crystals, however, are composed of an array of slightly misoriented crystal blocks (mosaic character) (Figure A.11). The ranges of geometric perfection are quite small. Even crystals that show some primary extinction exhibit mosaic character to some degree, and we may write

$$I \propto |F|^m \qquad (A.5)$$

Generally, the mosaic blocks are small, and $m$ is effectively 2.

Another process which leads to attenuation of the X-ray beam by a crystal set at the Bragg angle is known as secondary extinction. It may be encountered in single-crystal X-ray studies, and the magnitude of the effect can be appreciable. Consider a situation in which the first planes encountered by the X-ray beam reflect a high proportion of the incident beam. Parallel planes further in the crystal receive less incident intensity, and, hence, reflect less than might be expected. The effect is most noticeable with large crystals and intense (usually low-order) reflections. Crystals in which the mosaic blocks are highly misaligned have negligible secondary extinction, because only a small number of planes are in the reflecting position at a given time. Such crystals are termed *ideally imperfect*; this condition can be developed by subjecting the crystals to the thermal shock of dipping them in

---

* This $\pi/2$ phase change is usually neglected since it arises for all reflections.

liquid air. The effect of secondary extinction on the intensity of a reflection can be brought into the least-squares refinement (page 289) as an additional variable, the extinction parameter $\zeta$. The quantity minimized in the refinement of the atomic parameters is then

$$\sum_{hkl} w\left(|F_o| - \frac{1}{K\zeta}|F_c|\right)^2 \tag{A.6}$$

## A.5.2 Intensity of Reflected Beam

The real or imperfect crystal will reflect X-rays over a small angular range centered on a Bragg angle $\theta$. We need to determine the total energy of a diffracted beam $\mathscr{E}(hkl)$ as the crystal, which is completely bathed in an X-ray beam of incident intensity $I_0$, passes through the reflecting range.

At a given angle $\theta$, let the power of the reflected beam be $d\mathscr{E}(hkl)/dt$. The greater the value of $I_0$, the greater the power. Hence,

$$d\mathscr{E}(hkl)/dt = R(\theta)I_0 \tag{A.7}$$

where $R(\theta)$ is the reflecting power. Figure A.12 shows a typical curve of $R(\theta)$ against $\theta$. The area under the curve is called the integrated reflection $J(hkl)$:

$$J(hkl) = \int_{-\delta\theta_0}^{\delta\theta_0} R(\theta)\, d\theta \tag{A.8}$$

Using (A.7), we obtain

$$J(hkl) = (1/I_o) \int_{-\delta\theta_0}^{\delta\theta_0} [d\mathscr{E}(hkl)/dt]\, d\theta \tag{A.9}$$

If the crystal is rotating with angular velocity $\omega\ (= d\theta/dt)$,

$$J(hkl) = \omega\mathscr{E}(hkl)/I_0 \tag{A.10}$$

where $\mathscr{E}(hkl)$ is the total energy of the diffracted beam for one pass of the crystal through the reflecting range, $\pm\delta\theta_0$. Since intensity is a measure of

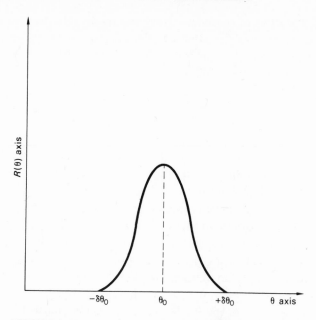

FIGURE A.12. Variation of reflecting power $R(\theta)$ with $\theta$ arising from "mosaic" character: $\theta_0$ is the ideal Bragg angle, and $\pm\delta\theta_0$ represent the limits of reflection.

energy per unit time, we have

$$\mathscr{E}(hkl) = I_0(hkl)t \tag{A.11}$$

and, from (4.50), we obtain

$$\mathscr{E}(hkl) \propto K^2 C(hkl)|F_o(hkl)|^2 \tag{A.12}$$

where $C(hkl)$ includes correcting factors for absorption and extinction, and for the Lorentz and polarization effects (page 351). Because of the proportionality between energy and intensity (A.11), although we are actually measuring the energy of the diffracted beam, we usually speak of the corresponding intensity.

## A.5.3  Intensity Measurements

X-ray intensities are measured either from the blackening of photographic film emulsion or by direct quantum counting.

Film Measurements

The optical density $D$ of a uniformly blackened area of an X-ray diffraction spot on a photographic film is given by

$$D = \log(i_0/i) \tag{A.13}$$

where $i_0$ is the intensity of light hitting the spot and $i$ is the intensity of light transmitted by it. $D$ is proportional to the intensity of the X-ray beam $I_0$ for values of $D$ less than about 1. In practice, this means spots which are just visible to those of a medium-dark grey on the film.

An intensity scale can be prepared by allowing a reflected beam from a crystal to strike a film for different numbers of times and according each spot a value in proportion to this number; Figure A.13 shows one such scale. Intensities may be measured by visual comparison with the scale, and, with care, the average deviation of intensity from the true value would be about 15%.

In place of the scale and the human eye, a photometric device may be used to estimate the blackening. In this method, the background intensity is measured and subtracted from the peak intensity. This process is carried out automatically in the visual method. Carefully photometered intensities would have an average deviation of less than 10%.

The accuracy of film measurements can be enhanced if an integrating mechanism is used in conjunction with either a Weissenberg or a precession camera in recording intensities. In this method, a diffraction spot (Figure A.14a) is allowed to strike the film successively over a grid of points (Figure A.14b). Each point acts as a center for building up the spot. The results of this process are a central plateau of uniform intensity in each spot and a series of spots of similar, regular shape: Figure A.15 illustrates, diagrammatically, the building up of the plateau, and Figure A.16 shows a Weissenberg photograph comparing the normal and integrating methods with the same crystal.

The average deviation in intensity measurements from carefully photometered, integrated Weissenberg photographs is about 5%. The general

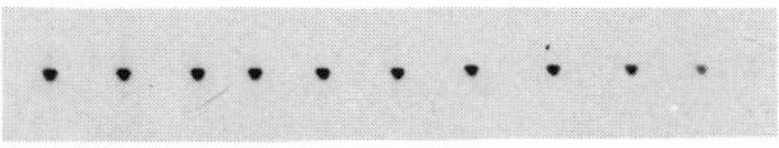

FIGURE A.13. Sketch of a crystal-intensity scale.

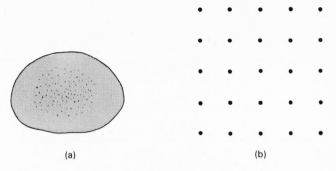

FIGURE A.14.  Spot integration: (a) typical diffraction spot, (b) $5 \times 5$
grid of points.

subject of accuracy in photographic measurements has been discussed
exhaustively by Jeffery.*

## Diffractometer Measurements

The principle of a four-circle, single-crystal diffractometer is shown in
Figure A.17. The incident and diffracted beams are maintained in the
horizontal plane, and the scintillation counter $C$, which counts the diffracted
photons, rotates about an axis normal to this plane. The crystal is brought
into the desired reflecting position by the correct setting of the angles of the
circles designated by $\chi$, $\phi$ and $\Omega$.

The intensity of a reflection peak (Figure A.12) is measured by allowing
the crystal to move on the $\Omega$ circle by amounts $\pm \delta\theta_0$ while following this
movement with the counter moving along the $2\theta$ circle at twice the rate of
the crystal rotation. The best recorded precision of diffractometer intensity
measurements lies in the range of 1–3%. Routine work may be carried out
with speed at a 5–6% level.

### A.5.4   Intensity Corrections

From (A.11) and (A.12), we see that certain corrections are necessary
in order to convert measured intensities into values of $|F|^2$. We shall write

$$I_0(hkl) \propto ALp|F_o(hkl)|_{\mathrm{rel}}^2 \qquad \text{(A.14)}$$

* See Bibliography, Chapter 3.

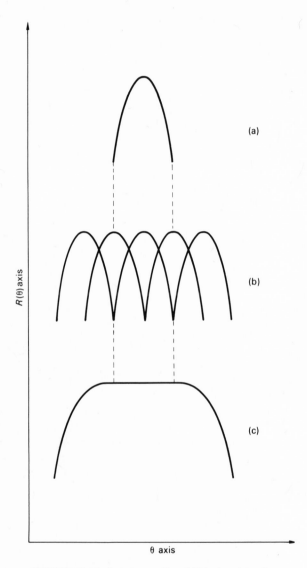

FIGURE A.15.  Spot integration: (a) ideal peak profile,
(b) superposition, by translation, of five profiles, (c)
integrated profile showing a central plateau.

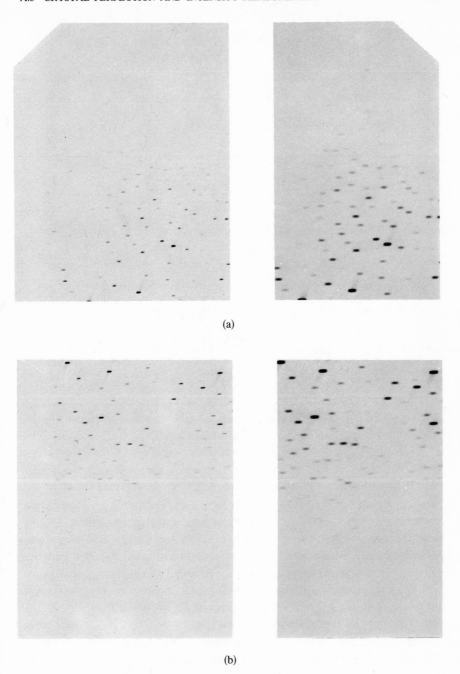

(a)

(b)

FIGURE A.16.  Weissenberg photographs: (a) normal, (b) integrated.

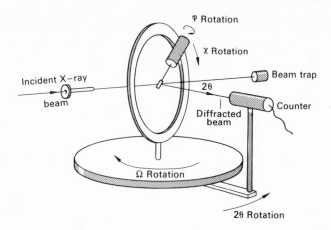

FIGURE A.17. Typical four-circle diffractometer; the counter rotates about the $2\theta$ axis in one plane, and the crystal is oriented about the three axes $\phi$, $\chi$, and $\Omega$ so that the incident and reflected beams lie in the horizontal ($2\theta$) plane. (Reproduced from *An Introduction to X-ray Crystallography* by M. M. Woolfson, with the permission of the Cambridge University Press, London.)

and

$$|F_o(hkl)| = K|F(hkl)|_{\text{rel}} \tag{A.15}$$

where $A$ is an absorption factor (including extinction for the purpose of this discussion), $L$ is the Lorentz factor, $p$ is the polarization factor, and $K$ is the scale factor which places $|F|$ values onto an absolute scale, represented by $|F_o|$; it includes, implicitly, the proportionality constant of (A.12). Absorption corrections may often be neglected, especially with small, approximately equidimensional crystals containing light atoms, and we shall not consider it further. The Lorentz factor expresses the fact that, for a constant angular velocity of rotation of the crystal, different reciprocal lattice points pass through the sphere of reflection at different rates and thus have different times-of-reflection opportunity. The form of the $L$ factor depends upon the experimental arrangement. For both zero-level photographs taken with the X-ray beam normal to the rotation axis and four-circle diffractometer measurements, $L$ has the simple form of $1/\sin 2\theta$.

The radiation from a normal X-ray tube is unpolarized, but after reflection from a crystal the beam is polarized. The fraction of energy lost in this process is dependent only on the Bragg angle:

$$p = (1 + \cos^2 2\theta)/2 \tag{A.16}$$

Application of the $L$ and $p$ factors, where absorption and secondary extinction are negligible, is essential in order to bring the $|F|^2$ data onto a correct relative scale. The scale factor $K$ can be determined approximately by Wilson's method (page 245) and refined as a parameter in a least-squares analysis.

## A.6  Transformations

The main purpose of this appendix is to obtain a relationship between the indices of a given plane referred to two different unit cells in one and the same lattice. However, several other useful equations will emerge in the discussion.

In Figure A.18, a centered unit cell $(\mathbf{A}, \mathbf{B})$ and a primitive unit cell $(\mathbf{a}, \mathbf{b})$ are shown; for simplicity, only two dimensions are considered. From the geometry of the diagram,

$$\mathbf{A} = \mathbf{a} - \mathbf{b} \tag{A.17}$$

$$\mathbf{B} = \mathbf{a} + \mathbf{b} \tag{A.18}$$

$$\mathbf{a} = \mathbf{A}/2 + \mathbf{B}/2 \tag{A.19}$$

$$\mathbf{b} = -\mathbf{A}/2 + \mathbf{B}/2 \tag{A.20}$$

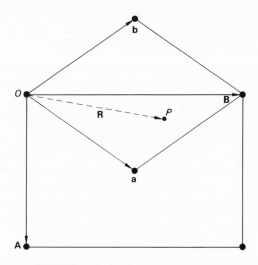

FIGURE A.18.  Unit-cell transformations.

We have encountered this type of transformation before, in our study of lattices (page 79).

The point $P$ may be represented by fractional coordinates $X$, $Y$ in the centered unit cell and by $x$, $y$ in the primitive cell. Since $\mathbf{OP}$ is invariant under unit cell transformation,

$$\mathbf{R} = X\mathbf{A} + Y\mathbf{B} = x\mathbf{a} + y\mathbf{b} \tag{A.21}$$

Substituting for $\mathbf{A}$ and $\mathbf{B}$ from (A.17) and (A.18), we obtain

$$(X + Y)\mathbf{a} + (-X + Y)\mathbf{b} = x\mathbf{a} + y\mathbf{b} \tag{A.22}$$

whence

$$x = X + Y \tag{A.23}$$

$$y = -X + Y \tag{A.24}$$

Similarly, it may be shown that

$$X = x/2 - y/2 \tag{A.25}$$

$$Y = x/2 + y/2 \tag{A.26}$$

The vector to the reciprocal lattice point $hk$ is given, from (2.15), by

$$\mathbf{d}^*(hk) = h\mathbf{a}^* + k\mathbf{b}^* \tag{A.27}$$

and that to the same point, but represented by $HK$, is

$$\mathbf{d}^*(HK) = H\mathbf{a}^* + K\mathbf{b}^* \tag{A.28}$$

The scalar $\mathbf{d}^* \cdot \mathbf{R}$ is invariant with respect to unit cell transformation, since it represents the path difference between that point and the origin[†] (see page 144). Hence, evaluating $\mathbf{d}^* \cdot \mathbf{R}$ with respect to both unit cells and using the properties of the reciprocal lattice discussed on pages 68–72, we obtain

$$hx + ky = HX + KY \tag{A.29}$$

[†] The full significance of this statement can be appreciated after studying Chapter 4.

Substituting for $x$ and $y$ from (A.23) and (A.24), we find

$$(h - k)X + (h + k)Y = HX + KY \qquad (A.30)$$

Hence,

$$H = h - k \qquad (A.31)$$

$$K = h + k \qquad (A.32)$$

which is the same form of transformation as that for the unit cell, given by (A.17) and (A.18). Generalization of this treatment to three dimensions and oblique unit cells is straightforward, if a little time consuming.

## A.7   Comments on Some Orthorhombic and Monoclinic Space Groups

### A.7.1   Orthorhombic Space Groups

In Chapter 2, we looked briefly at the problem of choosing the positions of the symmetry planes in the space groups of class $mmm$ $\left(\dfrac{2}{m} \dfrac{2}{m} \dfrac{2}{m}\right)$ with respect to a center of symmetry at the origin of the unit cell. We give now some simple rules whereby this task can be accomplished readily, while still making use implicitly of the ideas already discussed, including the relative orientations of the symmetry elements given by the space-group symbol itself (see Tables 1.5 and 2.5).

Half-Translation Rule

*Location of Symmetry Planes.* Consider space group *Pnna*; the translations associated with the three symmetry planes are $(b + c)/2$, $(c + a)/2$, and $a/2$, respectively. If they are summed, the result ($T$) is $(a + b/2 + c)$. We disregard the *whole* translations $a$ and $c$, because they refer to neighboring unit cells; thus, $T$ becomes $b/2$, and the center of symmetry is displaced by $T/2$, or $b/4$, from the point of intersection of the three symmetry planes $n$, $n$ and $a$. As a second example, consider *Pmma*. The only translation is $a/2$; thus, $T = a/2$, and the center of symmetry is displaced by $a/4$ from *mma*.

Space group *Imma* may be formed from *Pmma* by introducing the body-centering translation $\frac{1}{2}$, $\frac{1}{2}$, $\frac{1}{2}$ (Fig. 6.18b). Alternatively, the half-translation rule may be applied to the complete space-group symbol. In all, *Imma* contains the translations $(a+b+c)/2$ and $a/2$, and $T = a + (b+c)/2$, or $(b+c)/2$; hence, the center of symmetry is displaced by $(b+c)/4$ from *mma*. This center of symmetry is one of a second set of eight introduced, by the body-centering translation, at $\frac{1}{4}$, $\frac{1}{4}$, $\frac{1}{4}$ (half the *I* translation) from a *Pmma* center of symmetry. This alternative setting is given in the *International Tables for X-Ray Crystallography*, Vol. I*; it corresponds to that in Figure 6.18b with the origin shifted to the center of symmetry at $\frac{1}{4}$, $\frac{1}{4}$, $\frac{1}{4}$. Space groups based on *A*, *B*, *C*, and *F* unit cells similarly introduce additional sets of centers of symmetry. The reader may care to apply these rules to space group *Pnma* and then check the result with Figure 2.36.

*Type and Location of Symmetry Axes.* The quantity *T*, reduced as above to contain half-translations only, readily gives the types of twofold axes parallel to *a*, *b*, and *c*. Thus, if *T* contains an $a/2$ component, then $2_x$ (parallel to $a$) $\equiv 2_1$, otherwise $2_x \equiv 2$. Similarly for $2_y$ and $2_z$, with reference to the $b/2$ and $c/2$ components. Thus, in *Pnna*, $T = b/2$, and so $2_x \equiv 2$, $2_y \equiv 2_1$, and $2_z \equiv 2$. In *Pmma*, $T = a/2$; hence, $2_x \equiv 2_1$, $2_y \equiv 2$, and $2_z \equiv 2$.

The location of each twofold axis may be obtained from the symbol of the symmetry plane perpendicular to it, being displaced by half the corresponding glide translation (if any). Thus, in *Pnna*, we find 2 along $[x, \frac{1}{4}, \frac{1}{4}]$, $2_1$ along $[\frac{1}{4}, y, \frac{1}{4}]$, and another 2 along $[\frac{1}{4}, 0, z]$. In *Pmma*, $2_1$ is along $[x, 0, 0]$, 2 is along $[0, y, 0]$, and another 2 is along $[\frac{1}{4}, 0, z]$. The reader may care to continue the study of *Pmma*, and then check the result, again against Figure 2.36.

## General Equivalent Positions

Once we know the positions of the symmetry elements in a space-group pattern, the coordinates of the general equivalent positions in the unit cell follow readily.

Consider again *Pmma*. From the above analysis, we may write

$\bar{1}$ at 0, 0, 0 (choice of origin)

$m_x \parallel (\frac{1}{4}, y, z)$,     $m_y \parallel (x, 0, z)$,     $a \parallel (x, y, 0)$

---

* N. F. M. Henry and K. Lonsdale (Editors), *International Tables for X-Ray Crystallography*, Vol. I, Birmingham, Kynoch Press.

Taking a point $x$, $y$, $z$ across the three symmetry planes in turn, we have (from Figure 2.33)

$$x, y, z \xrightarrow{\ m_x\ } \tfrac{1}{2}-x, y, z$$

$$\xrightarrow{\ m_y\ } x, \bar{y}, z$$

$$\xrightarrow{\ a\ } \tfrac{1}{2}+x, y, \bar{z}$$

These four points are now operated on by $\bar{1}$ to give the total of eight equivalent positions for *Pmma*:

$$\pm\{x, y, z; \quad \tfrac{1}{2}-x, y, z; \quad x, \bar{y}, z; \quad \tfrac{1}{2}+x, y, \bar{z}\}$$

The reader may now like to complete the example of *Pnma*.

A similar analysis may be carried out for the space groups in the $mm2$ class, with respect to origins on 2 or $2_1$ (consider, for example, Figure 4.13), although we have not discussed specifically these space groups in this book.

### A.7.2 Monoclinic Space Groups

In the monoclinic space groups of class $2/m$, a $2_1$ axis, with a translational component of $b/2$, shifts the center of symmetry by $b/4$ with respect to the point of intersection of $2_1$ with $m$ (Figure S.13b). In $P2/c$, the center of symmetry is shifted by $c/4$ with respect to $2/c$, and in $P2_1/c$ the corresponding shift is $(b+c)/4$ (Figure 2.32).

# Solutions

## Chapter 1

**1.1.** (1, 3.366).

**1.2.** (a) $(1\bar{2}0)$. (b) (164). (c) $(00\bar{1})$. (d) $(3\bar{3}4)$. (e) $(0\bar{4}3)$. (f) $(\bar{4}2\bar{3})$.

**1.3.** (a) $[\bar{5}11]$. (b) $[3\bar{5}2]$. (c) [111]. (d) [110].

**1.4.** $(52\bar{3})$; $(52\bar{3})$ and $(\bar{5}2\bar{3})$ are parallel. It should be noted that a similar situation exists in Problem 1.3, but for coincident zone symbols. $[UVW]$ and $[\bar{U}\bar{V}\bar{W}]$.

**1.5.** (a) see Figure S.1.
(b) $c/a = \cot 29.37° = 1.7771$.
(c) In this example, the zone circles may be sketched in carefully, and the stereogram indexed without using a Wulff's net. Draw on the procedures used in Problems 1.3 and 1.4. (The center of the stereogram corresponds to 001, even though this face is not present on the crystal.) By making use of the axial ratio, the points of intersection of the

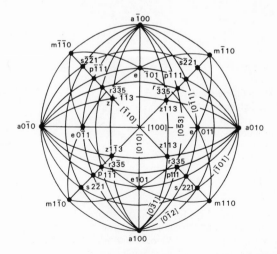

FIGURE S.1

361

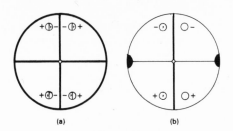

FIGURE S.2

zone circles with the $Y$ axis may be indexed, even though they do not all represent faces present. Reading from center to right, they are 001, 013, 035, 011, 021, and 010 (letter symbols indicate faces actually present). Hence, the zone symbols and poles may be deduced. Confirm the assignments of indices by means of the Weiss zone law.

**1.6.**  (a) *mmm*.  (b) $2/m$.  (c) 1.

**1.7.**  See Figure S.2.  (a) *mmm*          (b) $2/m$
                        $m.m.m \equiv \bar{1}$.      $2.m \equiv \bar{1}$.

**1.8.**

|         | $\{010\}$ | $\{\bar{1}10\}$ | $\{11\bar{3}\}$ |
|---------|-----------|-----------------|-----------------|
| $2/m$   | 2         | 4               | 4               |
| $\bar{4}2m$ | 4     | 4               | 8               |
| $m3$    | 6         | 12              | 24              |

**1.9.**  (a) 1.  (b) $m$.  (c) 2.  (d) $m$.  (e) 1.  (f) 2.  (g) 6.  (h) $6mm$.  (i) 3.  (j) $2mm$.

**1.10.**  (a) Hexachloroplatinate ion      $m3m$   $O_h$   $\{100\}$
    (b) Carbonate ion            $\bar{6}m2$   $D_{3h}$   $(10\bar{1}0\}$ or $\{01\bar{1}0\}$

    (c) Benzene                $\dfrac{6}{m}mm$   $D_{6h}$

    (d) Methane               $\bar{4}3m$   $T_d$   $\{111\}$ or $\{1\bar{1}1\}$
    (e) *cis*-1,2-Dichloroethylene   $mm2$   $C_{2v}$
    (f) Naphthalene          $mmm$   $D_{2h}$
    (g) Sulfate ion            $\bar{4}3m$   $T_d$
    (h) Tetrakismethylthiomethane   $\bar{4}$   $S_4$   $\{111\}$ or $\{1\bar{1}1\}$
        (tetramethylorthothiocarbonate)
    (i) Cyanuric triazide        $\bar{6}$   $C_{3h}$
    (j) Bromochlorofluoromethane   1   $C_1$

**1.11.**  2

# Chapter 2

**2.1.**  (a) (i) $4mm$, (ii) $6mm$.  (b) (i) Square, (ii) hexagonal.  (c) (i) Another square can be drawn as the conventional ($p$) unit cell.  (ii) The symmetry at each point is degraded to $2mm$. A rectangular net is produced, and may be described by a $p$ unit cell. The transformation equations for both examples are

$$\mathbf{a}' = \mathbf{a}/2 + \mathbf{b} + 2, \qquad \mathbf{b}' = -\mathbf{a}/2 + \mathbf{b}/2$$

*Note.* A regular hexagon of points with another point at its center is not a centered hexagonal unit cell; it represents three adjacent $p$ hexagonal unit cells in different orientations.

**2.2.** The $C$ unit cell may be obtained by the transformation $\mathbf{a}' = \mathbf{a}$, $\mathbf{b}' = \mathbf{b}$, $\mathbf{c}' = -\mathbf{a}/2 + \mathbf{c}/2$. The new dimensions are $c' = 5.763$ Å and $\beta' = 139°17'$; $a'$ and $b'$ remain as $a$ and $b$, respectively. $V_c(C \text{ cell}) = V_c(F \text{ cell})/2$.

**2.3.** (a) The symmetry is no longer tetragonal.
(b) The tetragonal symmetry is apparently restored, but the unit cell no longer represents a lattice because the points do not all have the same environment.
(c) A tetragonal $F$ unit cell is obtained, which is equivalent to $I$ under the transformation $\mathbf{a}' = \mathbf{a}/2 + \mathbf{b}/2$, $\mathbf{b}' = -\mathbf{a}/2 + \mathbf{b}/2$, $\mathbf{c}' = \mathbf{c}$.

**2.4.** 28.74 Å ($F$ cell); 28.64 Å.

**2.5.** It is not a new system because the symmetry of the unit cell is not higher than $\bar{1}$. It represents a special case of the triclinic system with $\gamma = 90°$.

**2.6.** (a) Plane group $c2mm$.

| | | | $(0, 0; \frac{1}{2}, \frac{1}{2})+$ | | | | Limiting conditions |
|---|---|---|---|---|---|---|---|
| 8 | (f) | 1 | $x, y$; | $x, \bar{y}$; | $\bar{x}, y$; | $\bar{x}, \bar{y}$ | $hk: h + k = 2n$ |
| 4 | (e) | $m$ | $0, y$; | $0, \bar{y}$ | | | — |
| 4 | (d) | $m$ | $x, 0$; | $\bar{x}, 0$ | | | — |
| 4 | (c) | 2 | $\frac{1}{4}, \frac{1}{4}$; | $\frac{1}{4}, \frac{3}{4}$ | | | As above + |
| | | | | | | | $hk: h = 2n, (k = 2n)$ |
| 2 | (b) | $2mm$ | $0, \frac{1}{2}$ | | | | — |
| 2 | (a) | $2mm$ | $0, 0$ | | | | — |

(b) Plane group $p2mg$. See Figures S.3 and S.4. If the symmetry elements are arranged with 2 at the intersection of $m$ and $g$, they do not form a group. Attempts to draw such an arrangement lead to continued halving of the "repeat" parallel to $g$.

**2.7** See Figures S.5 and S.6.

| | | | | | | |
|---|---|---|---|---|---|---|
| 4 | (e) | 1 | $x, y, z$; | $\bar{x}, \bar{y}, \bar{z}$; | $x, \frac{1}{2} - y, \frac{1}{2} + z$; | $hkl$: None |
| | | | $\bar{x}, \frac{1}{2} + y, \frac{1}{2} - z$ | | | $h0l: l = 2n$ |
| | | | | | | $0k0: k = 2n$ |
| 2 | (d) | $\bar{1}$ | $\frac{1}{2}, 0, \frac{1}{2}$; | $\frac{1}{2}, \frac{1}{2}, 0$ | | |
| 2 | (c) | $\bar{1}$ | $0, 0, \frac{1}{2}$; | $0, \frac{1}{2}, 0$ | | As above + |
| 2 | (b) | $\bar{1}$ | $\frac{1}{2}, 0, 0$; | $\frac{1}{2}, \frac{1}{2}, \frac{1}{2}$ | | $hkl: k + l = 2n$ |
| 2 | (a) | $\bar{1}$ | $0, 0, 0$; | $0, \frac{1}{2}, \frac{1}{2}$ | | |
| (100) | $p2gg$ | $b' = b, c' = c$ | | | | |
| (010) | $p2$ | $a' = a, c' = c/2$ | | | | |
| (001) | $p2gm$ | $a' = a, b' = b$ | | | | |

The two molecules lie with the center of their $C(1)-C(1)'$ bonds on any pair of special positions (a)–(d). The molecule is therefore centrosymmetric and planar.

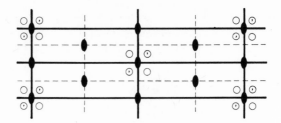

FIGURE S.3

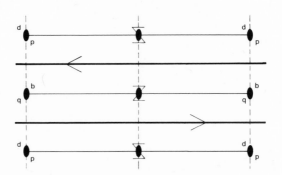

FIGURE S.4

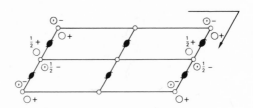

FIGURE S.5

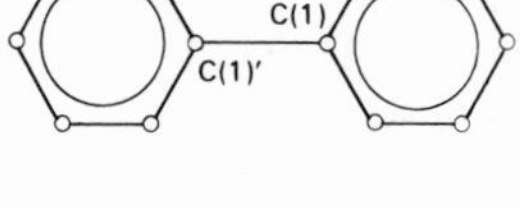

FIGURE S.6

**2.8.** Each pair of positions forms two vectors, between the origin and the points $\pm\{(x_1-x_2),$ $(y_1-y_2), (z_1-z_2)\}$: one vector at each of the locations

$$2x, 2y, 2z; \quad 2\bar{x}, 2\bar{y}, 2\bar{z}; \quad 2x, 2\bar{y}, 2z; \quad 2\bar{x}, 2y, 2\bar{z}$$

and two vectors at each of the locations

$$2x, \tfrac{1}{2}, \tfrac{1}{2}+2z; \quad 0, \tfrac{1}{2}+2y, \tfrac{1}{2}; \quad 2\bar{x}, \tfrac{1}{2}, \tfrac{1}{2}-2z; \quad 0, \tfrac{1}{2}-2y, \tfrac{1}{2}$$

Note: $-(2x, \tfrac{1}{2}, \tfrac{1}{2}+2z) \equiv 2\bar{x}, \tfrac{1}{2}, \tfrac{1}{2}-2z.$

**2.9.**

$$x, y, z \xrightarrow{\ -b\ } 2p - x, -\tfrac{1}{2}+y, z$$
$$\downarrow{-\bar{1}}$$
$$\bar{x}, \bar{y}, \bar{z} \qquad\qquad \downarrow{-a}$$

$$\left\{-1 \text{ (or 0)} + 2p - x, 2q - y, 2r - z \xleftarrow{\ -n\ } -\tfrac{1}{2}+2p - x, 2q +\tfrac{1}{2}- y, z\right.$$

The points $\bar{x}, \bar{y}, \bar{z}$ and $2p - x, 2q - y, 2r - z$ are one and the same; hence, by comparing coordinates, $p = q = r = 0$.

**\*2.10.** See Figure S.7. General equivalent positions:

$$x, y, z; \qquad x, y, \bar{z}; \quad \tfrac{1}{2}-x, \tfrac{1}{2}-y, z; \quad \tfrac{1}{2}-x, \tfrac{1}{2}-y, \bar{z}$$

$$\tfrac{1}{2}+x, \bar{y}, x; \quad \tfrac{1}{2}+x, \bar{y}, \bar{z}; \quad \bar{x}, \tfrac{1}{2}+y, z; \quad \bar{x}, \tfrac{1}{2}+y, \bar{z}$$

Centers of symmetry:

$$\tfrac{1}{4}, \tfrac{1}{4}, 0; \quad \tfrac{1}{4}, \tfrac{3}{4}, 0; \quad \tfrac{3}{4}, \tfrac{1}{4}, 0; \quad \tfrac{3}{4}, \tfrac{3}{4}, 0$$
$$\tfrac{1}{4}, \tfrac{1}{4}, \tfrac{1}{2}; \quad \tfrac{1}{4}, \tfrac{3}{4}, \tfrac{1}{2}; \quad \tfrac{3}{4}, \tfrac{1}{4}, \tfrac{1}{2}; \quad \tfrac{3}{4}, \tfrac{3}{4}, \tfrac{1}{2}$$

Change of origin: (i) subtract $\tfrac{1}{4}, \tfrac{1}{4}, 0$ from the above set of general equivalent positions, (ii) let $x_0 = x - \tfrac{1}{4}, y_0 = y - \tfrac{1}{4}, z_0 = z$, (iii) continue in this way, and finally drop the subscript:

$$\pm(x, y, z; \quad \bar{x}, \bar{y}, z; \quad \tfrac{1}{2}+x, \tfrac{1}{2}-y, \bar{z}; \quad \tfrac{1}{2}-x, \tfrac{1}{2}+y, \bar{z})$$

This result may be confirmed by redrawing the space-group diagram with the origin on $\bar{1}$.

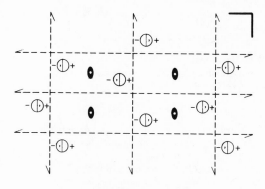

FIGURE S.7

**2.11.** Two unit cells of space group $Pn$ are shown on the (010) plane (see Figure S.8). In the transformation to $Pc$, only the $c$ axis changed:

$$\mathbf{c}'(Pc) = -\mathbf{a}(Pn) + \mathbf{c}(Pn)$$

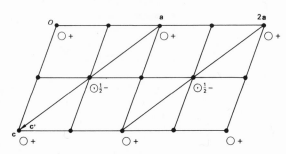

Hence, $Pn \equiv Pc$. By interchanging the labels of the $X$ and $Z$ axes (which are not constrained by the twofold symmetry axis), we see that $Pc \equiv Pa$. However, because of the translations of $\frac{1}{2}$ along $a$ and $b$ in $Cm$, from the centering of the unit cell, $Ca \not\equiv Cc$, although $Cc \equiv Cn$. We have $Ca \equiv Cm$, and the usual symbol for this space group is $Cm$. If the $X$ and $Z$ axes are interchanged in $Cc$, the equivalent symbol is $Aa$.

**2.12.** $P2/c$    (a) $2/m$; monoclinic.
                 (b) Primitive unit cell, $c$-glide plane $\perp b$, twofold axis $\| b$.
                 (c) $h0l$: $l = 2n$.

       $Pca2_1$    (a) $mm2$; orthorhombic.
                 (b) Primitive unit cell, $c$-glide plane $\perp a$, $a$-glide plane $\perp b$, $2_1$ axis $\| c$.

       $Cmcm$    (a) $mmm$; orthorhombic.
                 (b) $C$-face-centered unit cell, $m$ plane $\perp a$, $c$-glide plane $\perp b$, $m$ plane $\perp c$.
                 (c) $hkl$: $h + k = 2n$; $h0l$: $l = 2n$.

       $P\bar{4}2_1c$    (a) $\bar{4}2m$; tetragonal.
                 (b) Primitive unit cell, $\bar{4}$ axis $\| c$, $2_1$ axes $\| a$ and $b$, $c$-glide planes $\perp [110]$ and $[1\bar{1}0]$.
                 (c) $hhl$: $l = 2n$; $h00$: $h = 2n$; $0k0$: $(k = 2n)$.

       $P6_322$    (a) $622$; hexagonal.
                 (b) Primitive unit cell, $6_3$ axis $\| c$, twofold axes $\| a$, $b$, and $u$, twofold axes $30°$ to $a$, $b$, and $u$, in the (0001) plane.
                 (c) $000l$: $l = 2n$.

       $Pa3$    (a) $m3$; cubic.
                 (b) Primitive unit cell, $a$-glide plane $\perp c$, $b$-glide plane $\perp a$, $c$-glide plane $\perp b$ (the glide planes are equivalent under the cubic symmetry), threefold axes $\| [111], [1\bar{1}1], [\bar{1}11]$, and $[\bar{1}\bar{1}1]$.
                 (c) $0kl$: $k = 2n$; $h0l$: $(l = 2n)$; $hk0$; $(h = 2n)$.

**2.13.** Plane group $p2$; the unit-cell repeat along $b$ is halved, and $\gamma$ has the particular value of $90°$.

# Chapter 3

**3.1** (a) Tetragonal crystal; optic axis parallel to the needle axis ($c$) of the crystal.

(b) Section extinguished for any rotation in the $ab$ plane.

(c) Horizontal $m$ line. Symmetric oscillation photograph with $a$, $b$, or $\langle 110 \rangle$ parallel to the beam at the center of the oscillation would have $2mm$ symmetry ($m$ lines horizontal and vertical).

**3.2.** (a) Orthorhombic. (b) The edges of the brick. (c) Horizontal $m$ line. (d) $2mm$ ($m$ lines horizontal and vertical).

**3.3.** (a) Monoclinic, or possibly orthorhombic.

(b) If monoclinic, $y \| p$. If orthorhombic, $p \| x$, $y$, or $z$.

(c) (i) Mount the crystal perpendicular to $p$, either about $q$ or $r$, and take a Laue photograph with the X-ray beam parallel to $p$. If monoclinic, twofold symmetry would be observed. If orthorhombic, $2mm$, but with the $m$ lines in general directions on the film which define the directions of the crystallographic axes normal to $p$. If the crystal is rotated so that X-rays are perpendicular to $p$, a vertical $m$ line would appear on the Laue photograph of either a monoclinic or an orthorhombic crystal.

(ii) Use the same crystal mounting as in (i) and take symmetric oscillation photographs with the X-ray beam parallel or perpendicular to $p$ at the center of the oscillation. The rest of the answer is as in (i).

**3.4.** $a = 9.00$, $b = 6.00$, $c = 5.00$ Å.

$a^* = 0.167$, $b^* = 0.250$, $c^* = 0.300$ RU.

$2 \sin \theta (146) > 2.0$. Each photograph would have a horizontal $m$ line, conclusive of orthorhombic symmetry if the crystal is known to be biaxial; otherwise, tests for higher symmetry would have to be carried out.

**3.5.** (a) $a = 4.322$, $c = 7.506$ Å.

(b) $n_{max} = 4$.

(c) No symmetry in (i). Horizontal $m$ line in (ii).

(d) The photographs would be identical because of the fourfold axis of oscillation. (See Figure S.9)

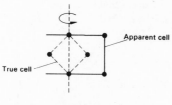

FIGURE S.9

**3.6.** Remembering that the $\beta$ angle is, conventionally, oblique, and that in the monoclinic system $\beta = 180° - \beta^*$, $\beta^* = 86°$ and $\beta = 94°$.

# Chapter 4

**4.1.** The coordinates show that the structure is centrosymmetric. Hence, $A'(hk)$ is given by (4.62) with $l = 0$, $B'(hkl) = 0$, and the structure factors are real $[F(hk) = A'(hk)]$:

$$F(5, 0) = 2(-g_P + g_O), \qquad F(0, 5) = 2(g_P - g_O)$$

$$F(5, 5) = 2(-g_P - g_O), \qquad F(5, 10) = 2(-g_P + g_O)$$

For $g_P = 2g_O$, $\phi(0, 5) = 0$ and $\phi(5, 0) = \phi(5, 5) = \phi(5, 10) = \pi$.

**4.2.** The structure is centrosymmetric.

$$A(hkl) = 4 \cos 2\pi[ky + (h + k + l)/4] \cos 2\pi(h + k)/4$$

|  | $y = 0.10$ | $y = 0.15$ |
|---|---|---|
| $|F_c(020)|$ | 86.5 | 86.5 |
| $|F_c(110)|$ | 258.9 | 188.1 |

Hence, 0.10 is the better value for $y$, as far as one can judge from these two reflections.

**4.3.** The shortest U—U distance is between $0, y, \frac{1}{4}$ and $0, \bar{y}, \frac{3}{4}$ and has the value 2.76 Å.

**4.4.** (a) $P2_1, P2_1/m$. (b) $Pa, P2/a$. (c) $Cc, C2/c$. (d) $P2, Pm, P2/m$.

**4.5.** (a) $P2_12_12$.
(b) $Pbm2, Pbmm$.
(c) $Ibm2 \ (Icm2); Ib2m \ (Ic2m); Ibmm \ (Icmm)$
$hkl: h + k + l = 2n$
$0kl: k = 2n, (l = 2n),$ or $l = 2n, (k = 2n)$
$h0l: (h + l = 2n)$
$hk0: (h + k = 2n)$
$h00: (h = 2n)$
$0k0: (k = 2n)$
$00l: (l = 2n)$.

**4.6.** (a)  (i)  $h0l: h = 2n; 0k0: k = 2n$. No other independent conditions.
(ii)  $h0l: l = 2n$. No other independent conditions.
(iii)  $hkl: h + k = 2n$. No other independent conditions.
(iv)  $h00: h = 2n$. No other conditions.
(v)  $0kl: l = 2n; h0l: l = 2n$. No other independent conditions.
(vi)  $hkl: h + k + l = 2n; h0l: h = 2n$. No other independent conditions.

Space groups with the same conditions:  (i) None.  (ii) $P2/c$.  (iii) $Cm, C2/m$.  (iv) None.  (v) $Pccm$.  (vi) $Ima2, I2am$.
(b)  $hkl$: None; $h0l: h + l = 2n; 0k0: k = 2n$.
(c)  $C2/c, C222$.

# Chapter 5

**5.1.** $A(hkl) = 4 \cos 2\pi[0.2h + 0.1l + (k + l)/4] \cos 2\pi(l/4)$. Systematically absent reflections are $hkl$ for $l$ odd. The $c$ dimension appertaining to $P2_1/c$ should be halved, because the true cell contains two atoms in space group $P2_1$. This problem illustrates the consequences of siting atoms on glide planes. Although this answer applies to a hypothetical structure containing a single atomic species, in a mixed-atom structure an atom may, by chance, be situated on a translational symmetry element. See Figure S.10.

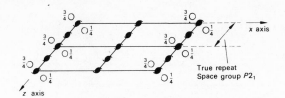

FIGURE S.10

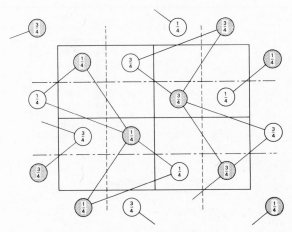

FIGURE S.11

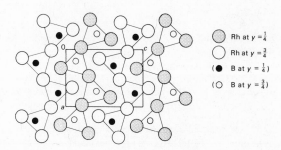

FIGURE S.12

**\*5.2.** There are eight Rh atoms in the unit cell. The separation of atoms related across any $m$ plane is $\frac{1}{2} - 2y$, which is less than $b/2$ and thus, prohibited. The Rh atoms must therefore lie in two sets of special positions, with either $\bar{1}$ or $m$ symmetry. The positions on $\bar{1}$ may be eliminated, again by spatial considerations. Hence, we have (see Figures S.11 and S.12)[†]

$$4\,\text{Rh}(1):\ \pm\{x_1, \tfrac{1}{4}, z_1;\ \tfrac{1}{2}+x_1, \tfrac{1}{4}, \tfrac{1}{2}-z_1\}$$

$$4\,\text{Rh}(2):\ \pm\{x_2, \tfrac{1}{4}, z_2;\ \tfrac{1}{2}+x_2, \tfrac{1}{4}, \tfrac{1}{2}-z_2\}$$

† R. Mooney and A. J. E. Welch, *Acta Crystallographica* **7**, 49 (1954).

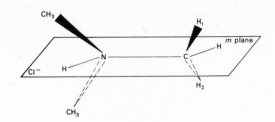

FIGURE S.13a

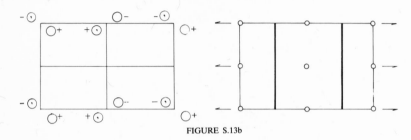

FIGURE S.13b

Origin at $\bar{1}$; unique axis $b$

Limiting conditions

4  $f$  1   $x, y, z$;  $\bar{x}, \bar{y}, \bar{z}$;  $\bar{x}, \frac{1}{2}+y, z$;  $x, \frac{1}{2}-y, z$.

$hkl$:  None
$h0l$:  None
$0k0$:  $k = 2n$

| 2 | $e$ | $m$ | $x, \frac{1}{4}, z$; | $\bar{x}, \frac{3}{4}, \bar{z}$ |
|---|---|---|---|---|
| 2 | $d$ | $\bar{1}$ | $\frac{1}{2}, 0, \frac{1}{2}$; | $\frac{1}{2}, \frac{1}{2}, \frac{1}{2}$. |
| 2 | $c$ | $\bar{1}$ | $0, 0, \frac{1}{2}$; | $0, \frac{1}{2}, \frac{1}{2}$. |
| 2 | $b$ | $\bar{1}$ | $\frac{1}{2}, 0, 0$; | $\frac{1}{2}, \frac{1}{2}, 0$. |
| 2 | $a$ | $\bar{1}$ | $0, 0, 0$; | $0, \frac{1}{2}, 0$. |

As above +
$hkl$:  $k = 2n$

Symmetry of special projections

(001) $pgm$;  $a' = a, b' = b$     (100) $pmg$;  $b' = b, c' = c$     (010) $p2$;  $c' = c, a' = a$

**5.3.** Space group $P2_1/m$. Molecular symmetry cannot be $\bar{1}$, so it must be $m$. (a) Cl lie on $m$ planes; (b) N lie on $m$ planes; (c) two C on $m$ planes, and four other C probably in general positions; (d) 16 H in general positions, two H (in NH groups) on $m$ planes, and two H (from the $CH_3$ that have their C on $m$ planes) on $m$ planes. This arrangement is shown schematically in Figure S.13a. The groups $CH_3$, $H_1$, and $H_2$ lie above and below the $m$ plane. (The alternative space group, $P2_1$, was considered, but the structure analysis† confirmed the assumption of $P2_1/m$. The diagram of space group $P2_1/m$ shown in Figure S.13b is reproduced from the *International Tables for X-Ray Crystallography*, Vol. I, edited by N. F. M. Henry and K. Lonsdale, with the permission of the International Union of Crystallography.)

† J. Lindgren and I. Olovsson, *Acta Crystallographica* **B24**, 554 (1968).

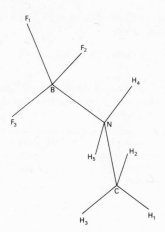

FIGURE S.14a

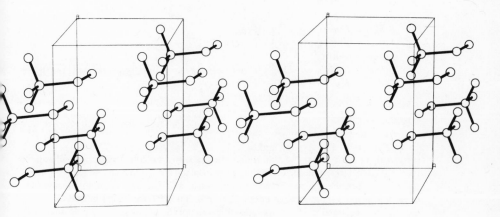

FIGURE S.14b

**5.4**

|       |         |             |           |              | $X_{Cl} = 0.23$ | | $X_{Cl} = 0.24$ | |
|-------|---------|-------------|-----------|--------------|---------|-----------|---------|-----------|
| *hhh* | $|F_o|$ | $g_{Pt}$ | $g_K$ | $g_{Cl}$ | $|F_c|$ | $K_1|F_o|$ | $|F_c|$ | $K_2|F_o|$ |
| 111   | 491     | 73.5        | 17.5      | 15.5         | 341     | 315       | 317     | 329       |
| 222   | 223     | 66.5        | 14.5      | 13.0         | 152     | 143       | 160     | 150       |
| 333   | 281     | 59.5        | 12.0      | 10.5         | 145     | 180       | 191     | 189       |
|       |         |             |           |              | $K_1$   | $R_1$     | $K_2$   | $R_2$     |
|       |         |             |           |              | 0.641   | 0.11      | 0.671   | 0.036     |

Clearly, $x_{Cl} = 0.24$ is the preferred value. Pt—Cl $= 2.34$ Å. For sketch and point group, see Problem (and Solution) 1.10(a).

**5.5.** $A_U(hkl) = 4 \cos 2\pi[hx_U - (h+k)/4] \cos 2\pi[ky_U + (h+k+l)/4]$.     $x_U \approx \frac{1}{8}$,  $y_U \approx 0.20$ (mean of $\frac{1}{4}$, $\frac{3}{16}$, and $\frac{1}{6}$).

**5.6.** Since $Z = 2$, the molecules lie either on $\bar{1}$ or $m$. Chemical knowledge eliminates $\bar{1}$. The $m$ planes are at $\pm(x, \frac{1}{4}, z)$, and the C, N, and B atoms must lie on these planes. Since the shortest distance between $m$ planes is 3.64 Å, $F_1$, B, N, C, and $H_1$ (see Figure S.14) lie on one $m$ plane. Hence, the remaining F atoms and the four H atoms must be placed symmetrically across the same $m$ plane. The conclusions were borne out by the structure analysis.* Figure S.14 shows a stereoscopic pair of packing diagrams for $CH_3NH_2 \cdot BF_3$. $F_1$, B, N, C, and $H_1$ lie on a mirror plane; the $F_2$, $F_3$, $H_4$, $H_5$, and $H_2$, $H_3$, atom pairs are related across the same $m$ plane.

# Chapter 6

**6.1.** (a)  $|F(hkl)| = |F(\bar{h}k\bar{l})|$
$|F(0kl)| = |F(0\bar{k}\bar{l})|$
$|F(h0l)| = |F(\bar{h}0\bar{l})|$

(b)  $|F(hkl)| = |F(\bar{h}k\bar{l})| = |F(h\bar{k}l)| = |F(\bar{h}\bar{k}\bar{l})|$
$|F(0kl)| = |F(0\bar{k}\bar{l})| = |F(0\bar{k}l)| = |F(0k\bar{l})|$
$|F(h0l)| = |F(\bar{h}0\bar{l})|$

(c)  $|F(hkl)| = |F(\bar{h}k\bar{l})| = |F(\bar{h}kl)| = |F(h\bar{h}l)| = |F(hk\bar{l})| = |F(h\bar{k}\bar{l})| = |F(\bar{h}k\bar{l})| = |F(\bar{h}kl)|$
$|F(0kl)| = |F(0\bar{k}\bar{l})| = |F(0\bar{k}l)| = |F(0k\bar{l})|$
$|F(h0l)| = |F(\bar{h}0\bar{l})| = |F(\bar{h}0l)| = |F(h0\bar{l})|$

**6.2.** (a)  $Pa$; $[\frac{1}{2}, v, 0]$.
$P2/a$: $[\frac{1}{2}, v, 0]$ and $(u, 0, w)$.
$P222_1$; $(0, v, w)$, $(u, 0, w)$, and $(u, v, \frac{1}{2})$.

(b)  $(u, 0, w)$ is the Harker section for a structure with a twofold axis along $b$, whereas $[0, v, 0]$ is the Harker line corresponding to an $m$ plane normal to $b$ and passing through the origin. Since the crystal is noncentrosymmetric, and assuming no other concentrations of peaks, the space group is either $P2$ or $Pm$. If it is $P2$, then there must be chance coincidences between the $y$ coordinates of atoms not related by symmetry. If it is $Pm$, then the chance coincidences must be between both the $x$ and the $z$ coordinates of atoms not related by symmetry.

**6.3.** (a)  $P2_1/n$ (a nonstandard setting of $P2_1/c$; see Problem 2.11 for a similar relationship between $Pc$ and $Pn$).

(b) Vectors:   1:   $\pm\{\frac{1}{2}, \frac{1}{2} + 2y, \frac{1}{2}\}$         double weight
2:   $\pm\{\frac{1}{2} + 2x, \frac{1}{2}, \frac{1}{2} + 2z\}$      double weight
3:   $\pm\{2x, 2y, 2z\}$               single weight
4:   $\pm\{2x, 2\bar{y}, 2z\}$               single weight

Section $v = \frac{1}{2}$: type 2 vector.
Section $v = 0.092$: type 1 vector.
Section $v = 0.408$: type 3 or 4 vector.

4 S:   $\pm\{0.182, 0.204, 0.226\}$   and   $\pm\{0.682, 0.296, 0.726\}$

* S. Geller and J. L. Hoard, *Acta Crystallographica* **3**, 121 (1950).

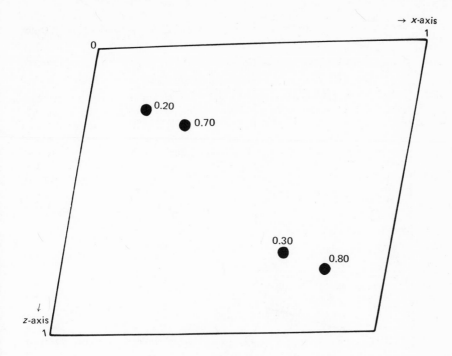

You may have selected one of the other seven centers of symmetry, in which case the coordinates determined may be transformed accordingly. The positions are plotted in Figure S.15. Differences in the third decimal places of the coordinates determined from the maps in Problems 6.3 and 6.4 are not significant.

**6.4.** (a) The sulfur atom $x$ and $z$ coordinates are S(0.264, 0.141), S'(−0.264, −0.141).

(b) Plot the position −S on tracing paper and copy the Patterson map (excluding the origin peak) with its origin over −S (Figure S.16a). On another tracing, carry out the same procedure with respect to −S' (Figure S.16b). Superimpose the two tracings (Figure S.16c). Atoms are located where both maps have positive areas.

**6.5.** (a) $P(v)$ shows three nonorigin peaks. If the highest is assumed to arise from Hf atoms at $\pm\{0, y_{Hf}, \frac{1}{4}\}$, then $y_{Hf} = 0.11$. The other two peaks may be Hf—Si vectors; the difference in their height is due partly to the proximity of the peak of lesser height to the low vector density around the origin peak—an example of poor resolution—and partly due to the $y$ value of one Si atom.

(b) The signs are, in order and omitting 012,0 and 016,0, $+ - - + + -$. $\rho(y)$ shows a large peak at 0.107, which is a better value for $y_{Hf}$, and smaller peaks at 0.05, 0.17, and 0.25. The values 0.05 and 0.25 give vectors for Hf—Si which coincide with peaks on $P(v)$. We conclude that these values are the approximate $y$ coordinates for Si, and that the peak at 0.17 is spurious, arising from both the small-number of data and experimental errors therein.

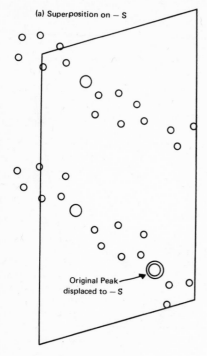

(a) Superposition on − S

Original Peak
displaced to − S

FIGURE S.16a

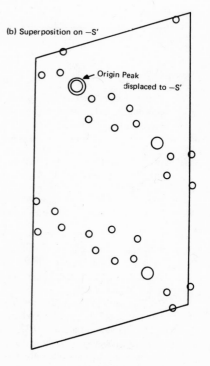

(b) Superposition on −S′

Origin Peak
displaced to −S′

FIGURE S.16b

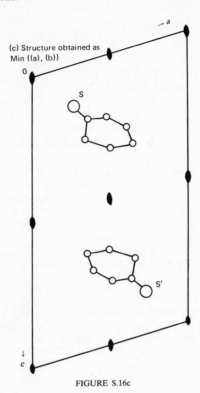

(c) Structure obtained as
Min ((a), (b))

FIGURE S.16c

*6.6. Since the sites of the replaceable atoms are the same in each derivative, and the space group is centrosymmetric, we may write $F(M_1) = F(M_2) + 4(f_{M_1} - f_{M_2})$

(a)

| NH$_4$ | K | Rb | Tl | |
|---|---|---|---|---|
| − | − | + | + | * Indeterminate because $\|F\|$ is unobserved. |
| * | + | + | + | |
| + | + | + | + | |
| − | † | + | + | † Indeterminate because $\|F\|$ is small. |
| + | + | + | + | Omit from the electron density synthesis. |
| + | + | + | + | |
| − | − | * | + | |
| * | + | + | + | |

(b) Peaks at 0 and $\frac{1}{2}$ represent K and Al, respectively. The peak at 0.35 is due, presumably, to the S atom.

(c) The effect of the isomorphous replacement can be noted first from the increases in $|F(555)|$ and $|F(666)|$ and the decrease in $|F(333)|$. These changes are not in accord with the findings in (b). Comparison of the electron density plots shows that $x_{S/Se}$ must be 0.19. The peak at 0.35 arises, in fact, from a superposition of oxygen atoms in projection, and it is not altered appreciably by the isomorphous replacement. Aluminum, at 0.5, is not represented strongly in these projections.

## Chapter 7

**7.1.** $705, 6\bar{1}7, 8\bar{1}4$. $42\bar{6}$ is a structure invariant, 203 is linearly related to $8\bar{1}4$ and $6\bar{1}7$, and 432 has a low $|E|$ value. Alternative sets are $705, 203, 8\bar{1}4$ and $705, 203, 6\bar{1}7$. A vector triplet exists between $8\bar{1}4, 42\bar{6}$, and $4\bar{3}2$.

**7.2.** $|F(hkl)| = |F(\bar{h}\bar{k}\bar{l})| = |F(h\bar{k}l)| = |F(\bar{h}k\bar{l})|$
$k = 2n$:  $\phi(hkl) = -\phi(\bar{h}\bar{k}\bar{l}) = -\phi(h\bar{k}l) = \phi(\bar{h}k\bar{l})$
$k = 2n+1$:  $\phi(hkl) = -\phi(\bar{h}\bar{k}\bar{l}) = \pi - \phi(h\bar{k}l) = \pi + \phi(\bar{h}k\bar{l})$

**7.3.** Set (b) would be chosen. There is a redundancy in set (a) among 041, $\bar{1}62$, and $\bar{1}23$, because $F(041) = F(04\bar{1})$ in this space group. In space group $C2/c$, $h + k$ must be even, Hence, reflections 012 and 162 would not be found. The origin could be fixed by 223 and $13\bar{7}$ because there are only four parity groups for a $C$-centered unit cell.

**7.4.** From (7.32) and (7.33), $K = 4.0 \pm 0.4$ and $B = 6.6 \pm 0.3\ \text{Å}^2$. (You were not expected to derive the standard errors in these quantities; they are listed in order to give some idea of the precision of the results obtained by the Wilson plot.) The rms displacement $\overline{(u^2)} = 0.28\ \text{Å}$.

**7.5.** The shortest distance is between points like $\frac{1}{4}, y, z$ and $\frac{3}{4}, \bar{y}, z$. Hence, from (7,41), $d^2(\text{Cl} \cdots \text{Cl}) = a^2/4 + 4y^2b^2$, or $d(\text{Cl} \cdots \text{Cl}) = 4.64\ \text{Å}$. Using (7.44), $[2d\sigma(d)]^2 = [2a\sigma(a)/4]^2 + [8y^2b\sigma(b)]^2 + [8yb^2\sigma(y)]^2$, whence $\sigma(d) = 0.026\ \text{Å}$.

**7.6.** $|F(010)| = 149$, $|F(0\bar{1}0)| = 145$, $\phi(010) = 55°$, $\phi(0\bar{1}0) = -49°$

## Chapter 8

**8.1.** The I—I vector lies at $2x, \frac{1}{2}, 2z$. Hence, by measurement, $x = 0.423$ and $z = 0.145$, with respect to the origin $O$.

| $hkl$ | $(\sin\theta)/\lambda$ | $2f_\text{I}$ | $\cos 2\pi[(0.423h) + (0.146l)]$ | $F_\text{I}$ | $|F_o|$ |
|---|---|---|---|---|---|
| 001 | 0.0261 | 105 | 0.608 | 64 | 40 |
| 0014 | 0.365 | 66 | 0.962 | 63 | 37 |
| 300 | 0.207 | 82 | −0.119 | −10 | 35 |
| 106 | 0.176 | 86 | −0.303 | −26 | 33 |

The signs of 001, 0014, and 106 are probably +, +, and −, respectively. The magnitude of $F_\text{I}(300)$ is a small fraction of $|F_o(300)|$, and the negative sign is unreliable. Note that small variations in your values for $F_\text{I}$ are acceptable; they would probably indicate differences in the graphical interpolation of $f_\text{I}$.

**8.2.** A simplified $\Sigma_2$ listing follows:

| h | k | h−k | $|E(\mathbf{h})\|E(\mathbf{k})\|E(\mathbf{h}-\mathbf{k})|$ |
|------|------|-------|------|
| 0018 | 081 | 0817 | 9.5 |
| 001 | 024 | 035 | 3.0 |
| | 026 | 035 | 0.3 |
| 021 | 038 | 059 | 3.6 |
| | 0310 | 059 | 3.2 |
| 024 | 035 | 059 | 9.6 |
| 038 | 059 | 0817 | 7.2 |
| | 081 | 011,7 | 6.0 |
| | 081 | 011,9 | 10.2 |
| 0310 | 059 | 081 | 7.9 |
| | 081 | 011,9 | 9.2 |

In space group $P2_1/a$, $s(hkl) = s(\overline{h}\overline{k}\overline{l}) = (-1)^{h+k}s(h\overline{k}l)$, which means that $s(hk\overline{l}) = (-1)^{h+k}s(\overline{h}kl)$. The origin may be specified by $s(081) = +$ and $s(011,9) = +$.

Sign determination

| h | k | h−k | Conclusion |
|------|------|-------|------|
| 011,9 | 081 | 038 | $s(038) = +$ |
| 011,9 | 081 | 0310 | $s(0310) = +$ |
| 038 | 081 | 011,7 | $s(011,7) = +$ |
| 0310 | 081 | 059 | $s(059) = -$ |
| 059 | 038 | 0817 | $s(0817) = -$ |
| 038 | 059 | 021 | $s(021) = -$ |
| 0310 | 059 | 021 | $s(021) = -$ |
| 0817 | 081 | 0018 | $s(0018) = -$ |
| | | Let | $s(035) = A$ |
| 059 | 035 | 024 | $s(024) = -A$ |
| 035 | 024 | 011 | $s(011) = -$ |
| 035 | 011 | 026 | $s(026) = A$ |

Note that $s(026) = A$ has a low probability compared with those of the other conclusions. To obtain most nearly equal numbers of plus and minus signs, $A = +$ would be chosen. Hence, finally, $s(035) = +$ and $s(024) = -$; $s(026)$ is $+$, but with low probability.

# Index